Crop Protection and Fungicides

Crop Protection and Fungicides

Editor: Chris Frost

www.callistoreference.com

Callisto Reference,
118-35 Queens Blvd., Suite 400,
Forest Hills, NY 11375, USA

Visit us on the World Wide Web at:
www.callistoreference.com

ISBN: 978-1-63239-793-5 (Hardback)

The publisher's policy is to use permanent paper from mills that operate a sustainable forestry policy. Furthermore, the publisher ensures that the text paper and cover boards used have met acceptable environmental accreditation standards.

Printed in the United States of America.

Cataloging-in-publication Data

Crop protection and fungicides / edited by Chris Frost.
   p. cm.
Includes bibliographical references and index.
ISBN 978-1-63239-793-5
1. Fungicides. 2. Plants, Protection of. 3. Plants--Effect of Fungicides on. 4. Crops--Nutrition. I. Frost, Chris.
SB951.3 .C76 2017
632.952--dc23

Table of Contents

Preface

This book discusses the fundamental as well as modern approaches of using fungicides for crop production. It strives to provide a fair idea about this discipline and to help develop a better understanding of the latest advances within this field. Crop protection refers to study and practice of protecting crops from harmful pests, fungi, weeds and diseases. Fungicides are an important part of crop protection science, along with pesticides, herbicides, etc. This book elucidates new techniques and their applications in a multidisciplinary approach. It presents researches and studies performed by experts across the globe. This text is a compilation of chapters that discuss the most vital concepts and emerging trends in this field. It is appropriate for students seeking detailed information in this area as well as for experts.

It is often said that books are a boon to mankind. They document every progress and pass on the knowledge from one generation to the other. They play a crucial role in our lives. Thus I was both excited and nervous while editing this book. I was pleased by the thought of being able to make a mark but I was also nervous to do it right because the future of students depends upon it. Hence, I took a few months to research further into the discipline, revise my knowledge and also explore some more aspects. Post this process, I began with the editing of this book.

I thank my publisher with all my heart for considering me worthy of this unparalleled opportunity and for showing unwavering faith in my skills. I would also like to thank the editorial team who worked closely with me at every step and contributed immensely towards the successful completion of this book. Last but not the least, I wish to thank my friends and colleagues for their support.

Editor

A Single *Streptomyces* Symbiont Makes Multiple Antifungals to Support the Fungus Farming Ant *Acromyrmex octospinosus*

Ryan F. Seipke[1]*, Jörg Barke[1], Charles Brearley[1], Lionel Hill[2], Douglas W. Yu[1,3], Rebecca J. M. Goss[4], Matthew I. Hutchings[1]*

1 School of Biological Sciences, University of East Anglia, Norwich Research Park, Norwich, United Kingdom, 2 Metabolic Biology, John Innes Centre, Norwich Research Park, Norwich, United Kingdom, 3 State Key Laboratory of Genetic Resources, and Evolution, Ecology, Conservation, and Environment Center, Kunming Institute of Zoology, Chinese Academy of Sciences, Kunming, China, 4 School of Chemistry, University of East Anglia, Norwich Research Park, Norwich, United Kingdom

Abstract

Attine ants are dependent on a cultivated fungus for food and use antibiotics produced by symbiotic Actinobacteria as weedkillers in their fungus gardens. Actinobacterial species belonging to the genera *Pseudonocardia*, *Streptomyces* and *Amycolatopsis* have been isolated from attine ant nests and shown to confer protection against a range of microfungal weeds. In previous work on the higher attine *Acromyrmex octospinosus* we isolated a *Streptomyces* strain that produces candicidin, consistent with another report that attine ants use *Streptomyces*-produced candicidin in their fungiculture. Here we report the genome analysis of this *Streptomyces* strain and identify multiple antibiotic biosynthetic pathways. We demonstrate, using gene disruptions and mass spectrometry, that this single strain has the capacity to make candicidin and multiple antimycin compounds. Although antimycins have been known for >60 years we report the sequence of the biosynthetic gene cluster for the first time. Crucially, disrupting the candicidin and antimycin gene clusters in the same strain had no effect on bioactivity against a co-evolved nest pathogen called *Escovopsis* that has been identified in ~30% of attine ant nests. Since the *Streptomyces* strain has strong bioactivity against *Escovopsis* we conclude that it must make additional antifungal(s) to inhibit *Escovopsis*. However, candicidin and antimycins likely offer protection against other microfungal weeds that infect the attine fungal gardens. Thus, we propose that the selection of this biosynthetically prolific strain from the natural environment provides *A. octospinosus* with broad spectrum activity against *Escovopsis* and other microfungal weeds.

Editor: Jae-Hyuk Yu, University of Wisconsin – Madison, United States of America

Funding: This work was funded by the Medical Research Council Milstein Award (grant G0801721) to MIH, RJMG and DWY. The funders had no role in study design, data collection and analysis, decision to publish, or preparation of the manuscript.

Competing Interests: The authors have declared that no competing interests exist.

* E-mail: r.seipke@uea.ac.uk (RFS); m.hutchings@uea.ac.uk (MIH)

Introduction

Insect fungiculture has been best studied in the attine ants (subfamily Myrmicinae, tribe Attini) whose common ancestor is estimated to have evolved agriculture around 50 million years ago [1]. Attine ants are so dependent on the cultivation of fungus that when a daughter queen leaves to found a new nest she carries a piece of the cultivar fungus in her mouth in order to establish a culture of that fungus in her new nest [1]. Intriguingly, attine ants have also evolved a mutualism with Actinobacteria that produce antibiotics that the ants use as weedkillers to keep their fungal gardens free of other microbes [2,3,4]. The relationship between Actinobacteria, fungal cultivar and attine ant has been intensely studied in the branch of higher attines known as the leaf-cutting ants (genera *Atta* and *Acromyrmex*) which harvest fresh vegetation to feed to their highly specialised fungal cultivar, *Leucoagaricus gongylophorus* [1]. The fungus has evolved lipid and carbohydrate rich hyphae known as gongylidia which the ants harvest and use as food [5]. Pathogens of the fungal garden, most notably fungi of the genus *Escovopsis*, if left

unchecked, can destroy a fungal garden and lead to the collapse of the colony within weeks [6,7].

Two overlapping but conflicting theories have been put forward to explain the evolution of mutualism between attine ants and Actinobacteria. The first suggests co-evolution of attine ants and Actinobacteria belonging to the genus *Pseudonocardia*. This theory suggests that the fungus garden pathogen *Escovopsis* has also co-evolved and that *Pseudonocardia* and *Escovopsis* are engaged in an evolutionary arms race in which the bacteria evolve compounds that specifically target *Escovopsis* but do not inhibit the growth of the fungal cultivar [5]. The second model suggests that attine ants select antifungal-producing Actinobacteria from the environment and is consistent with the identification of additional Actinobacterial genera on leaf-cutting ants, including *Streptomyces* and *Amycolatopsis* species [3,4,8,9]. However, these theories are not mutually exclusive and evidence suggests attine ants co-evolve with *Pseudonocardia* bacteria and still select other antifungal producing bacteria from the soil, perhaps to prevent evolution of resistance in the fungal pathogens [9,10]. Indeed, there is good evidence, both direct and indirect, that leaf-cutting ants use multiple antifungals produced by multiple Actinobacteria during the cultivation of their fungal gardens suggesting that both models do apply, at least in the higher attine genus *Acromyrmex* [3,11]. Recent work demonstrated that *Acromyrmex octospinosus* are associated with a *Pseudonocardia* strain that may have co-evolved and a *Streptomyces* strain that was most likely acquired from the environment relatively recently [9].

One intriguing question that still needs to be addressed in the environmental recruitment model concerns how the ants select beneficial bacteria. Actinobacteria are well known producers of useful secondary metabolites, including around 60% of all known antibiotics [12] and it seems likely that this production capability is key to their success as mutualists. Clearly the production of antifungals makes them useful to the ants and we hypothesise that production of multiple antifungals with different targets by single Actinobacterial species would make them more attractive to the ants as mutualists. In this work we carry out a more in depth analysis of the antifungals made by one of the strains associated with the leaf-cutting ant *A. octospinosus*, a species of *Streptomyces* which has been proposed to support fungus growing ants through production of the polyene antifungal candicidin [4,9]. We report the genome sequence and analysis of this strain indicating its capacity to make numerous antibiotics, including at least three antifungal compounds. Curiously, this strain makes both of the antifungals that have been reported in the *Streptomyces* attine ant mutualists but neither the candicidin or antimycins are obligatory for the inhibition of the co-evolved nest pathogen *E. weberi*. We propose that additional and as yet unknown antifungal(s) made by this strain specifically target *Escovopsis* and that candicidin and antimycins offer protection against other microfungal weeds. We propose that the ability of this single species to make multiple antifungal compounds makes it an attractive acquisition for the ants and their fungiculture.

Results

Genome sequencing and analysis

To determine the antibiotic biosynthetic capability of *Streptomyces* S4 a combination of shotgun, 3 kbp and 8 kbp paired end libraries were constructed and 454 pyrosequenced to generate >335 Mbp of sequence that was assembled into 12 scaffolds containing 211 large contigs. The genome consists of one ~7.5 Mbp linear chromosome, which is within the size range reported for genomes of other sequenced streptomycetes, as well as one linear plasmid (~180 kbp) and one circular plasmid (~2 kbp) [13]. The *Streptomyces* S4 genome was annotated by a combination of manual and automated methods and multiple biosynthetic gene clusters predicted to produce secondary metabolites were identified. Table 1 summarizes characteristics of predicted natural product biosynthetic gene clusters in *Streptomyces* S4. As expected based on our previous work, the *Streptomyces* S4 genome contained a candicidin biosynthetic gene cluster that shares 98% nucleotide sequence identity with the candicidin biosynthetic gene cluster from *Streptomyces* sp. FR-008 [14]. Other biosynthetic gene clusters of note are a non-ribosomal peptide synthetase (NRPS) biosynthetic gene cluster that is predicted to direct the biosynthesis of an antibacterial similar to mannopeptimycin, as well as a NRPS biosynthetic gene cluster whose predicted product is a gramicidin-like antibacterial. We note these antibacterials could be useful in eliminating competition during colonisation of the ant cuticle. *Streptomyces* S4 also contains a type II polyketide synthase (PKS) biosynthetic gene cluster that shares 100% nucleotide identity to that of the fredericamycin biosynthetic gene cluster characterized in *S. griseus* [15]. Fredericamycin is mostly known for its antitumor properties [16]. Additionally, there are six functionally unassigned biosynthetic gene clusters (three NRPS and three hybrid NRPS/PKS) that are not similar to gene clusters with known products. It

Table 1. Putative secondary metabolites encoded by *Streptomyces* S4.

Predicted biosynthetic system	Genome coordinates	Predicted metabolite or close relative	Biological properties
Hopene / squalene synthase	scaffold08: 588141–598581	Hopanoids	Membrane stabilizers
NRPS-independent siderophore synthetase	scaffold05: 959198–972403	Desferrioxamine	Siderophore
NRPS-independent siderophore synthetase	scaffold08: 1448607–1457963	Unknown	Unknown
Ectoine synthase	scaffold05: 68880–72152	Ectoine	Osmolyte
Phytoene / polyprenyl synthase	scaffold06: 410147–419826	Carotenoids	Pigment
Terpene synthase	scaffold08: 1719586–1721871	Geosmin	Unknown
Type III PKS	scaffold06: 295706–300701	1,3,6,8-tetrahydroxynaphthalene	Pigment
Type I PKS	scaffold06: 115150–253654	Candicidin	Antifungal
Type I PKS / Type III PKS	scaffold05:1001127–1064995	Kendomycin	Anticancer
Type II PKS	scaffold08: 3878554–3911349	Fredericamycin	Anticancer
Hybrid NRPS / PKS	scaffold06: 81953–106578	Unknown	Unknown
Hybrid NRPS / PKS	scaffold06: 7264–45109	Unknown	Unknown
Hybrid NRPS / PKS	scaffold08: 503983–520001	Unknown	Unknown
NRPS	scaffold08: 4240081–4309220	Gramicidin	Antibacterial
NRPS	scaffold08: 3002155–3042863	Mannopeptimycin	Antibacterial
NRPS	scaffold06: 65083–81878	Unknown	Unknown
NRPS	scaffold08: 276268–301035	Unknown	Unknown
NRPS	scaffold08: 3930113–3950474	Unknown	Unknown

NRPS, non-ribosomal peptide synthetase, PKS, polyketide synthase.

is likely that at least some these clusters encode secondary metabolites with antibacterial or antifungal activity.

Mutagenesis of the candicidin biosynthetic gene cluster does not abolish antifungal activity

Evidence that candicidin is not the sole antifungal generated by *Streptomyces* S4 is demonstrated by the retained antifungal activity of a mutant strain deficient in the biosynthesis of this compound. The candicidin biosynthetic gene cluster was disrupted by deletion of the polyketide synthase gene, *fscC*, which encodes the candicidin biosynthetic modules 6–10 [14]. LC-MS analysis of butanol-extracted culture supernatants from the wild-type strain revealed a molecular ion (m/z 1109.6) consistent with that of candicidin D and showed characteristic polyene absorption bands in its UV spectrum, with absorbance maxima at 360, 380, 403 nm (Fig. 1). As predicted, the molecular ion m/z 1109.6 was not detected in the Δ*fscC* mutant, indicating that candicidin production is abolished in this strain (Fig. 1). Bioassays of the isogenic wild-type and Δ*fscC* strains against *C. albicans* and the nest pathogen

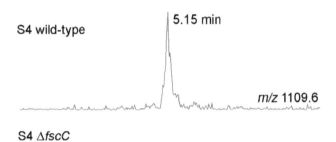

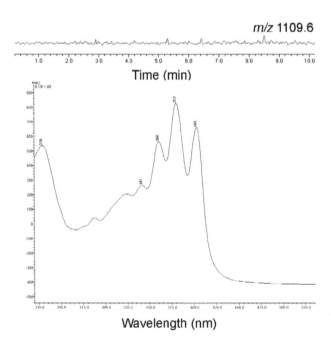

Figure 1. Deletion of the candicidin biosynthetic gene, *fscC* abolishes production of candicidin. LC-MS was used to analyze supernatant from *Streptomyces* S4 wild-type and S4 Δ*fscC*. The extracted ion chromatogram for candicidin (m/z 1109.6) is shown and confirmed that only S4 wild-type and not the Δ*fscC* mutant produced candicidin. The UV visible spectra for the peak at RT 5.15 min displays absorption characteristics consistent with polyene compounds is also shown (bottom).

Escovopsis weberi demonstrated that loss of candicidin has no effect on the antifungal bioactivity of *Streptomyces* S4 (Fig. 2). This result suggests that *Streptomyces* S4 makes at least one additional antifungal compound that has not been identified previously, most likely encoded by one of the other biosynthetic gene clusters identified in this work.

Identification of the antimycin biosynthetic gene cluster

While this work was in progress another group reported that antimycins are produced by a number of the other *Streptomyces* strains associated with attine ant nests [17]. Antimycins inhibit the respiratory chain and are known to have antifungal activity. We investigated whether *Streptomyces* S4 is making antimycin compounds in addition to candicidin and hypothesised that antimycins could potentially account for the retained bioactivity against *Escovopsis* observed for the *Streptomyces* S4 *fscC* mutant. LC-MS analysis of culture supernatants of the wild-type strain identified eight compounds with m/z that match those reported for antimycins A1–A4 (Fig. 3). To determine if any of the eight compounds could be antimycins, we co-injected commercially available antimycin standards A1–A4 with our wild-type extract, which revealed that four of the eight compounds possess the same retention time as the antimycin A1–A4 standards (Fig. 3). Four of the eight S4 compounds were identical to the commercially available antimycin standards A1–A4 both in terms of UV absorbance profile and LC retention time. Whilst the remaining four S4 compounds possess the same UV absorbance characteristics as the antimycin standards (Fig. S1), and the same m/z parent ions as those of the standards, they exhibit different retention times (Fig. 3); further experiments are being carried out to identify and characterize these four previously unreported compounds.

To our knowledge the gene cluster that encodes the antimycin biosynthetic pathway has not been identified, despite these compounds first being isolated over 60 years ago [18]. The structure of antimycin suggests that it may be synthesized, at least in part, by an NRPS, and that threonine may be utilized as a substrate (Fig. 3). The Basic Local Alignment Search Tool [19] revealed a region of the *Streptomyces* S4 genome with 57% amino acid sequence identity to the threonine adenylation domain from the daptomycin biosynthetic protein, DptA [20]. This region of homology enabled us to identify a hybrid NRPS/PKS biosynthetic gene cluster that displays significant amino acid identity to a hybrid NRPS/PKS biosynthetic gene cluster present in both *S. albus* and *S. ambofaciens* and potentially encodes for the biosynthesis of antimycins (Fig. 4). Table 2 displays the proposed functions of proteins present in the hybrid NRPS/PKS cluster. In order to determine if this biosynthetic gene cluster can direct the production of antimycin, we disrupted the hybrid NRPS/PKS gene, *antC* and assessed antifungal activity against *C. albicans* in a plate bioassay. The Δ*antC* mutant displayed dramatically reduced antifungal activity against *C. albicans* compared to that of the wild-type strain (Fig. 2). This strongly suggested that the product of this cluster was an antifungal compound and is consistent with the hypothesis that this cluster could potentially mediate the biosynthesis of antimycins, compounds known to possess strong antifungal activity against *C. albicans* (Fig. 2). Confirmation that the hybrid NRPS/PKS encoded by *Streptomyces* S4 directs the biosynthesis of antimycins was obtained by comparing the LC-MS profiles of the wild-type, Δ*fscC* and Δ*antC* mutant strains. Extracted ion chromatograms revealed that the Δ*antC* mutant does not produce the eight antimycins and that the Δ*fscC* mutant retained the ability to produce these antimycin compounds (Fig. 3).

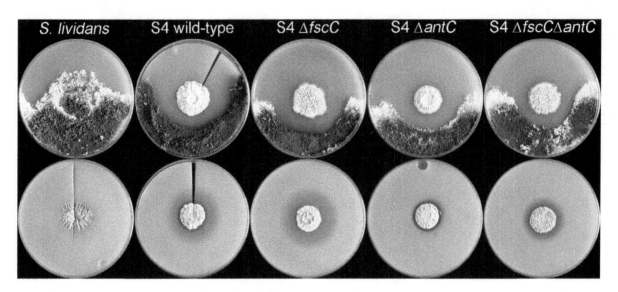

Figure 2. Antifungal bioactivity of the non-antifungal-producing strain *Streptomyces lividans*, *Streptomyces* S4 wild-type and mutant strains. Bioassays with *Streptomyces lividans*, S4 wild-type, S4 ΔfscC, S4 ΔantC and S4 ΔfscC ΔantC against *Escovopsis weberi* (top panel) and *C. albicans* (bottom panel) demonstrate that deletion of *fscC* does not abolish antifungal activity and that deletion of *antC* only reduces antifungal activity against *C. albicans* and not *E. weberi*. The S4 ΔfscC ΔantC double mutant does not display reduced antifungal activity against *E. weberi* suggesting the presence of an additional antifungal compound that is responsible for the phenotype observed during in vitro bioassays.

A *Streptomyces* S4 mutant which cannot make candicidin or antimycins still has antifungal activity against *E. weberi*

Our bioassays against *E. weberi* with the ΔfscC and ΔantC mutant strains of *Streptomyces* S4 did not show a reduction in bioactivity against this nest parasite indicating that key antifungal compound(s) affording protection against the natural fungal pathogen still remain to be identified (Fig. 2). To determine whether additional antifungals are made by *Streptomyces* S4 we generated a new mutant in which the *fscC* and *antC* genes were disrupted in the same strain and assessed the bioactivity of this strain against *C. albicans* and *E. weberi*. As we predicted, the antifungal activity of the ΔfscC ΔantC double mutant against *E. weberi* was comparable to that of the wild-type strain, confirming that additional antifungal compound(s) made by *Streptomyces* S4 account for the majority of the antifungal activity observed against *E. weberi* in vitro. The additional antifungal(s) may be encoded by one of the five functionally unassigned biosynthetic gene clusters identified in our genome analysis (Table 1) or by a gene cluster not identified in our analysis.

Discussion

Although the attine ant-fungal mutualism has been studied for more than a century, the antibiotic-producing Actinobacterial mutualists were only discovered ~15 years ago and it is only very recently that scientists have started to address the nature of the antibiotics being produced by these bacteria [2,4,9,17]. It has been hypothesised, although not proven, that these antibiotics are used by the ants to kill off contaminated parts of the garden and / or to suppress the growth of fungal pathogens including co-evolved pathogens in the genus *Escovopsis* and many other microfungal weeds [6,7]. Recent studies have shown that strains belonging to two key genera are typically associated with attine ants, species of *Pseudonocardia*, which have been suggested to have co-evolved with the ants and to be transmitted vertically by the queens, and species of *Streptomyces* which have been suggested to be more recently acquired from the environment [3,4,8,9]. Two antifungals have

been identified from proposed mutualist species in each genus and both inhibit the nest pathogen *Escovopsis* in vitro. *Pseudonocardia* associated with the lower attine *A. dentigerum* makes dentigerumycin and *Pseudonocardia* associated with the higher attine *A. octospinosus* makes nystatin P1 [2,9]. *Streptomyces* mutualists associated with higher attines of the genus *Acromyrmex* are known to produce the well-known antifungals candicidin and antimycins and it has been suggested that these compounds account for the bioactivity of this *Streptomyces* strain against *Escovopsis* [4,9,17]. To date almost all of this work has been carried out through isolation and mass spectrometry analysis of the antifungal compounds, although we used genome scanning of a *Pseudonocardia* mutualist to identify a nystatin-like biosynthetic gene cluster and its product which we named nystatin P1 [9].

In this work we have undertaken the first in-depth genome sequence analysis of a proposed attine ant mutualist, in this case a candicidin-producing *Streptomyces* strain isolated from *A. octospinosus* garden worker ants collected in Trinidad [9]. Genome sequencing and analysis identified 17 gene clusters that are predicted to encode for known or unknown secondary metabolites, including the known gene cluster for candicidin biosynthesis. Following the discovery of antimycin production by *Streptomyces* strains isolated from attine ants in a separate study [17] we identified a gene cluster encoding a pathway that is consistent with antimycin biosynthesis. Surprisingly, despite antimycins first being isolated and characterised >60 years ago the antimycin biosynthetic pathway was not known. We identified eight compounds which we assigned as antimycins and then identified and disrupted the hybrid NRPS/PKS gene cluster which we predicted to encode antimycin biosynthesis. The production of the eight antimycin compounds was abolished in the mutant strain providing strong evidence that this gene cluster does indeed encode the antimycin biosynthetic pathway.

In bioassays of the wild-type *Streptomyces* S4 strain alongside strains which cannot make candicidin, antimycins or either of these antifungal compounds we found that whilst antimycin- and candicidin-deficient strains had reduced activity against the

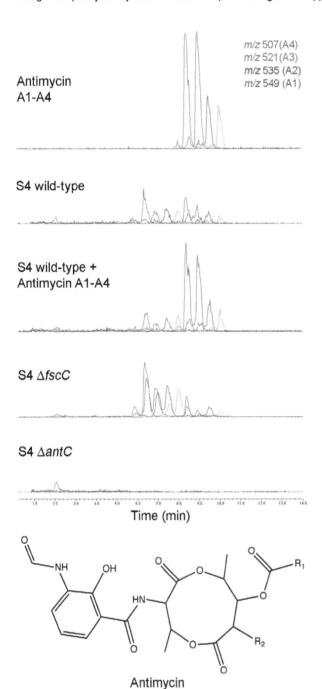

Figure 3. LC/MS analysis of *Streptomyces* S4 wild-type and mutant strains compared to antimycin standards. The extracted ion chromatograms for antimcyins A1–A4 are shown. Eight compounds consistent with the mass of antimycin A1–A4 were produced by S4 wild-type and S4 Δ*fscC*, but were not produced by the Δ*antC* mutant. Co-injection of antimycin A1–A4 with the S4 wild-type extract demonstrated that antimycin A1–A4 have the same retention time as four of the eight compounds produced by S4 wild-type. The UV visible spectra and ESI positive mode mass spectra for antimycin A1–A4 and the eight antimycin compounds produced by S4 wild-type are shown in Fig. S1 and Fig. S2, respectively. Antimycin A1: $R_1 = CH(CH_3)CH_2CH_3$, $R_2 = (CH_2)_5CH_3$. Antimycin A2: $R_1 = CH(CH_3)_2$, $R_2 = (CH_2)_5CH_3$. Antimycin A3: $R_1 = CH(CH_3)CH_2CH_3$, $R_2 = (CH_2)_3CH_3$. Antimycin A4: $R1 = CH(CH_3)_2$ $R_2 = (CH_2)_3CH_3$.

human pathogen *C. albicans* their activity against the nest pathogen *E. weberi* was unaltered. This is curious as the *Streptomyces* S4 strain and *E. weberi* are thought to have co-evolved in this mutualism [21]. This suggests that despite previous research demonstrating that purified candicidin and antimycin preparations inhibit the growth of *Escovopsis* in vitro neither compound is responsible for the activity observed in bioassays where *Streptomyces* S4 is challenged with *E. weberi*. We conclude that these compounds potentially inhibit the growth of other microfungal weeds found in the ant-fungus gardens while additional and currently unknown antifungal(s) produced by *Streptomyces* S4 have stronger activity against *Escovopsis*. We also propose that the combination of antifungals produced by this single *Streptomyces* strain coupled with the antifungal(s) produced by a *Pseudonocardia* strain isolated from the same nest provides a broad spectrum of antifungal activity that is used by the ants to farm their fungus. Furthermore, the antibacterials made by *Streptomyces* S4 potentially help it to outcompete other bacteria for the ant host.

Our data suggest that the *Streptomyces* strains isolated by other researchers from attine ants are likely to make additional antifungals since they appear to be closely related to *Streptomyces* S4. It will be important to re-examine the biosynthetic capability of these strains in order to fully understand the chemical basis of their interactions with attine ants and their fungal cultivar. This reflects a common problem in the field of natural product antibiotic discovery, in which the reisolation of known compounds hampers the discovery of new antibiotics. The approach we have outlined here is time consuming and technically challenging, but it is perhaps the only way to determine the entire biosynthetic capability of an antibiotic-producing strain particularly if some of the antibiotics being made, and their biosynthetic gene clusters, are new to science. Future work will be aimed at determining the products of the five unassigned biosynthetic gene clusters in *Streptomyces* S4 and identifying the additional antifungal compound(s) made by this strain. This is likely to involve significant challenges if, as we predict, these are novel secondary metabolites.

In conclusion, although good progress has been made recently we are still a long way from understanding the chemical basis of the symbioses between antibiotic-producing Actinobacteria and their attine ant hosts. We hope that our study will stimulate further research in this area and the identification of additional antifungal and antibacterial compounds in this system.

Materials and Methods

Growth media and strains

Streptomyces strains were routinely grown on soya flour mannitol (SFM) agar plates or in liquid TSB/YEME while *E. coli* strains were grown on Lysogeny both- Lennox (LB) [22]. Media was supplemented with antibiotics as required at the following concentrations: carbenicillin (100 µg/ml), hygromycin B (50 µg/ml), nalidixic acid (25 µg/ml), apramycin (50 µg/ml). S4 was isolated and identified by 16S rDNA sequencing in a previous study (GenBank accession HM179229). Antifungal bioassays with *C. albicans* and *E. weberi* were carried out as described previously [9]. Strains and plasmids are described in Table 3.

Construction of *Streptomyces* S4 mutant strains

In order to create the Δ*fscC* mutant, two 3 kb knockout arms were PCR amplified using GoTaq Polymersae (Promega) with oligonucleotide primers RFS78 and RFS79 (upstream arm) and RFS80 and RFS81 (downstream arm), respectively. Oligonucleotide primers (Integrated DNA Technologies) were engineered at their 5′ end to contain restriction sites for cloning (Table S1). The

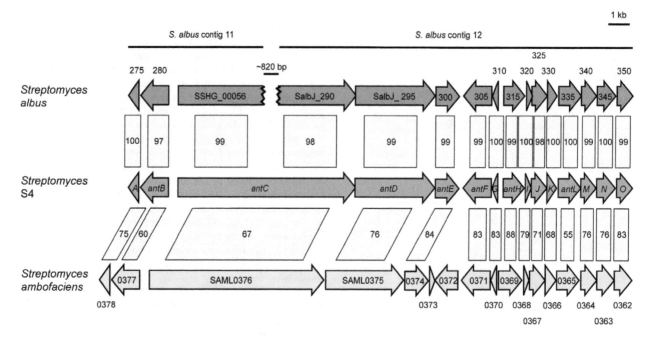

Figure 4. Gene schematic of the *Streptomyces* S4 antimycin biosynthetic gene cluster and comparison to putative antimycin clusters in *Streptomyces albus* and *Streptomyces ambofaciens*. The percent amino acid homology shared between S4 proteins and proteins in *S. albus* and *S. ambofaciens* is indicated in the shaded boxes. The draft genomic sequence of *S. albus* is incomplete and the sequence for the putative antimycin biosynthetic gene cluster is split over contig 11 and contig 12 with an estimated gap of ~820 bp in the *antC* gene. The *Streptomyces* S4 antimycin biosynthetic gene cluster is located on scaffold06 at coordinates 81953–106578. The partial genome sequences of *Streptomyces* S4, *S. albus* J1074, and *S. ambofaciens* ATCC 23877 are available under accession numbers CADY00000000, ABYC00000000 and AM238663, respectively. The gene names for *S. albus* have been shortened to eliminate the first nine numbers of the gene name (e.g. SalbJ_290 = SalbJ_010100000290).

resulting PCR products were cloned into pGEMT-EZ (Promega) and sequenced to verify their identity. The upstream arm was released from pGEMT-EZ with HindIII and BamHI and cloned into pKC1132 (which contained the RK2 conjugal origin of transfer and as well as an apramycin resistance gene [22,23] cut with the same enzymes to result in pKC1132-Up. Next, the

downstream arm was released from pGEMT-EZ with BamHI and EcoRI and cloned into pKC1132-Up cut with the same enzymes to result in pKC1132-UpDn. Finally, a hygromycin B resistance cassette was PCR-amplified from pIJ10700 [24] using oligonucleotides RFS94 and RFS95 engineered to contain BamHI sites at their 5′ end. The hygromycin resistance cassette was cloned into pGEMT-EZ and subsequently released by BamHI digestion and cloned into pKC1132-UpDn cut with the same enzyme to result in pKC1132-UpHygDn.

The plasmid pKC1132-UpHygDn was electroporated into *E. coli* strain ET12567/pUZ8002 [25] and transferred to *Streptomyces* S4 by cross-genera conjugation as previously described [22]. Transconjugants were selected for apramycin resistance. An apramycin-resistant transconjugant was obtained and subsequently replica plated to obtain hygromycin-resistant and apramycin-sensitive colonies, a phenotype indicating that the *fscC* gene had been entirely replaced by the hygromycin resistance cassette and that the plasmid backbone was no longer present. Loss of the pKC1132-UpHygDn plasmid backbone and mutagenesis of the *fscC* gene in the Δ*fscC* strain was confirmed by PCR.

In order to disrupt the *antC* gene a ~1.5 kb internal fragment of the *antC* gene was PCR amplified using oligonucleotide primers RFS121 and RFS122 which were engineered to contain BamHI and EcoRI restriction sites at their 5′ end, respectfully (Table S1). The resulting PCR product was sequenced to verify its identity and cloned into pGEMT-EZ to result in pGEMT-Ant. The apramycin resistance cassette containing a conjugal origin of transfer (*aac(3)*IV+oriT) was isolated from pIJ773 as a BamHI fragment and cloned into the BamHI site (provided by RFS121) in pGEMT-Ant to result in pGEMT-AntApr. The pGEMT-AntApr plasmid was electroporated into ET12567/pUZ8002 and mobilized to S4 wild-type and S4 Δ*fscC* by conjugation. Transconju-

Table 2. Proposed functions of proteins encoded by the antimycin biosynthetic gene cluster.

Streptomyces S4 protein	Proposed function
AntA	Sigma factor
AntB	Condensation domain
AntC	Non-ribosomal peptide synthetase
AntD	Polyketide synthase
AntE	Dehydrogenase
AntF	Acyl-CoA ligase
AntG	Thiolation domain
AntH	Phenylacetate dioxygenase
AntI	Phenylacetate dioxygenase
AntJ	Phenylacetate dioxygenase
AntK	Phenylacetate dioxygenase
AntL	Oxidoreductase
AntM	Dehydrogenase
AntN	Tryptophan 2,3-dioxygenase
AntO	Lipase

Table 3. Strains and plasmids used in this study.

Strain or plasmid	Genotype or comments	Source or reference
Streptomyces S4	Wild type	[9]
Streptomyces S4 ΔfscC	S4 fscC null mutant	This study
Streptomyces S4 ΔantC	S4 antC disruption mutant	This study
Streptomyces S4 ΔfscC ΔantC	S4 fscC and antC double knockout strain	This study
Candida albicans	Candidia albicans CA-6	[26]
Escovopsis weberi	Escovopsis weberi (CBS 11060)	[9]
E. coli ET12567	Non-methylating host for transfer of DNA into Streptomyces spp. (dam, dcm, hsdM)	[25]
E. coli TOP10	Host for routine cloning procedures	Invitrogen
Plasmids		
pGEMT-EZ	Cloning vector for PCR products; Amp^R	Promega
pIJ773	Source of the aac(3)IV+oriT apramycin resistance resistance marker	[27]
pIJ10700	PCR template for hygR cassette	[24]
pUZ8002	Encodes conjugation machinery for mobilization of plasmids from E. coli to Streptomyces; Kan^R	[25]
pKC1132	Suicide vector used for constructing gene deletions in Streptomyces spp. Apr^R and contains conjugal origin of transfer	[23]
pKC1132-Up	Derivative of pKC1132 containing the fscC upstream knockout arm cloned into the HindIII and BamHI restriction sites	This study
pKC1132-UpDn	Derivative of pKC1132-Up containing the fscC downstream knockout arm cloned into the BamHI and EcoRI restriction sites	This study
pKC1132-UpHygDn	Derivative of pKC1132-UpDn containing the hygromycin resistance cassette from pIJ10700 cloned into the BamHI site	This study
pGEMT-Ant	Derivative of pGEMT-EZ containing the a 1.5 kb fragment of the antC gene, Amp^R	This study
pGEMT-AntApr	Derivative of pGEMT-Ant containing the aac(3)IV+oriT apramycin resistance gene from pIJ773 cloned into the BamHI site provided by RFS121	This study

Amp^R, ampicillin resistance, Apr^R, apramycin resistance, Hyg^R, hygromycin resistance, Kan^R, kanamycin resistance, oriT, origin of transfer.

gants were selected for apramycin resistance, a phenotype indicating that the disruption plasmid has crossed into the chromosome. Disruption of the *antC* gene was confirmed by PCR amplification using oligonucleotide primers (RFS147 and M13F, and RFS148 and M13R) targeting the DNA sequence upstream and downstream of the expected site of integration.

LC-MS analysis

Streptomyces S4 wild-type and mutant strains were cultivated in mannitol-soya flour liquid medium in a 250 ml flask shaking at 270 rpm. For analysis of candidicin production, cultures were harvested after 10 days of growth, bacterial cells were removed by centrifugation and the supernatant of three biological replicates was combined. Fifty milliliters of the combined supernatant was extracted three times with an equal volume of butanol. Butanol extracts were combined and evaporated to dryness under vacuum and the residue was resuspended in 0.5 ml of 50% aqueous methanol. For analysis of antimycin production, cultures were harvested after 4 days of growth, bacterial cells were removed by centrifugation and the supernatant of two biological replicates was combined and extracted with XAD16 resin (Sigma). Following extraction, the resin was washed twice with ten milliliters of deionized water and *Streptomyces* S4 metabolites were eluted from the resin with one milliliter of 100% methanol. Prior to LC-MS analysis the methanol elution was diluted with water to a final methanol content of 50%. Antimycin A1–A4 standards were purchased from Sigma Aldrich. Immediately prior to LC-MS analysis samples were spun in a microcentrifuge at maximum speed for 5 minutes to remove insoluble material. Only the

supernatant (10 µl) was used for injection into a Shimadzu single quadrupole LC-MS-2010A mass spectrometer equipped with Prominence HPLC system as described previously [9]. For co-injection of antimycin A1-A4 with S4 wild-type extract 5 µl of standard and 5 µl of wild-type extract were mixed immediately prior to injection into the LC-MS.

Supporting Information

Figure S1   UV absorbance spectra for antimycin A1–A4 and eight antimycin compounds produced by *Streptomyces* S4. The UV absorbance spectra is shown for A) antimycin A4 (RT = 8.40), B) antimycin A3 (RT = 8.97), C) antimycin A2 (RT = 9.50), D) antimycin A1 (RT = 10.00), E) S4 metabolite 1 (RT = 6.38), F) S4 metabolite 2 (RT = 6.87), G) S4 metabolite 3 (RT = 7.43), H) S4 metabolite 4 (RT = 8.00), I) S4 antimycin A4 (RT = 8.40), J) S4 antimycin A3 (RT = 8.97), K) S4 antimycin A2 (RT = 9.50), L) S4 antimycin A1 (RT = 10.00).

Figure S2   Mass spectra for antimycin A1–A4 and eight antimycin compounds produced by *Streptomyces* S4. The ESI positive mode detection mass spectra is shown for A) antimycin A4 (RT = 8.40), B) antimycin A3 (RT = 8.97), C) antimycin A2 (RT = 9.50), D) antimycin A1 (RT = 10.00), E) S4 metabolite 1 (RT = 6.38), F) S4 metabolite 2 (RT = 6.87), G) S4 metabolite 3 (RT = 7.43), H) S4 metabolite 4 (RT = 8.00), I) S4 antimycin A4 (RT = 8.40), J) S4 antimycin A3 (RT = 8.97), K) S4 antimycin A2 (RT = 9.50), L) S4 antimycin A1 (RT = 10.00).

Table S1 Oligonucleotide primers used in this study.

Acknowledgments

We thank Sabine Grüschow and Mark Philo for assistance with LC-MS and useful discussions, Mark Buttner and Maureen Bibb for providing pKC1132 and The Genome Analysis Centre for sequencing and assembly of the *Streptomyces* S4 genome.

Author Contributions

Conceived and designed the experiments: RFS JB MIH. Performed the experiments: RFS JB LH. Analyzed the data: RFS JB LH. Contributed reagents/materials/analysis tools: CB DWY RJMG MIH. Wrote the paper: RFS RJMG MIH.

References

1. Schultz TR, Brady SG (2008) Major evolutionary transitions in ant agriculture. Proc Natl Acad U S A 105: 5435–5440.
2. Oh DC, Poulsen M, Currie CR, Clardy J (2009) Dentigerumycin: a bacterial mediator of an ant-fungus symbiosis. Nat Chem Biol 5: 391–393.
3. Sen R, Ishak HD, Estrada E, Dowd SE, Hong E, et al. (2009) Generalized antifungal activity and 454-screening of *Pseudonocardia* and *Amycolatopsis* bacteria in nests of fungus-growing ants. Proc Natl Acad U S A 106: 17805–17810.
4. Haeder S, Wirth R, Herz H, Spiteller D (2009) Candicidin-producing *Streptomyces* support leaf-cutting ants to protect their fungus garden against the pathogenic fungus *Escovopsis*. Proc Natl Acad U S A 106: 4742–4746.
5. Currie CR (2001) A community of ants, fungi, and bacteria: a multilateral approach to studying symbiosis. Annu Rev Microbiol 55: 357–380.
6. Reynolds HT, Currie CR (2004) Pathogenicity of *Escovopsis weberi*: the parasite of the attine ant-microbe symbiosis direclty consumes the ant-cultivated fungus. Mycologia 96: 955–959.
7. Rodrigues A, Bacci M, Mueller UG, Ortiz A, Pagnocca FC (2008) Microfungal 'weeds' in the leafcutter ant symbiosis. Microb Ecol 56: 604–614.
8. Kost C, Lakatos T, Bottcher I, Arendholz W-R, Redenbach M, et al. (2007) Non-specific association between filamentous bacteria and fungus-growing ants. Naturwissenschaften 94: 821–828.
9. Barke J, Seipke RF, Grüschow S, Heavens D, Drou N, et al. (2010) A mixed community of actinomycetes produce multiple antibiotics for the fungus farming ant *Acromyrmex octospinosus*. BMC Biol 8: 109.
10. Barke J, Seipke RF, Yu DW, Hutchings MI (2011) A mutualistic microbiome: how do fungus-growing ants select their antibiotic-producing bacteria. Commun Integr Biol 4: 41–43.
11. Mueller UG, Dash D, Rabeling C, Rodrigues A (2008) Coevolution between attine ants and actinomycete bacteria: a reevaluation. Evolution 62: 2894–2912.
12. Challis GL, Hopwood DA (2003) Synergy and contingency as driving forces for the evolution of multiple secondary metabolite production by *Streptomyces* species. Proc Natl Acad U S A 100: 14555–14561.
13. Seipke RF, Crossman L, Drou D, Heavens D, Bibb MJ, et al. (2011) Draft genome sequence of *Streptomyces* S4, a symbiont of the leafcutter ant *Acromyrmex octospinosus*. J Bacteriol;doi:10.1128/JB.05275-11.
14. Chen S, Huang X, Zhou X, Bai L, He J, et al. (2003) Organizational and mutational analysis of a complete FR-008/candicidin gene cluster encoding a structurally related polyene complex. Chem Biol 10: 1065–1076.
15. Wendt-Pienkowski E, Huang Y, Zhang J, Li B, Jiang H, et al. (2005) Cloning, sequencing, analysis, and heterologous expression of the fredericamycin biosynthetic gene cluster from *Streptomyces griseus*. J Am Chem Soc 127: 16442–16452.
16. Pandey RC, Toussaint MW, Stroshane RM, Kalita CC, Aszalos AA, et al. (1981) Fredericamycin A, a new antitumor antibiotic I. production, isolation and physicochemical properties. J Antibiot 34: 1389–1401.
17. Schoenian I, Spiteller M, Ghaste M, Wirth R, Herz H, et al. (2011) Chemical basis of the synergism and antagonism in microbial communities in the nests of leaf-cutting ants. Proc Natl Acad U S A 108: 1955–1960.
18. Dunshee BR, Leben C, Keitt GW, Strong FM (1949) The isolation and properties of antimycin A. J Am Chem Soc 71: 2436–2437.
19. Altschul SF, Gish W, Miller W, Myers EW, Lipman DJ (1990) Basic local alignment search tool. J Mol Biol 215: 403–410.
20. Miao V, Coeffet-LeGal M-F, Brian P, Brost R, Penn J, et al. (2005) Daptomycin biosynthesis in *Streptomyces roseosporus*: cloning and analysis of the gene cluster and revision of peptide stereochemistry. Microbiol 151: 1507–1523.
21. Currie CR, Wong B, Stuart AE, Schultz TR, Rehner A, et al. (2003) Ancient tripartite coevolution in the attine ant-microbe symbiosis. Science 299: 386–388.
22. Kieser T, Bibb MJ, Buttner MJ, Chater KF, Hopwood DA (2000) Practical *Streptomyces* Genetics. Norwich, UK: The John Innes Foundation. 613 p.
23. Bierman M, Logan R, O'Brien K, Seno ET, Rao RN, et al. (1992) Plasmid cloning vectors for the conjugal transfer of DNA from *Escherichia coli* to *Streptomyces* spp. Gene 116: 43–49.
24. Gust B, Chandra G, Jakimowicz D, Yuqing T, Bruton CJ, et al. (2004) Lambda Red-mediated genetic manipulation of antibiotic-producing *Streptomyces*. Adv Appl Microbiol 54: 107–128.
25. MacNeil DJ, Gewain KM, Ruby CL, Dezeny G, Gibbons PH, et al. (1992) Analysis of *Streptomyces avermitilis* genes required for avermectin biosynthesis utilizing a novel integrative vector. Gene 111: 61–68.
26. Maconi P, Bistoni F, Boncio A, Boncio L, Bersiani A, et al. (1976) Utilizzazione di una soluzione salina ipertonica di cloruro di potassio (3M KCl) per l'estrazione di antigeni solubili da Candida albicalns. Ann Sclavo 18: 61–66.
27. Gust B, Challis GL, Fowler K, Kieser T, Chater KF (2003) PCR-targeted *Streptomyces* gene replacement identifies a protein domain needed for biosynthesis of the sesquiterpene soil odor geosmin. Proc Natl Acad U S A 100: 1541–1546.

2

Soil Moisture and Fungi Affect Seed Survival in California Grassland Annual Plants

Erin A. Mordecai*

Department of Ecology, Evolution, and Marine Biology, University of California, Santa Barbara, Santa Barbara, California, United States of America

Abstract

Survival of seeds in the seed bank is important for the population dynamics of many plant species, yet the environmental factors that control seed survival at a landscape level remain poorly understood. These factors may include soil moisture, vegetation cover, soil type, and soil pathogens. Because many soil fungi respond to moisture and host species, fungi may mediate environmental drivers of seed survival. Here, I measure patterns of seed survival in California annual grassland plants across 15 species in three experiments. First, I surveyed seed survival for eight species at 18 grasslands and coastal sage scrub sites ranging across coastal and inland Santa Barbara County, California. Species differed in seed survival, and soil moisture and geographic location had the strongest influence on survival. Grasslands had higher survival than coastal sage scrub sites for some species. Second, I used a fungicide addition and exotic grass thatch removal experiment in the field to tease apart the relative impact of fungi, thatch, and their interaction in an invaded grassland. Seed survival was lower in the winter (wet season) than in the summer (dry season), but fungicide improved winter survival. Seed survival varied between species but did not depend on thatch. Third, I manipulated water and fungicide in the laboratory to directly examine the relationship between water, fungi, and survival. Seed survival declined from dry to single watered to continuously watered treatments. Fungicide slightly improved seed survival when seeds were watered once but not continually. Together, these experiments demonstrate an important role of soil moisture, potentially mediated by fungal pathogens, in driving seed survival.

Editor: Jack Anthony Gilbert, Argonne National Laboratory, United States of America

Funding: Funding was provided by the Mildred E. Mathias Foundation and United States Department of Agriculture (USDA) National Research Initiative Grant 2005-02252 (http://www.csrees.usda.gov/funding rfas/nri_rfa.html). The funders had no role in study design, data collection and analysis, decision to publish, or preparation of the manuscript.

Competing Interests: The author has declared that no competing interests exist.

* E-mail: Mordecai@lifesci.ucsb.edu

Introduction

Most plant life cycles begin with seeds: to complete the life cycle, seeds must germinate, survive, mature, and produce new seeds. The environmental factors that control the beginning of the life cycle—seed survival in the soil and germination—are often overlooked in ecological studies. Yet many important processes that affect plant population dynamics and community composition depend on seed survival and germination [1]. For example, seed survival in the soil can buffer plant populations against environmental variability [2,3]. Seed survival and germination cues are important mechanisms by which plants respond to environmental fluctuations. Germination cues often correspond with the environmental conditions most favorable for seedling growth, survival, and/or avoidance of competition [4]. Seed banking, the storage of ungerminated seed in the soil between growing seasons, can buffer population losses due to competition or poor growing conditions. In variable climates such as coastal California, seed survival and germination vary substantially between years, affecting both the dynamics of individual populations and the outcome of competition between species [5].

Understanding the influence of environmental conditions on seed survival and germination is critical not only for population and community ecology but also for restoration and conservation [6]. Removing unwanted species requires understanding the seed bank and controls over its longevity. Conversely, for native species that are the target of restoration efforts, sites and species with long-lived seeds may not need to be sown. Thus, understanding how environmental characteristics influence seed survival will clarify the positive and negative effects of seed banks on conservation and restoration.

Fungal pathogens reduce seed survival in many habitats [7–13]. Soil fungi respond to moisture, plant litter, and other soil characteristics [10,14]. Although soil fungi are ubiquitous, their impact on survival depends on the combination of fungus and seed species [11,15]. Soil fungi can kill seeds, persist as harmless commensals, or even protect seeds [13]. Of the fungal pathogens, some are generalists while others are closely associated with particular species of seeds or plants in the surrounding community [16]. Pathogens may force a race for survival between the seed and the fungus [17], generating selective pressure that influences the timing of germination. Because of these properties, some soil pathogens may be useful as microbial biocontrol agents for weed seed banks [18]. Because ungerminated seeds are not yet dependent on nutrients and water, much of the influence of the environment on seed survival may be mediated by fungal pathogens, for example if wet conditions promote fungal pathogens that kill seeds.

In light of the importance of seed dynamics for a range of ecological applications, it is important to understand variability in seed survival in the soil across species, space, and time, and how this variation depends on abiotic and biotic factors. In particular, controls over seed survival at a landscape level are not well understood. To address this gap, I conducted a one-year seed survival survey across eight annual plant species in Santa Barbara County, California. The region has a Mediterranean climate, and rainfall varies substantially in quantity and timing between years [5]. Seed survival was measured at 18 sites that differ in vegetation cover (grassland or coastal sage scrub), proximity to the coast (inland or coastal), and soil moisture. To test the extent to which soil fungi drove the results from this field survey, I conducted two experiments that manipulated environmental conditions and fungi. California grasslands are heavily invaded by annual grasses that deposit thick layers of thatch, which may modify soil microclimate, promote fungal growth, or inoculate seeds on the ground—all of which could influence fungal attack on seeds. Therefore, I first measured seasonal variation in responses to exotic annual grass cover and soil fungi by removing grass litter and adding fungicide across winter and summer seasons for six species. I tested for effects of thatch removal and interactions between thatch and fungicide. Second, to examine the relationship between soil moisture and fungal pathogens more directly, I applied watering and fungicide treatments in the laboratory for seven annual plant species.

I hypothesized that seed survival would be lower in wetter environments and in the winter (the rainy season) as opposed to the summer, due in part to increased fungal growth in moist soil. I also expected exotic grass thatch to influence seed survival by modifying soil microclimate and thereby promoting fungal growth. Because seed species differ in size, seed coat, shape, phylogenetic origin, and other factors, I expected the species to vary substantially both in seed survival and in their responses to environmental drivers.

Methods

Field survey of seed survival

To survey seed survival across geographic location and plant community type, I buried mesh bags containing seeds of eight species at 18 sites in Santa Barbara County, California. The climate is Mediterranean with a strong coastal influence. Winters are cool and wet and summers and are hot and dry. Temperatures are more moderate year-round in the coastal sites due to low-hanging clouds and fog occurring throughout the summer. The inland sites are hotter and drier in the summer. Annual precipitation is highly variable, usually coming in a series of heavy rainstorms from December through March.

The species used in the experiment are a mixture of native and exotic annual grasses and forbs that occur in southern California grasslands (Table 1). The seeds used in this survey were purchased from S & S Seed Company in Carpinteria, California, and all genotypes were from Santa Barbara County. Each seed bag consisted of a mesh strip with a separate compartment for each species, each containing 50 seeds. The compartments were 7.5 cm wide by 10 cm tall with a 2.5 cm strip separating each compartment from the next, to prevent contamination across species. Seed bags were buried on December 14, 2008, prior to the first rain of the season, and recovered on October 24, 2009, at the end of the growing season and prior to the onset of the next season's rain.

The sites in the study represent a mixture of coastal and inland geography and grassland and coastal sage scrub (CSS) vegetation. Coastal sites were located in the region of Santa Barbara County between the coastal mountain range and the ocean, while inland sites were located north (inland) of the coastal range. Grassland sites were dominated by exotic grasses and contained few shrubs. Coastal sage scrub sites (characterized by vegetation, not geographic location), contained mostly sagebrush, sage, and other shrubs, along with a few understory grasses and forbs. There were a total of five coastal CSS, four coastal grassland, five inland CSS, and four inland grassland sites (sample sizes were uneven due to loss of several of the original sites). All necessary permits were obtained for this and the following field experiment. Ten seed bags were buried at each site within a ten-meter area, for a total of 180 bags. At least 152 intact samples were recovered for each species.

I measured soil moisture on March 18, 2009, in the middle of the growing season at approximately peak soil moisture but not

Table 1. Description of species used in the study.

Species	Experiment used	Plant type	Provenance
Clarkia purpurea	field survey	Forb	Native
Lupinus bicolor	field survey	Forb	Native
Hemizonia fasciculate	field survey	Forb	Native
Amsinkia intermedia	field survey	Forb	Native
Phacelia distans	field survey	Forb	Native
Brassica nigra	field survey	Forb	Exotic
Erodium cicutarium	field survey	Forb	Exotic
Vulpia microstachys	field survey, field experiment, lab experiment	Grass	Native
Plantago erecta	field experiment, lab experiment	Forb	Native
Chaenactis glabriuscula	field experiment, lab experiment	Forb	Native
Salvia columbariae	field experiment, lab experiment	Forb	Native
Chorizanthe palmerii	field experiment, lab experiment	Forb	Native
Avena barbata	field experiment, lab experiment	Grass	Exotic
Hordeum vulgare	field experiment, lab experiment	Grass	Exotic
Bromus hordeaceus	lab experiment	Grass	Exotic

Native species are native to California grasslands; exotic species are widespread invaders from Europe. All species are annuals. For brevity, throughout the paper I refer to the species by their genus only.

within two weeks of a rain event. At each site I collected two soil cores and measured soil moisture by weight, comparing wet weight of soil to dry weight after 3 days in the drying oven at 60°C. I calculated percent soil moisture from each sample gravimetrically, and averaged the two values for each site.

Due to the large number of seeds in the study, full seed viability analysis was not feasible. Instead, I treated the number of seeds that germinated either in the field or in the lab as a proxy for survival. The correlation between germination and survival was 0.95 in a laboratory experiment in which I tested both (discussed below), suggesting that germination in the laboratory is a good proxy for seed survival. To assess germination I cut open the bags and placed them on wetted germination paper (Versa-Pak, Anchor Paper Company) in clear plastic germination boxes. The seeds were held at 15°C for five days, then placed on the lab bench at room temperature for three weeks [5]. Germinated seeds were counted and removed daily.

I analyzed the effects of vegetation type, geographic location, and soil moisture on seed germinability using binomial generalized linear mixed models (GLMMs). The fraction of seeds that germinated out of the total number of seeds in each bag was the response variable, so that each bag produced a single data point per species [19]. The basic model structure is:

$$\log\left(\frac{p}{1-p}\right) = \alpha + \beta_1 \times VegetationType + \beta_2 \times Location +$$
$$\beta_3 \times Moisture + \beta_4 \times Location \times Moisture + \quad (1)$$
$$\beta_5 \times Location \times Site + \beta_6 \times (Moisture)^2 +$$
$$RandomEffects$$

where the left-hand side is the logit-transformation of p, the proportion of seeds that germinated from each bag. α is the intercept, or the average response at the baseline condition, and the β's represent the effect of each environmental variable relative to the baseline value. For example, in the models presented in Table 2 the baseline condition is coastal location, CSS vegetation, and soil moisture equal to the average across sites. β_1 represents the effect of grasslands, β_2 the effect of inland location, β_3 the effect of deviations from mean moisture, β_4 the interactive effect of deviations in moisture at inland sites, β_5 the interactive effect of

inland grasslands, and β_6 the squared effect of deviations in mean soil moisture. Moisture squared was included to test for unimodal effects of soil moisture on seed survival. Deviations in soil moisture were used (i.e. the mean was subtracted), rather than absolute values, to improve model fitting. The categorical variables are coded as 0 for the baseline value and 1 for alternative values, so that for non-baseline values the model includes the intercept and the additional coefficient.

I first examined trends across species by analyzing all data together with species as a random effect in a mixed model. I then fit models to each species separately to pinpoint how species differed in their responses. To avoid pseudoreplication and control for the nested structure, I fit GLMMs with random effects of bag nested within site. This accounts for the fact that seed germination was measured at the bag level (i.e. ten bags per site) but predictor variables were measured at the site level. Including bag-level random effects also controls for overdispersion, i.e. higher variance than expected for binomial data [20]. I compared nested models containing subsets of the fixed effects using Akaike Information Criterion (AIC). In order to isolate differences among species, I fit GLMMs with the same fixed effects and nested random effects for each species separately. I selected the best model for each species using AIC. All models were implemented in R (R Core Development Team; r-project.org, version 2.11.1) using "glmer" in the "lme4" package.

Field experiment on the effects of fungi and thatch

To measure the effect of fungi and exotic grass thatch on seed survival, I added fungicide and removed thatch over buried seed bags in Sedgwick Reserve in Santa Ynez, California. The site, located inland in Santa Barbara County on Figueroa Mountain, was one of the sites included in the field seed survival survey. The fungicide and thatch experiment was conducted in an invaded serpentine grassland with mixed forbs. Before the annual grass invasion, the area was probably dominated by native perennial bunchgrasses and annual forbs. Because exotic annual grasses create much thicker layers of dead plant material (thatch) than the native plants, and thatch modifies the soil microclimate potentially to the benefit of soil pathogens, I expected thatch to mediate seed survival.

I buried 64 seed bags in the grassland. Seed bags were constructed as described in the field survey, except that each

Table 2. Estimates of the regression parameters from the best fit models for the field survey.

	intercept		inland		moisture		grassland		inland×moisture		(moisture)²	
All species	−1.95	***	0.18		−0.48		0.176	.	−14.3	***	-	
Clarkia	−0.27		0.071		−2.14		-		−42.83	***	-	
Lupinus	−3.74	***	0.74	***	-		-		-		-	
Hemizonia	−1.09	***	−0.09		0.943		0.592	***	−21.14	***	-	
Amsinkia	−2.02	***	−0.21		−1.99		0.299	*	−10.75	**	-	
Brassica	−1.92	***	-		−0.63		-		-		−207.5	***
Vulpia	−2.2	***	−0.11		4.347	*	0.282	*	−25.91	***	-	
Phacelia	−0.99	***	-		-		-		-		-	
Erodium	−2.95	***	0.843	***	−7.56	**	-		-		-	

Significance codes: 0 '***' 0.001 '**' 0.01 '*' 0.05 '.' 0.1 ' ' 1.
The intercept (α from equation 1) represents the baseline scenario of coastal location, CSS vegetation, and average soil moisture. The subsequent coefficients (β's from equation 1) represent change in the intercept relative to the baseline. The moisture coefficients represent responses to deviations from the mean value. The first row lists the estimates of fixed effects from the best fit GLMM for all species combined. The subsequent rows list the best GLMM for each species separately. The dash indicates a factor that was not included in the best-fit model.

contained 20 seeds of each species (listed in Table 1). Because this experiment focused on a single study site, I chose a set of species appropriate to this specific site. All seeds were collected from the site. The seed bags received factorial combinations of thatch absent/present and fungicide added/not added, for a total of 16 replicates per treatment combination. Following burial, thatch absent treatments were left with bare soil, while thatch present treatments had the displaced thatch replaced. Because this thatch was no longer rooted into the soil, these conditions may differ from naturally undisturbed thatch. I applied fungicide to seeds designated to that treatment by placing seeds in the seed bags after shaking them in vials with powdered Captan fungicide (Southern Ag).

I conducted the seed bag study once in the summer and fall, and again in the winter and spring. The summer seed bags were buried on July 9, 2008 and retrieved on October 28, 2008, before any rain had fallen. The winter seed bags were buried on December 12, 2008 and retrieved on July 28, 2009. Following retrieval, seeds were removed from seed bags and tested for germinability as described in the preceding section.

To examine general patterns of seed survival I fit binomial GLMMs with season, fungicide, and thatch as fixed effects and species and bag as random effects. The fraction of seeds in each bag that germinated was the response variable. The models follow the same basic structure as equation (1) for the field survival survey. I tested for season-by-fungicide and thatch-by-fungicide interactions. The bag-level random effect was included to correct for overdispersion. I compared nested models using AIC.

To compare responses between species, I removed the random effects and fitting generalized linear models (GLMs) for each species using the same fixed effects. It was not necessary to fit bag-level random effects for each species (as in the field survey) because the treatments were applied individually to each bag, and thus the experiment was not pseudoreplicated. To account for over-dispersion I used quasi-binomial GLMs, which fit an additional dispersion parameter, c_{hat} [21]. Because AIC is not defined for quasi-binomial models, I compared model fits by quasi-AIC (qAIC). qAIC = (residual deviance/c_{hat})+2k, where c_{hat} is the dispersion parameter on the most complex model and k is one plus

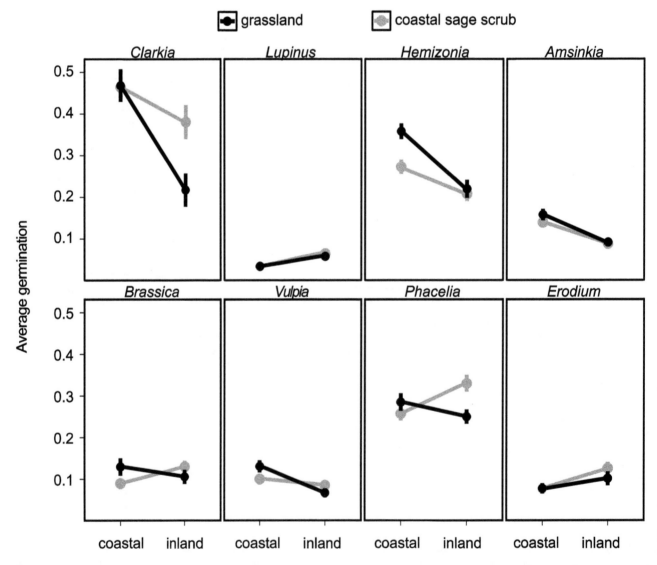

Figure 1. Average germination fraction by species, from the field survey. Different points represent each combination of location (coastal vs. inland) and vegetation type (coastal sage scrub, gray points vs. grassland, black points) for each species. Error bars are ±1 standard error.

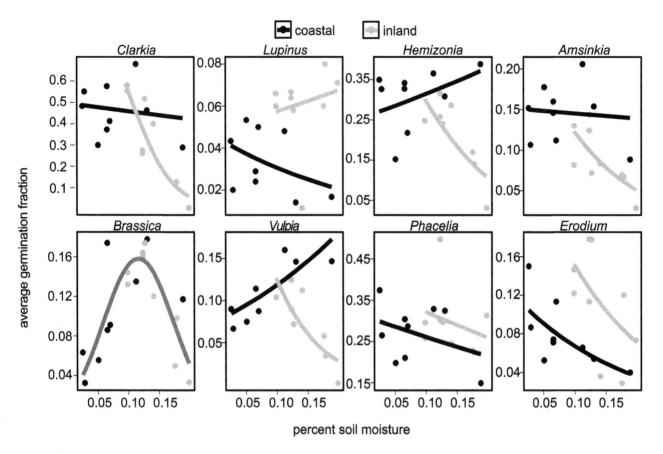

Figure 2. Average germination fraction by species as a function of soil moisture, from the field survey. Points are the average germination fraction pooled by site. Black points are coastal sites and gray points are inland sites. Lines are the fitted soil moisture by location models for each species except *Brassica*, which shows the (moisture)² model because it was a better fit.

the number of parameters estimated [21]. GLMs were implemented in R using "glm" in the "stats" package.

Laboratory experiment on the effects of water and fungicide

I further explored the relationship between soil moisture, fungi, and seed survival by manipulating fungicide and water in a laboratory experiment. I constructed mesocosms with seeds buried in field-collected soil, and applied fungicide to half of the samples following the methods described above. Each sample was either never watered, watered once at the beginning of the experiment, or watered twice per week throughout the experiment. Although percent soil moisture was not measured in this experiment, the samples that were never watered or watered only once were very dry at the end of the experiment, similar to conditions in the field during the summer drought period. Continuously watered samples remained moist throughout the study, with conditions similar to field soils during the rainy season in the winter and spring. Fungicide treatments were crossed with watering treatments for six treatment combinations. I used all the species from the field experiment plus one additional exotic grass, which was not available during the field experiment (Table 1).

The soil used in the experiment was collected from Sedgwick Reserve at the site of the field experiment. The soil was homogenized and sieved to 4 mm. Twenty seeds of a single species were sandwiched between two layers of mesh and buried within soil in a 6-cm diameter aluminum weighing boat. There

were six replicates of each treatment for each species. The replicates were placed on trays in a growth chamber set to cycle between 11°C, 22°C, and 31°C daily with a 12-hour light-dark cycle. The experiment ran from August to December 2008.

At the end of the experiment, the seeds were removed and germinated as described above. Seeds that did not germinate after six days were treated with 350 ml of a 400 ppm gibberellic acid solution overnight to stimulate germination [5]. To test for viability of ungerminated seed, those that did not germinate were cut in half and soaked in 1% tetrazolium solution for one hour, staining live embryo pink. The tetrazolium treatment effectively allowed me to distinguish between live and dead seeds based on stained or unstained embryo, respectively. The number of viable seeds is equal to the number of seeds that germinated naturally or with gibberellic acid plus the number that stained with tetrazolium. I used the fraction of viable seeds in each replicate at the end of the experiment as the response variable in the analyses.

To evaluate the influence of the treatments on seed survival, I followed the data analysis methods outlined under the field experiment, with fungicide, water treatment, and their interaction as fixed effects and species and experimental block (tray) as random effects. I also fit quasi-binomial GLMs to each species individually.

Results

Water was an important predictor of seed survival across all three parts of the study. In the field survey, inland vs. coastal

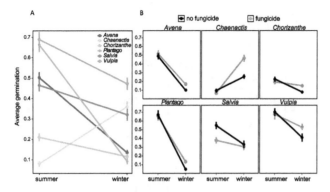

Figure 3. Average germination fraction from summer or winter with or without fungicide, from the field experiment. Differences in germination between summer and winter for all species (with both fungicide treatments combined), (a), plotted by fungicide treatment for each species separately, (b). In (b), black dots are treatments with fungicide, and gray dots are without fungicide. Error bars are ±1 standard error.

location interacted with soil moisture to affect seed survival (Figures 1, 2; Table 2). Overall, soil moisture reduced survival, but this decline was steeper inland than in coastal sites (Figure 2; Table 2, All Species model). Grasslands had slightly higher survival than coastal sage scrub sites (Figure 1; Table 2), suggesting that thatch may improve survival to some degree.

Soil moisture affected seed survival in the survey for all species individually except *Phacelia* and *Lupinus* (Table 2). Survival declined with moisture at inland sites for all the remaining species, but declined less steeply or even increased with moisture at coastal sites (Figure 2). *Brassica* survival peaked at intermediate moisture (Figure 2). These results suggest that seed responses to soil moisture are mediated by other unmeasured factors that vary geographically, such as soil type or other climate variables. Geographic location itself (independent of soil moisture) influenced survival for some species: *Lupinus* and *Erodium* had higher survival inland while *Clarkia*, *Hemizonia*, and *Amsinkia* had higher survival near the coast (Figure 1).

Some of the environmental drivers that may underlie the results of the field survey were highlighted in the field experiment. First, seed survival was lower in the winter than in the summer, a pattern which held for all species except *Chaenactis* (Figure 3; Table 3).

Because winter is the rainy season and summer had no precipitation, this again suggests that soil moisture influenced seed survival. Fungicide improved winter survival but did not affect survival in the summer (Figure 3; Table 3). This effect was significant for *Chaenactis*, *Chorizanthe*, and *Plantago*, and the pattern also held for *Avena* and *Vulpia*. Fungi therefore contribute to over-winter mortality of seeds, and may be partly responsible for the decline in seed survival with soil moisture in the field survey. Although grasslands had higher survival than coastal sage scrub in the field survey, removal of grass thatch did not significantly affect survival in the field experiment (Table 3). This may be because thatch was no longer rooted into the soil, because thatch had already altered the soil biota and microclimate prior to removal, or because thatch does not strongly affect seed survival.

The laboratory experiment further confirmed the negative effect of water on seed survival and suggested a possible role of fungi in this decline. Seed survival was highest in samples that were never watered, (Figure 4; Table 4). As expected, survival declined as watering increased from never to once to continuously. However, *Bromus* and *Salvia* deviated from this pattern, having the lowest survival when watered only once (Figure 4). Fungicide slightly improved survival in samples that were watered once, but not in samples that were watered continuously (Table 4, All Species model). However, this improved survival with fungicide in samples watered once was significant only for *Chorizanthe* in the single-species models (Table 4), suggesting that fungal seed mortality is variable across species. Fungicide may either have killed beneficial mutualist fungi or have a toxic effect on seeds, which could explain the slight decrease in survival for *Chorizanthe* and *Salvia* with fungicide in the continuous watering treatments. Because fungicide treatments were applied only once at the beginning of the study, fungicide may have washed away in the continuous watering treatment.

Discussion

In this study, I investigated patterns and drivers of seed survival in California annual plant communities with a field survey, a field experiment, and a laboratory experiment and fifteen species. The most notable result was the variation in seed survival across species, location, and season. Soil moisture was related to seed survival in all three experiments. In the year-long survey of seed survival across coastal and inland grasslands and coastal sage scrub habitats, the interaction between soil moisture and geographic location explained significant variation in survival for most species

Table 3. Estimates of regression parameters from the best fit models for the field experiment.

	intercept		thatch	winter		no fungicide		winter×no fungicide	
All species	−0.5		-	−0.69	***	0.205		−0.87	***
Avena	0.011		-	−1.87	***	-		-	
Chaenactis	−2.66	***	-	2.52	***	0.399		−1.32	**
Chorizanthe	−1.42	***	-	−0.31		0.19		−0.91	*
Plantago	0.644	***	-	−2.5	***	0.07		−1.16	*
Salvia	−0.34	*	-	−0.62	***	0.396	*	-	
Vulpia	0.794	***	-	−0.91	***	-		-	

Significance codes: 0 '***' 0.001 '**' 0.01 '*' 0.05 '.' 0.1 ' ' 1.
The intercept represents the mean for the baseline scenario of thatch removed, summer, and fungicide added. The subsequent coefficients represent the change relative to the baseline. The first row lists the parameter estimates for the GLMM with species- and bag- level random effects. The subsequent rows list the best fit quasibinomial GLM for each species separately. The dash indicates a factor that was not included in the best-fit model.

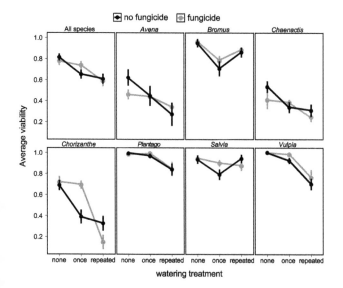

Figure 4. Average seed viability as a function of watering and fungicide treatments, from the laboratory experiment. The top left panel is all species combined, and the rest are individual species. Moisture treatments are dry, pulse, and wet, and fungicide treatments are with (gray points) and without (black points) fungicide. Error bars are ±1 standard error.

(Table 2; Figure 2). Some of the effect of soil moisture on seed survival may be mediated by soil fungi. In the field experiment at one of the survey sites, fungicide application modestly improved survival for most species in the wet winter months but not in the dry summer (Table 3; Figure 3). Similarly, seed survival in the laboratory declined in treatments that received water either once or throughout the experiment (Table 4; Figure 4). Fungicide improved survival in watered treatments in some cases, but the effect was not consistent across species (Table 4; Figure 4). Taken together, these experiments provide evidence for soil moisture-mediated seed survival, which may be partly due to soil-borne fungi. These results agree with the findings of previous studies in

which seed survival varied by species, soil moisture, and habitat type, with part of these effects due to fungal pathogens [10,14].

Although species varied considerably in seed survival and responses to environmental drivers, soil moisture appeared to play a role in all three studies. That species differ in seed survival and in responses to moisture and fungi is not surprising given the range of seed characteristics (e.g. size, shape, seed coat) as well as the range of specific and non-specific soil pathogens that naturally vary across the landscape. Although the effect of fungicide was relatively weak and variable across species, the estimated effect sizes are conservative given that a single dose of non-specific fungicide was applied only once in both of the experiments. In particular, in the winter trial in the field experiment and the wet treatments in the lab experiment some of the fungicide probably washed away. In addition, the fungicide may have killed beneficial fungi such as mychorrizae, so the fungicide treatments demonstrate the net effect of fungi in the system. This study also does not account for other water-dependent pathogens in the soil such as bacteria and *Pythium* species, which could affect seed survival.

The field survey did not support a strong role of vegetation type in mediating seed survival, although there was a significant effect of vegetation type for three species. This suggests that patterns of season- and fungal-mediated seed survival are usually a function of the seed species, not the surrounding vegetation. Additionally, vegetation types (e.g. grasslands versus coastal sage scrub) do not appear to affect seed survival consistently, for example by promoting fungal growth on all seeds. Instead, species responded idiosyncratically to vegetation type, geographic location, and, to some extent, soil moisture.

Although consistent with the results of the field and laboratory experiments, the field survey provides only a snapshot of the relationship between soil moisture and seed survival taken at one point during the rainy season (not within two weeks of a rain event). Because soil moisture was measured near its peak, this may reflect the time of year when soil moisture has the strongest effect on survival. Unmeasured aspects of the soil moisture regime such as differences in summer soil moisture may also affect seed survival. Some of the residual variation in survival across locations may be due to unmeasured variation in soil moisture.

In the field survey, I expected coastal locations to be wetter and therefore have lower seed survival, but soil moisture was actually

Table 4. Estimates of regression parameters from the best fit models for the laboratory experiment.

	intercept		watered once		continuous watering		no fungicide	watered once×no fungicide		continuous watering×no fungicide
All species combined	2.05	***	−0.44	.	−1.49	***	0.27	−0.90	**	−0.20
Avena	0.15		−0.37		−0.97	**	-	-		-
Bromus	2.98	***	−1.93	***	−1.08	*	-	-		-
Chaenactis	−0.12		−0.44	.	−0.84	**	-	-		-
Chorizanthe	0.92	**	−0.13		−2.59	***	−0.09	−1.19	*	1.00
Plantago	4.31	***	−0.48		−2.69	**	-	-		-
Salvia	2.63	***	−0.93	.	−0.35		-	-		-
Vulpia	21.75	***	−18.73	***	−20.76	***	-	-		-

Significance codes: 0 '***' 0.001 '**' 0.01 '*' 0.05 '.' 0.1 ' ' 1.
The intercept represents the mean for the baseline scenario of never watered with fungicide added. The subsequent parameters represent the change relative to the baseline. The first row lists the parameter estimates for the GLMM with species- and observation- level random effects. The subsequent rows list the best fit quasi-binomial GLM for each species separately. The dash indicates a factor that was not included in the best-fit model. Species listed in bold followed the survival pattern never watered > watered once > continuous watering.

lower in coastal locations, despite foggier and more humid conditions. This was likely due to the differences in soil type. Coastal soils were looser and more sandy, whereas inland soils contained more clay and held more water. These soil type differences may partly explain the steeper decline in seed survival with soil moisture at inland locations.

This study showed considerable variation in seed survival across species, with some evidence for a soil moisture- mediated effect of fungal pathogens. A more comprehensive pathogen removal study that also manipulated soil moisture in the field would greatly clarify the role of moisture-mediated fungal pathogens in seed survival. Fungicide could be applied repeatedly throughout the year while controlling for the effects on soil moisture using water-only controls. Sterilization of field soils (e.g. by autoclaving) would elucidate the importance of the soil biome as a whole, including fungi, bacteria, nematodes, and other organisms. Finally, direct manipulation of soil moisture in the field across geographic locations (ideally crossed with a fungicide manipulation) would clarify the role of soil moisture and its impact on pathogenic fungi. However, given the variability among species, site, and season in this study, researchers interested in controlling seed banks for particular locations or species will likely need to study species-specific drivers of seed survival at a local scale.

Seed survival is important for plant species in variable environments, such as the Mediterranean climate in southern California, where rainfall can vary six-fold between years [22].

Banking ungerminated seeds in the soil is a bet hedging mechanism that aids population recovery following years of poor seed production due to competition or environmental factors [23]. Seed banking is also important for some mechanisms of species diversity maintenance that rely on species partitioning the environment through space and time [24,25]. Understanding the processes that control seed survival as well as the variability in seed survival across species, space, and time is important for understanding these population and community level effects of seed banking. Nonetheless, to date, ecologists are only beginning to understand the types of habitats in which species can efficiently bank seeds. This study supports the notion that ecologists might expect wetter sites to have generally lower seed survival, and that fungal pathogens are likely contributors to this effect. On the other hand, this study suggests that arid sites and drought years may have longer-lived seed banks.

Acknowledgments

Marie Madaras, Carolyn Finney, Eli Berger, and Natalie Nounou helped with data collection.

Author Contributions

Conceived and designed the experiments: EAM. Performed the experiments: EAM. Analyzed the data: EAM. Contributed reagents/materials/analysis tools: EAM. Wrote the paper: EAM.

References

1. Fenner M (2000) Seeds: the ecology of regeneration in plant communities. CABI. 432 p.
2. Pake C, Venable D (1996) Seed banks in desert annuals: Implications for persistence and coexistence in variable environments. Ecology 77: 1427–1435.
3. Venable DL (2007) Bet hedging in a guild of desert annuals. Ecology 88: 1086–1090.
4. Baskin C, Baskin J (2001) Seeds: ecology, biogeography, and evolution of dormancy and germination. Academic Press San Diego. p.
5. Levine JM, HilleRisLambers J (2009) The importance of niches for the maintenance of species diversity. Nature 461: 254–257.
6. Khurana E, Singh J s (2001) Ecology of seed and seedling growth for conservation and restoration of tropical dry forest: a review. Environmental Conservation 28: 39–52. doi:10.1017/S0376892901000042.
7. Kirkpatrick BL, Bazzaz FA (1979) Influence of certain fungi on seed germination and seedling survival of four colonizing annuals. Journal of Applied Ecology 16: 515–527.
8. Crist T, Friese C (1993) The impact of fungi on soil seeds: implications for plants and granivores in a semiarid shrub-steppe. Ecology: 2231–2239.
9. Gilbert GS (2002) Evolutionary ecology of plant diseases in natural ecosystems. Annual Reviews in Phytopathology 40: 13–43.
10. Schafer M, Kotanen PM (2003) The influence of soil moisture on losses of buried seeds to fungi. Acta Oecologica 24: 255–263.
11. Schafer M, Kotanen P (2004) Impacts of naturally-occurring soil fungi on seeds of meadow plants. Plant Ecology 175: 19–35.
12. O'Hanlon-Manners D, Kotanen P (2006) Losses of seeds of temperate trees to soil fungi: effects of habitat and host ecology. Plant Ecology 187: 49–58.
13. Terborgh J (2012) Enemies Maintain Hyperdiverse Tropical Forests. The American Naturalist 179: 303–314. doi:10.1086/664183.

14. Blaney CS, Kotanen PM (2001) Effects of fungal pathogens on seeds of native and exotic plants: a test using congeneric pairs. Journal of Applied Ecology 38: 1104–1113.
15. Orrock J, Damschen E (2005) Fungi-mediated mortality of seeds of two old-field plant species. The Journal of the Torrey Botanical Society 132: 613–617.
16. Gallery RE, Dalling JW, Arnold AE (2007) Diversity, host affinity, and distribution of seed-infecting fungi: a case study with Cecropia. Ecology 88: 582–588.
17. Beckstead J, Meyer SE, Molder CJ, Smith C (2007) A race for survival: can Bromus tectorum seeds escape Pyrenophora semeniperda-caused mortality by germinating quickly? Annals of Botany 99: 907–914.
18. Kremer RJ (1993) Management of weed seed banks with microorganisms. Ecological Applications 3: 42–52. doi:10.2307/1941791.
19. Wang X, Comita LS, Hao Z, Davies SJ, Ye J, et al. (2012) Local-scale drivers of tree survival in a temperate forest. PLoS ONE 7: e29469. doi:10.1371/journal.pone.0029469.
20. Elston DA, Moss R, Boulinier T, Arrowsmith C, Lambin X (2001) Analysis of aggregation, a worked example: Numbers of ticks on red grouse chicks. Parasitology 122: 563–569. doi:10.1017/S0031182001007740.
21. Burnham KP, Anderson DR (2002) Model selection and multimodel inference: a practical information-theoretic approach. Springer. 512 p.
22. Levine JM, McEachern AK, Cowan C (2008) Rainfall effects on rare annual plants. Journal of Ecology 96: 795–806. doi:10.1111/j.1365-2745.2008.01375.x.
23. Brown JS, Venable DL (1986) Evolutionary ecology of seed-bank annuals in temporally varying environments. The American Naturalist 127: 31–47.
24. Chesson P (2000) Mechanisms of maintenance of species diversity. Annual Reviews in Ecology and Systematics 31: 343–366.
25. Angert AL, Huxman TE, Chesson P, Venable DL (2009) Functional tradeoffs determine species coexistence via the storage effect. Proceedings of the National Academy of Sciences 106: 11641.

3

Unlikely Remedy: Fungicide Clears Infection from Pathogenic Fungus in Larval Southern Leopard Frogs (*Lithobates sphenocephalus*)

Shane M. Hanlon[1]*, **Jacob L. Kerby**[2], **Matthew J. Parris**[1]

1 Department of Biological Sciences, University of Memphis, Memphis, Tennessee, United States of America, **2** Department of Biology, University of South Dakota, Vermillion, South Dakota, United States of America

Abstract

Amphibians are often exposed to a wide variety of perturbations. Two of these, pesticides and pathogens, are linked to declines in both amphibian health and population viability. Many studies have examined the separate effects of such perturbations; however, few have examined the effects of simultaneous exposure of both to amphibians. In this study, we exposed larval southern leopard frog tadpoles (*Lithobates sphenocephalus*) to the chytrid fungus *Batrachochytrium dendrobatidis* and the fungicide thiophanate-methyl (TM) at 0.6 mg/L under laboratory conditions. The experiment was continued until all larvae completed metamorphosis or died. Overall, TM facilitated increases in tadpole mass and length. Additionally, individuals exposed to both TM and *Bd* were heavier and larger, compared to all other treatments. TM also cleared *Bd* in infected larvae. We conclude that TM affects larval anurans to facilitate growth and development while clearing *Bd* infection. Our findings highlight the need for more research into multiple perturbations, specifically pesticides and disease, to further promote amphibian heath.

Editor: Matthew Charles Fisher, Imperial College Faculty of Medicine, United Kingdom

Funding: Bd analyses were conducted on equipment supplied by NSF grant MRI 0923419. The funders had no role in study design, data collection and analysis, decision to publish, or preparation of the manuscript.

Competing Interests: The authors have declared that no competing interests exist.

* E-mail: shanlon1@memphis.edu

Introduction

Anthropogenic perturbations, such as pesticides, often act as stressors for non-target organisms. Pesticides are common contaminants that enter aquatic systems through runoff, over-spray, or pesticide drift [1,2]. Thus, ecologists are charged with examining the manner in which contaminants affect non-target aquatic organisms such as amphibians.

Under realistic conditions, agrochemicals such as pesticides are applied multiple times throughout a growing season and non-target organisms commonly experience reoccurring exposure [3,4]. Depending upon the half-life of the specific chemical, the reapplication of pesticides may not allow for its natural breakdown into less harmful products before another exposure occurs. When pesticide dosages are lethal, the number of exposures is inconsequential. However, at sublethal levels, repeatedly exposed individuals may not have an opportunity to recover from an initial dose. Accordingly, ecologists have examined the differences between "pulse" (a single, initial dose) and "press" (multiple exposures over time) treatments for many years [5–7]. Many pulse experiments have illustrated the ability of an individual, community, or both to rebound from a single exposure [4,8]. One might conclude that using press experiments in which individuals are subjected to the reoccurring pressure of pesticide exposure to be the test that most closely reflects patterns of exposure experienced by most affected organisms [9].

One pesticide that is applied worldwide is the fungicide thiophanate-methyl (TM; http://water.usgs.gov/nawqa/pnsp/ usage/maps/show_map.php?year = 02&map = m5019). TM is a broad-spectrum fungicide that targets mycorrhizal fungi. TM has been marketed as the replacement to benomyl [10]; the most widely used fungicide in the United States until its discontinuation in 2001 [10].

TM is used heavily in the Mississippi River Basin. Bishop et al. [11] assessed fungicide usage in Tennessee and northern Mississippi and concluded that TM was used widely throughout both states. It is reasonable to assume that TM is entering aquatic environments as has been found for other pesticides. However, no research has examined the effects of TM on anurans. This may be due in part to the relatively short half-life of TM [12] and potential breakdown into inert products. However, a byproduct of TM breakdown, carbendazim, has been linked to adverse effects on amphibian larval growth, development, and survival [13]. Even so, with repeat pesticide exposure, runoff, and spraydrift [9,14], aquatic organisms with prolonged larval periods are likely exposed to TM multiple times through ontogeny, thus decreasing the probability of breakdown into inert components. Multiple studies have examined the effects of an LD$_{50}$ (the lethal dosage required to kill half a population) injected dose of TM [15–17]. However, this application method does not serve as a valid proxy for environmental exposure. Accordingly, tests where TM is applied to an organism's habitat (LC$_{50}$), as opposed to directly into the organism itself, provide a more relevant indication of adverse effects. While natural water sample analysis has not been conducted to determine environmental concentrations of TM, pilot studies reveal that possible LC_{50} values range from 7.5 to 10 mg/L (Hanlon, unpublished data).

Along with the aforementioned chemical factors, organisms are also exposed to additional pressures, such as pathogens. One pathogen that is causing rapid declines in amphibian populations is the emerging infectious disease chytridiomycosis, caused by the pathogenic fungus *Batrachochytrium dendrobatidis* (*Bd*). *Bd* infects keratinized tissues, such as anuran larval mouthparts, reducing their foraging capabilities [18–20]. While *Bd* does not generally cause mortality in larvae (as it does in adults), the fungus often impairs growth and developmental rates [21–24]. Venesky et al. [20] found that *Bd* altered larval mouthparts, resulting in *Hyla chrysoscelis* (Cope's treefrog) larvae foraging less efficiently than uninfected individuals. Additionally, Hanlon et al. (unpublished data) found that infected *Hyla versicolor* (gray treefrog) larvae spent significantly more time foraging than uninfected individuals. However, the authors observed no corresponding increases in growth and development with increased foraging. Together, these results indicate that while infected larvae may spend more time foraging than uninfected individuals, they are unable to fully compensate for the deficits in efficiency.

Although much is known about the independent effects of pesticides and *Bd* on amphibians, a limited number of studies have examined the possible interactive effects of these two perturbations. Currently, two research approaches are being developed to test such interactions: 1) testing the effects of contamination on disease independent of hosts, and 2) testing the effects of contamination on disease in amphibian hosts. Studies that examine possible interactions of contaminants and pathogens outside of a host mimic situations prior to or following host infection. *Bd* can persist within the environment (independent of an amphibian host) for up to seven weeks [25], allowing for the possibility of an interaction outside of hosts. Additionally, Hanlon and Parris [26] showed that the pesticides carbaryl, glyphosate, and TM killed *Bd* in culture independent of potential hosts. On the other hand, research such as our current study that examine the interactive effects of *Bd* and a contaminant upon a host mimic a post-infection scenario. Such situations indicate the possibility of interactive effects between *Bd* and pesticides, causing reduced foraging efficiency and likely life history consequences.

While studies have found negative interactive effects of pesticides and pathogens within hosts [27,28], studies testing for interactions between *Bd* and anti-fungal agents have yielded significantly different results. Many anti-fungal treatments kill *Bd* in culture [29–31]. Also, anti-fungals kill *Bd* in hosts [29,31,32]. However, the broader impacts of such treatments on non-target organisms are largely unknown and thus prevent addition of such chemicals to natural habitats. While the addition of these chemicals to natural environments is not possible, fungicidal pesticides are applied in great quantities across the United States. Accordingly, we examined the interactive effects of the fungicide TM and *Bd* on larval anurans under laboratory conditions. We predicted that *Bd* exposure would facilitate reductions in growth and TM alone at sublethal levels would have no effect on growth. Also, because TM has been shown to kill *Bd* in culture independent of hosts, we predicted that TM would clear *Bd* infection in individuals exposed to both *Bd* and TM.

Methods

Animal Collection and Husbandry

L. sphenocephalus eggs were collected from ponds within Shelby Farms Park in Shelby County, TN (35° 9′ 13″ N/89° 51′ 7″ W). On March 29, 2010, we collected 9 *L. sphenocephalus* clutches. Eggs

were transported to the laboratory at the University of Memphis, Memphis, Tennessee. After hatching, tadpoles were maintained in 8 L aquaria in 4 L of water at a density of 2 clutches/aquaria. All tanks were the same size and dimensions and filled with the same amount of water (one tank contained a single clutch and was filled with half the amount of water for control purposes). Upon reaching the free-swimming stage (Gosner 25 [33]), tadpoles were combined from the different clutches and redistributed into tanks where density, tank size, water volume, and amount of and type of food (Tetramin® fish food) was controlled for and standardized. Such steps were used to distribute potential genetic effects of the traits measured. Test subjects were then randomly selected from this stock and placed into 1.5 L plastic containers filled with 1 L of aged tap water. Throughout the experiments, tadpoles were maintained on a 12 h light: 12 h dark photoperiod at 19°C and fed every 3 days.

Batrachochytrium Dendrobatidis Inoculation

The *Bd* isolate used in our experiment was locally isolated from an infected adult *L. sphenocephalus* captured from the University of Memphis Biological Field Station at Meeman-Shelby State Park, Shelby County TN ([35°23′22.66″N 90°02′15.75″W]) in May 2010. The isolate was grown in the laboratory in tryptone broth (1.6% tryptone, 0.2% gelatin hydrolysate, and 0.4% lactose [TGhL]) according to standard protocol [34]. Stock cultures were transferred monthly and all *Bd* inoculates were taken from these cultures. This strain has resulted in successful infections in both laboratory and field experiments.

Bd zoospores were harvested by adding 10.0 mL of sterile water to cultures and collecting the zoospores that emerged from the zoosporangia after 45 minutes. At Gosner 25, tadpoles were split into two groups: *Bd*-exposed and non-exposed (control) groups. The *Bd*-exposed group (N = 20) was inoculated with *Bd* through exposure to water baths containing infectious concentrations of fungal zoospores. Tadpoles were placed in individual 50 mL water baths (3 individuals per 50 mL) and an infectious concentration of zoospores (320,000 zoospores/mL) was added to each bath for 48 hours. The non-exposed group followed the same protocol but the water was added to plates with TGhL alone, thereby the additional group (N = 20) was exposed to water baths with no *Bd* zoospores. This design simulates transmission by water, a possible mode of *Bd* transmission in natural environments [35], and has resulted in successful infections in previous studies [19,20]. After 48 hrs of exposure, *Bd*+ and *Bd*− subjects were removed from water baths and all subjects from each treatment group were placed into separate 8 L containers. After six days, 10 tadpoles for *Bd* and 10 for TM*Bd* treatments were haphazardly selected from the single *Bd*+ pool for the experiment. A similar process was carried out for subjects in control and TM treatments: were selected from the single *Bd*− pool.

Pesticide Application

The experimental design employed 4 treatments with 10 replicates per treatment. Treatments were as follows: *Bd*− control (water), *Bd*+ control, TM+*Bd*+, and TM+*Bd*−. Individuals in TM+*Bd*+ and TM+*Bd*− treatments were exposed to the same dosage of TM at a concentration of 0.6 mg/L. This concentration was chosen because it is lower than LC_{50} levels and represents a realistic estimate in situations with direct overspray [10]. Pesticide was mixed with aged tap water in bulk to achieve the respective concentration. At this point, test subjects were exposed to their respective treatment.

Water was changed every 3 days, at which time the pesticide was reapplied and individuals were fed. Pesticide concentrations were confirmed via high-pressure liquid chromatography through Pacific Agricultural Labs in Portland OR.

Measurement of Life History Traits

We were interested in the effects of Bd and TM on life history traits as larvae and metamorphs. Thus, we measured life history traits of larvae prior to metamorphosis. On day 60, all larvae were anesthetized and measures of mass and snout-vent length (SVL) were recorded. Larval measurements were taken once during ontogeny to reduce the possibility of stress-induced behavioral or morphological alterations or death from repeat anesthetization and handling. When larvae began to metamorphose (day 78), containers were monitored daily for metamorphic animals. Metamorphosis was defined by the emergence of one forelimb [33]. Upon tail resorption, animals were weighed and SVL measurements were recorded.

Batrachochytrium Dendrobatidis qPCR Confirmation

Infection status ($Bd+/-$) of all experimental animals was confirmed using real-time quantitative polymerase chain reaction (qPCR) following the method used by Boyle et al. [36]. DNA was extracted from cotton swabs of tadpole mouthparts taken immediately after life history measurements were taken (day 60). Swabbing tadpoles requires the removal of the tadpole from the aquatic environment by netting, holding the tadpole in hand, and twisting a swab around the tadpole's mouthparts; thus, swabbing was conducted at a single time point to reduce handling time and stress, potentially resulting in tadpole mortality. A different pair of nitrile gloves was used between each subject to prevent contamination. Moreover, the same exposure protocols have resulted in successful infections in previous experiments (e.g. [19,20]).

All samples were stored in 100% EtOH until qPCR analyses. Standards were obtained from CSIRO labs in Australia and were the same as those used in Boyle et al. [36]. The standards served as the positive controls and each plate contained a negative control (which tested negative on all plates). For calculations of prevalence, swabs were categorized as Bd-positive when zoospore equivalents were ≥ 1 (as used by [37,38]).

Statistical Analysis

Multivariate analysis of variance (MANOVA) was conducted to consider whether the Bd and TM had a significant effect on each dependent variable when the two treatments were considered simultaneously. We then used two-way analyses of variance (ANOVA) to test for an effect of Bd and TM on each response (larval mass, larval SVL, metamorphic mass, and metamorphic SVL).

Results

No tadpoles from our control, TM+$Bd-$, or TM+$Bd+$ treatments tested positive for Bd infection. All tadpoles from our $Bd+$ treatment tested positive for infection. From qPCR in the $Bd+$ group, the mean zoospore equivalents were 168.44 (± 18.44) with a range of 42.24 to 397.77.

MANOVA indicated that there was a significant effect of Bd ($F_{2,34} = 6.96$, P = 0.003), TM ($F_{2,34} = 9.53$, P<0.001), and TM×Bd ($F_{2,34} = 3.91$, P = 0.030) on larval mass and SVL when considered simultaneously. There was a significant effect of TM at day 60 on larval mass ($F_{3,35} = 18.63$, P = <0.001) and SVL ($F_{3,35} = 16.62$, P = <0.001). Individuals exposed to TM alone were heavier and larger compared to Bd and control treatments (Fig. 1). Bd also had a significant effect on larval mass ($F_{3,35} = 10.69$, P = 0.002) and SVL ($F_{3,35} = 7.71$, P = 0.009) with $Bd+$ individuals being larger and

longer than non-Bd subjects (Fig. 1). However, the presence of TM likely influenced these results. Additionally, there was a significant TM by Bd interaction on larval mass ($F_{3,35} = 6.37$, P = 0.016) and SVL ($F_{3,35} = 4.45$, P = 0.042. The TM×Bd interaction caused individuals to be heavier and larger compared to all other treatments (Fig. 1).

MANOVA indicated that there was a significant effect of TM ($F_{2,25} = 8.81$, P = 0.001), but not of Bd ($F_{2,25} = 0.88$, P = 0.428) or TM×Bd ($F_{2,25} = 0.06$, P = 0.946) on metamorphic mass and SVL when considered simultaneously. TM had a significant effect on all metamorphic features as well. TM affected mass at metamorphosis ($F_{3,25} = 14.18$, P = <0.009) and SVL at metamorphosis ($F_{3,25} = 10.02$, P = <0.001) (Fig. 2). Independent of Bd, individuals subjected to TM were heavier and larger (Fig. 2) compared to all other treatments. There was not a significant effect of Bd on metamorphic mass ($F_{3,25} = 2.47$, P = 0.128) or SVL ($F_{3,25} = 2.33$, 0.1388). The TM×Bd interaction had similar effects as Bd alone. The TM×Bd interaction did not affect metamorphic mass ($F_{3,25} = 0.11$, P = 0.7415) and metamorphic SVL ($F_{3,25} = 0.10$, P = 0.759).

Discussion

Numerous anthropogenic factors have been implicated in amphibian declines [39–41]. For example, amphibian trade and land use changes have facilitated the spread of Bd [42]. The emergence of Bd in areas of pesticide exposure in the forms of runoff, spraydrift, and direct overspray (personal observation) has complicated our understanding of the role of this disease in the declines. It is likely that these two factors interact to impact amphibians by altering behavior, morphology, and physiology. Although studies that have examined each perturbation have shown that the separate effects of pesticides and Bd are usually deleterious, the results of our study provide evidence to the contrary.

Overall, TM was advantageous to all measured traits. TM facilitated larval growth, as individuals were heavier and larger. To the best of our knowledge, ours is the first study that found a pesticide to promote such measures without obvious tradeoffs. Although other studies have found pesticides to benefit specific life history traits (e.g. growth or development), such benefits have invariably been accompanied by a tradeoff in which other trait(s) were negatively impacted. For example, Boone et al. [43] found that Woodhouse's toads (*Bufo woodhousii*) that were exposed to carbaryl experienced increased growth at the cost of a longer developmental period. Semlitch et al. [44] found similar patterns in Gray treefrogs (*Hyla versicolor*), in which larvae in low-density treatments exposed to the insecticide carbaryl completed metamorphosis sooner, but were smaller than those in high-density treatments. There likely are tradeoffs that exist from TM exposure, but we were not able to identify any of these in the factors typical with previous work.

It should be noted that while we controlled for developmental stage at the beginning of the experiment, starting mass and size were not measured. However, the use of development stage as a starting measure in tadpole experiments that assess growth and development through development is an experimental standard [45–52]. Additionally, in the presence of constant conditions, tadpole growth and development are closely correlated (for review see [53]). While studies have shown that tradeoffs occur between tadpole growth and development, individuals housed in identical conditions experience such tradeoffs together [54,55]. Because the subjects in our study were housed in identical conditions, we are

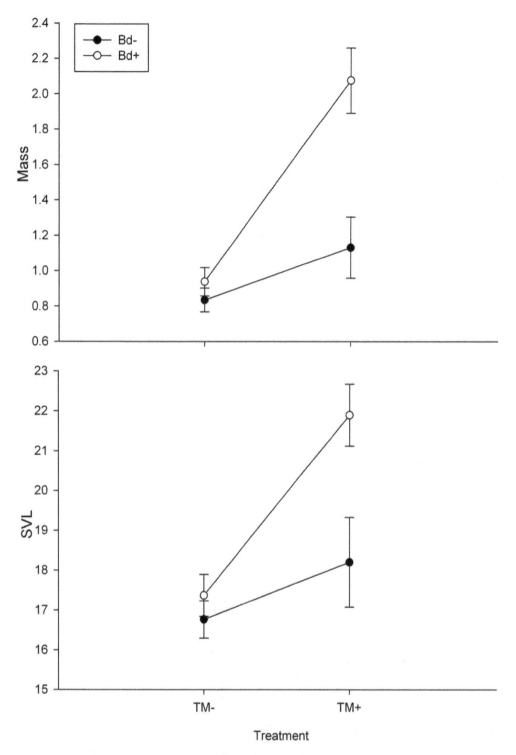

Figure 1. The effects of *Bd* and TM on larval mass and SVL at day 60. Asterisks (*) above plots indicate significant differences from TM− treatments (P<0.05). *indicates significant difference from TM− treatments while **indicates significant difference from TM− and *Bd*−TM+ treatments.

confident that initial mass and size were constant at the start by initiating the experiment with subjects at Gosner 25 [33].

Although we found no effect of *Bd* on any measured trait, qPCR revealed that exposure techniques were successful. Interestingly, larvae in *Bd* × TM treatments were heavier and larger than those in all other treatments. Additionally, larvae in this treatment that

were exposed to *Bd* at the onset of the experiment tested negative for *Bd* infection via qPCR immediately prior to metamorphosis. While we assert that TM was responsible for such observations, the possibility of alterative explanations cannot be ruled out. Because we did not swab tadpoles within one week of initial exposure, it is possible that those in TM*Bd* treatments were not

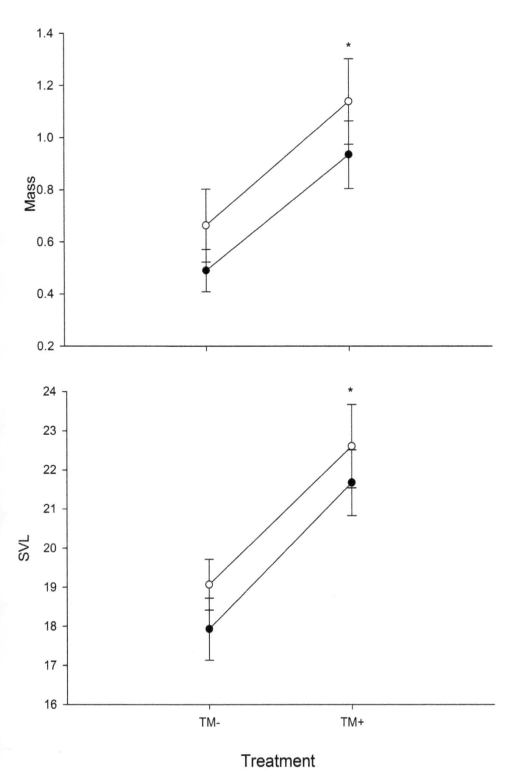

Figure 2. The effects of *Bd* and TM on metamorphlic mass and SVL. Asterisks (*) above plots indicate significant differences from TM− treatments (P<0.05).

initially infected. However, this is extremely unlikely. Subjects in solely *Bd+* treatments all tested positive for *Bd*. It is critical to reiterate that animals in both *Bd+* and TM*Bd+* treatments were derived from the same infection protocols. As previously stated, tadpoles in *Bd* and TM*Bd* treatments were selected from a single pool of *Bd* exposed individuals. Because of the 100% infection success in *Bd* groups and the 0% infection in TM*Bd* groups, as well as the expose to TM after *Bd* exposure, we conclude that our infection protocols were successful in TM*Bd* subjects and TM cleared *Bd* infection in hosts.

Perhaps the most surprising finding of our study was the facilitative qualities of TM. Whereas one might propose that this fungicide could be beneficial when combined with *Bd* (i.e., by controlling the progression of the infection), it is more difficult to understand the beneficial effects of TM when administered alone. To date, no other pesticide has been found to promote life history traits such as mass and size without any obvious costs. Thus, we hypothesize multiple pathways through which such facilitations may have occurred.

Contaminant-induced alterations in amphibian growth have been well documented (for review see [56]). Specifically, aquatic contaminants have been shown to alter amphibian physiology and subsequent growth [57–59]. The physiology of larval growth and development, ending with metamorphosis, is regulated primarily by the hypothalamus-pituitary-thyroid (HPT) axis. Amphibian growth and development is regulated through the production of thyroid hormones (TH [60,61]). Specifically, the thyroid hormone thyroxine (T_4) is converted to triiodothyronine (T_3) that acts on target tissues to promote growth and development [62]. While some argue that not all larval growth and development is controlled by TH and the HPT axis (e.g. growth hormone, prolactin [63]), other experiments manipulating TH in larvae have altered the timing of metamorphosis [58].

In our experiment, all larvae exposed to TM grew larger and weighed more than unexposed individuals. Given that the both growth in mass and size increased, we conclude that TM may have a direct effect on larval TH production, thereby increasing growth. Such effects have been observed in other environmental pollutants. The bactericide triclosan increased developmental and growth rates in larvae of bullfrogs (*Rana catesbiana* [64,65]) and common frogs (*Rana temporaria* [66]), respectively. Studies have also shown that the pesticide acetochlor increased developmental rates in northern leopard frogs (*Rana pipiens*) and African clawed frog larvae [67,68] by increasing T_3 and both TH receptor levels (α, β). TM is likely acting on individuals in a similar manner to increase growth in our current study.

In addition to the possible impacts of TM on hormone concentrations, the fungicide also cleared *Bd* infection in *Bd+* individuals. Unexpectedly, larvae exposed to *Bd* and TM were heavier and larger than those exposed to TM alone. One possible explanation for this observation could be adjusted from the "thrifty metabolism" hypothesis [69,70]. The hypothesis holds that if an individual is malnourished early in development, their metabolism will overcompensate and, as a side effect, induce adverse health effects later in life. In such studies, the metabolism of malnourished young overcompensated upon release from stressful circumstances, and this resulted in 'over-nutrition' (or catch-up weigh gain) with subsequent health issues such as diabetes and obesity. While we did not test the effects of *Bd* or TM on foraging abilities, is it possible that this hypothesis could apply to our data. We hypothesize that individuals exposed to both TM and *Bd* were forced to initially cope with the deleterious effects of *Bd*. However, upon the clearing of *Bd* by TM, these newly

uninfected individuals may have overcompensated by increasing feeding rates. This hypothesis parallels the "enemy release" hypothesis [28,71]. Usually pertaining to invasive species, the invasive organism is placed into a naïve habitat without any predators to control population sizes. Because of this 'release,' the introduced organism can thrive and usually become a pest in the absence of any predators. In our case, TM is "releasing" infected tadpoles from *Bd*; thus, allowing newly uninfected individuals to overcompensate in the absence of a previous health threat. This, combined with the beneficial effects of TM, might have resulted in growth that surpassed all other treatments.

While our study is the first to show significant promotions in life history traits of *Bd−* and pesticide-exposed amphibians, Gahl et al. [72] recently showed similar trends in *Bd−* and glyphosate-exposed frogs. In the study, exposure to *Bd* or glyphosate alone did not significantly alter growth. However, in both *Bd*-exposed and unexposed, a trend was observed with individuals exposed to glyphosate being larger and heavier than those not exposed to the pesticide. They cited three possibilities for their observations: possible direct inhibition of *Bd* in water, the addition of nutrients to the system from glyphosate, and glyphosate-induced immune responses that would operate to fight *Bd* infection. These results corroborate our current findings where in both *Bd*-exposed and unexposed anurans, the addition of TM into the system facilitated growth compared to subjects in treatments without pesticides.

The results of our current study were unexpected. While we predicted that TM would clear infection in *Bd* exposed individuals, we did not predict the overall morphologically beneficial properties of TM to both *Bd+* and *Bd−* individuals. We have offered reasonable speculations as to the mechanism whereby such benefits might occur, though we reiterate that they are just that – speculations. It is clear that further research must be conducted to elucidate the pathways through which TM is acting upon larvae to induce growth in both larvae and adults. Additionally, researchers must work to make clear possible effects of such substances (i.e., effects on reproduction and fitness). Only through such research will we truly be able to assess the effects of such contaminants and disease on amphibian health.

Acknowledgments

We thank Forrest Brem for assisting with the experiment and Emily Elderbrock, Steve Schoech, and Travis Wilcoxen for reviewing this manuscript. We also thank Michelle Boone, Christine Lehman, and Rick Relyea for providing suggestions on experiment design. Collection permits from TN were obtained prior to collecting the animals used in these experiments, and all experimental procedures were approved bye the University of Memphis IACUC.

Author Contributions

Conceived and designed the experiments: SMH MJP. Performed the experiments: SMH. Analyzed the data: SMH. Contributed reagents/materials/analysis tools: SMH JLK MJP. Wrote the paper: SMH MJP.

References

1. LeNoir JS, McConnell LL, Fellers GM, Cahill TM, Seiber JN (1999) Summertime transport of current use pesticides from California's central valley to the Sierra Nevada Mountain range, USA. Environmental Toxicology and Chemistry 18: 2715–2722.

2. Gillion RJ, Barbash JE, Crawford CG, Hamilton PA, Martin JD, et al. (2006) Pesticides in the nation's streams and ground water, 1992–2001. 172. United States Geological Survey, Circular 1291.

3. Howe G, Gillis R, Mowbray R (1998) Effect of chemical synergy and larval stage on the toxicity of atrazine and alachlor to amphibian larvae. Environmental Toxicology and Chemistry 17: 519–525.

4. Relyea RA and Diecks N (2008) An unforeseen chain of events: Lethal effects of pesticides on frogs at sublethal concentrations. Ecological Applications 18(7): 1728–1748.

5. Bender J (1994) Future Harvest: Pesticide-free Farming. University of Nebraska Press, Lincoln, NE.

6. Payne NJ (1998) Developments in aerial pesticide application methods for forestry. Crop Protection 17: 171–186.

7. Clements WH and Newman MC (2002) Community ecotoxicology. Wiley, Chichester, UK.

8. Jones DK, Hammond JI, Relyea RA (2011) Roundup and amphibians: The importance of concentration, application time, and stratification. Environmental Toxicology and Chemistry 29: 2016–2025.

9. Matthews GA (1992) Pesticide Application Methods. Harlow, UK: Longmans. 405 3rd ed.

10. Wilson GWT and Williamson MM (2008) Topsin-M: The new benomyl for mycorrhizal-suppression experiments. Mycologia 100: 548–554.

11. Bishop P, Sorochan J, Ownley BH, Samples T J, Windham AS, et al. (2008) Resistance of *Sclerotinia homoeocarpa* to iprodione, propiconazole, and thiophanate-methyl in Tennessee and northern Mississippi. Crop Science 48: 1615–1620.

12. Thomas TH (2008) Investigations into the cytokinin-like properties of benzimidazole-derived fungicides. Annals of Applied Biology 76(2): 237–241.

13. Yoon C-S, Jin J-H, Park J-H, Yeo C-Y, Kim S-J, et al. (2008) Toxic effects of carbendazim and *n*-butyl isocyanate, metabolites of the fungicide benomyl, on early development in the African clawed frog *Xenopus laevis*. Environmental Toxicology 23(1): 131–144.

14. Norris LA, Lorz HW, Gregory SV (1983) Influence of forest and range land management on anadromous fish habitat in Wester North America: Forest chemicals. Technical Report. PW-149. U.S. Department of Agriculture Forest Service, Portland, OR.

15. Capaldo A, Gay F, De Falco M, Virgillo F, Valiante S, et al. (2005) The newt *Triturus carnifex* as a model for model for monitoring the ecotoxic impact of the fungicide thiophanate methyl: Adverse effects on the adrenal gland. Comparative Biochemicstry and Physiology-Part C143: 86–93.

16. De Falco M, Sciarrillo R, Capaldo A, Russo T, Gay F, et al. (2007) The effects of the fungicide methyl thiophanate on adrenal gland morphophysiology of the lizard *Podarcis sicula*. Archives of Environmental Contamination and Toxicology 53: 241–248.

17. Sciarrillo R, De Falco M, Virgilio F, Laforgia V, Capaldo A, et al. (2008) Morphological and functional changes in the thyroid gland of methyl thiophanate- injected lizards, *Podarcis sicula*. Archives of Environmental Contamination and Toxicology 55: 254–261.

18. Fellers GM, Green ED, Longcore JE (2001) Oral chytridiomycosis in the mountain yellow-legged frog (*Rana muscosa*). Copeia 4: 945–953.

19. Venesky MD, Parris MJ, Storfer A (2009) Impacts of *Batrachochytrium dendrobatidis* infection on tadpole foraging performance. EcoHealth 6: 565–575.

20. Venesky MD, Wassersug RJ, Parris MJ (2010) Fungal pathogen changes the feeding kinematics of larval anurans. Journal of Parasitology 3: 552–557.

21. Parris MJ and Baud DR (2004) Interactive effects of a heavy metal and chytridiomycosis on gray treefrog larvae (*Hyla chrysoscelis*). Copeia 2: 334–350.

22. Parris MJ and Cornelius TO (2004) Fungal pathogen causes competitive and developmental stress in larval amphibian communities. Ecology 85(12): 3385–3395.

23. Garner TWJ, Walker S, Bosch J, Leech S, Rowcliffe JM, et al. (2009) Life history tradeoffs influence mortality associated with the amphibian pathogen *Batrachochytrium dendrobatidis*. Oikos 118: 783–791.

24. Smith KG, Weldon C, Conradie W, du Preez LH (2007) Relationship among size, development, and *Batrachochytrium dendrobatidis* infection in African tadpoles. Disease of Aquatic Organisms 74: 159–164.

25. Johnson ML and Speare R (2003) Survival of *Batrachochytrium dendrobatidis* in water: Quarantine and disease control implications. Emerging Infectious Diseases 9(8): 922–925.

26. Hanlon SM and Parris MJ (2012) The impact of pesticides on the pathogen Batrachochytrium dendrobatidis independent of potential hosts. Archives of Environmental Contamination and Toxicology DOI: 10.1007/s00244-011-9744-1.

27. Relyea RA and Mills N (2001) Predator-induced stress makes the pesticide carbaryl more deadly to gray treefrog tadpoles (*Hyla versicolor*). Proceedings of the National Academy of Sciences USA 98(5): 2491–2496.

28. Boone MD and Semlitsch RD (2003) Interactions of bullfrog tadpole predators and an insecticide: Predation release and facilitation. Oecologia 137: 610–616.

29. Johnson M, Berger L, Philips L, Speare R (2003) Fungicidal effects of chemical disinfectants, UV light, desiccation and heat on the amphibian chytrid, *Batrachochytrium dendrobatidis*. Diseases of Aquatic Organisms 57: 255–260.

30. Webb R, Berger L, Mendez D, Speare R (2005) MS-222 (tricane methane sulfonate) does not kill the amphibian chytrid fungus *Batrachochytrium dendrobatidis*. Diseases of Aquatic Organisms 68: 89–90.

31. Schimidt BR, Geiser C, Peyer N, Keller N, Rutte MY (2009) Assessing whether disinfectants against the fungus *Batrachochytrium dendrobatidis* have negative effects on tadpoles and zooplankton. Amphibia-Reptillia 30: 313–319.

32. Martel A, Van Rooij P, Vercauteren G, Baert K, Van Waeyenberghe L, et al. (2010) Developing a safe antifungal treatment protocol to eliminate Batrachochytrium dendrobatidis from amphibians. Medical Mycology 42(2): 143–149.

33. Gosner KL (I960) A simplified table for staging anuran embryos and larvae with notes on identification. Herpetologica 16: 183–190.

34. Longcore JE, Pessier AP, Nichols DK (1999) Batrachochytrium dendrobatidis gen. et sp. nov., a chytrid pathogenic to amphibians. Mycologia 91: 219–227.

35. Pessier AP, Nichols DK, Longcore JE, Fuller MS (1999) Cutaneous chytridiomycosis in poison dart frogs (*Dendrobates* spp.) and White's tree frogs (*Litoria caerulea*). Journal of Veterinary Diagnostic Investigation 11: 194–199.

36. Boyle DG, Boyle DB, Olsen V, Morgan JAT, Hyatt AD (2004) Rapid quantitative detection of chytridiomycosis (*Batrachochytrium dendrobatidis*) in amphibian samples using real-time Taqman PCR assay. Diseases of Aquatic Organisms 60: 141–148.

37. Hyatt AD, Boyle DG, Olsen V, Boyle DB, Berger L, et al. (2007) Diagnostic assays and sampling protocols for the detection of Batrachochytrium dendrobatidis. Diseases of Aquatic Organisms 73: 175–192.

38. Vredenburg VT, Knapp RA, Tunstall T, Briggs CJ (2010) Large-scale Amphibian Die-offs Driven by the Dynamics of an Emerging Infectious Disease. Proceedings of the National Academy of Sciences, USA 107: 9689–9694.

39. Berger L, Speare R, Daszak P, Green DE, Cunningham AA, et al. (1998) Chytrid- iomycosis causes amphibian mortality as- sociated with population declines in the rain forests of Australia and Central Amer- ica. Proceedings of the National Academy of Sciences, USA 95: 9031–36.

40. Daszak P, Cunningham AA, Hyatt AD (2003) Infectious disease and amphibian population declines. Diversity and Distributions 9: 141–150.

41. Hayes TB, Case P, Chui S, Chung D, Haeffele C, et al. (2006) Pesticide mixtures, endocrine disruption, and amphibian declines: Are we understanding the impact? Environmental Health Perspectives 114(1): 40–50.

42. Collins J and Crump M (2009) Extinction in Our Times: Global Amphibian Decline. New York, NY, Oxford University Press.

43. Boone, M D., and S. M James (2003) Interactions of an insecticide, herbicide, and natural Stressors in amphibian community mesocosms. Ecological Applications 13: 829–841.

44. Semlitsch RD, Bridges CM, Welch AM (2000) Genetic variation and a fitness tradeoff in the tolerance of gray treefrog (*Hyla versicolor*) tadpoles to the insecticide carbaryl. Oecologia, 125: 179–185.

45. Buskirk JV and Relyea RA (1998) Selection for phenotypic plasticity in rana sylvatica tadpoles. Biological Journal of the Linnean Society 65: 301–328.

46. Belden LK, Moore IT, Mason RT, Wingfield JC, and Blaustein A (2003) Survival, the hormonal stress response, and UV-B avoidance in cascade frog tadpoles (*Rana cascadae*) exposed to UV-B radiation. Functional Ecology 17: 409–416.

47. Parris MJ, Reese E, and Storfer A (2006) Antipredator behavior of chytridiomycosis-infected northern leopard frog (*Rana pipiens*) tadpoles. Canadian Journal of Zoology 84: 58–65.

48. Rohr JR, Swan A, Raffel TR, Hudson PJ (2009) Parasites, info-disruption, and the ecology of fear. Oecologia 159: 447–454.

49. Sadeh A, Truskanov N, Mangel M, and Blaustein L (2011) Compensatory development and costs of plasticity: Larval responses to desiccated conspecifics. PLoS ONE 6(1): e15602.

50. Sparling DW and Fellers GM (2009) Toxicity of two insecticides to California, USE, anurans and its relevance to declining amphibian populations. Environ Toxicol Chem 28(8): 1696–1703.

51. Steiner UK and Buskirk JV(2009) Predator-induced changes in metabolism cannot explain the growth/predation risk tradeoff. PLoS ONE 4(7): e6160.

52. Venesky MD, Kerby JL, Storfer A, Parris MJ (2011) Can differences in host behavior drive patterns of disease prevalence in tadpoles? PLoS ONE 6(9): e24991.

53. McDiarmid, R W. and R. A Altig (1999) Tadpoles: The biology of anuran larvae. Chicago, IL: University of Chicago Press.

54. Alford RA and Harris RN (1988) Effects of larval growth history on anuran metamorphosis. American Naturalist 131(1): 91–106.

55. Newman RA (1988) Adaptive plasticity in development of *Scaphiopus couchii* tadpoles in desert ponds. Evolution 42(2): 774–783.

56. Sparling DW, Linder G, Bishop C, and Krest S (2010) Ecotoxicology of Amphibians and Reptiles. Second Edition. Boca Raton, FL, SETAC/Taylor & Francis.

57. Denver RJ (1999) Evolution of the corticotropin-releasing hormone signaling system and its role in stress-induced phenotypic plasticity. Annals of the New York Academy of Sciences 897: 46–53.

58. Huang H, Cai L, Remo BF, Brown DD (2001) Timing of metamorphosis and the onset of the negative feedback loop between the thyroid gland and the pituitary is controlled by type II iodothyronine deiodinase in *Xenopus laevis*. Proceedings of the National Academy of Sciences, USA 98: 7348–7353.

59. Boone MD and Semlitsch RD (2002) Interactions of an insecticide with competition and pond drying in amphibian communities. Ecological Applications 12: 307–316.

60. Kikuyama S, Kawamura K, Tanaka S, Yamamoto K (1993) Aspects of amphibian metamorphosis: Hormonal control. International Review of Cytology 145: 105–148.

61. Rose CS (2005) Integrating ecology and developmental biology to explain the timing of frog metamorphosis. Trends in Ecology and Evolution 20, 129–135.

62. Denver RJ (1997) Environmental stress as a developmental cue: corticotropin-releasing hormone is a proximate mediator of adaptive phenotypic plasticity in amphibian metamorphosis. Hormones and Behavior 31: 169–79.

63. Hayes TM and Wu TM (1995) Interdependence of corticosterone and thyroid hormones in toad larvae (*Bufo boreas*). 2. Regulation of corticosterone and thyroid hormones. Journal of Experimental Zoology 271: 103–111.

64. Veldhoen N, Skirrow RC, Osachoff H, Wigmore H, Clapson DJ, et al. (2006) The bactericidal agent triclosan modulates thryoid hormone-dependent gene expression and disrupts postembryonic anuran development. Aquatic Toxicology 80: 217–227.

65. Fort DJ, Mathis MB, Hanson W, Fort CE, Navarro LT, et al. (2011) Triclosan and thyroid-mediated metamorphosis in anurans: differentiating growth effects from thyroid-driven metamorphosis in *Xenopus laevis*. Toxicological Sciences 121(12): 292–302.

66. Brande-Lavridsen N, Chistensen-Dalgaard J, Korsgaard B (2010) Effects of ethiylestradiol and the fungicide prochlorax on metamorphosis and thryoid gland morphology in *Rana temporaria*. Open Zoology Journal 3: 7–16.

67. Cheek AO, Ide CF, Bollinger JE, Rider CV, McLachlan JA (1999) Alteration of leopard frog (*Rana pipiens*) metamorphosis by the herbicide acetochlor. Archives of Environmental Contamination and Toxicology 37(1)70–77.

68. Crump.D, Werry K, Veldhoen N, Van Aggelen G, Helbing CC (2002) Exposure to the herbicide acetochlor alters thyroid hormone-dependent gene expression and metamorphosis in *Xenopus laevis*. Environmental Health Perspectives 110: 1199–1205.

69. Rolland-Cachera MF, Deheeger M, Maillot M, Bellisle F (2006) Early adiposity rebound: causes and consequences for obesity in children and adults. International Journal of Obesity 30: S11–S17.

70. Summermatter S, Mainieri D, Russell AP, Seydoux J, Montani JP, et al. (2008) Thrifty metabolism that favors fat storage after caloric restriction: a role for skeletal muscle phosphatidylinositol-3-kinase activity and AMP-activated protein kinase. The FASEB Journal 22: 774–785.

71. Colautti RI, Ricciardi A, Grigorovich, I A MacIsaac H J (2004), Is invasion success explained by the enemy release hypothesis? Ecology Letters, 7: 721–733.

72. Gahl Megan K, Bruce D Pauli, Jeff E Houlahan (2011) Effects of chytrid fungus and a glyphosate-based herbicide on survival and growth of wood frogs (*Lithobates sylvaticus*) Ecological Applications 21: 2521–2529.

4

The Two-Component Sensor Kinase TcsC and Its Role in Stress Resistance of the Human-Pathogenic Mold *Aspergillus fumigatus*

Allison McCormick[1], **Ilse D. Jacobsen**[2], **Marzena Broniszewska**[1], **Julia Beck**[1], **Jürgen Heesemann**[1,3], **Frank Ebel**[1]*

1 Max-von-Pettenkofer-Institut, Ludwig-Maximilians-University, Munich, Germany, 2 Department for Microbial Pathogenicity Mechanisms, Leibniz Institute for Natural Product Research and Infection Biology, Jena, Germany, 3 Center of Integrated Protein Science (Munich) at the Faculty of Medicine of the Ludwig-Maximilians-University, Munich, Germany

Abstract

Two-component signaling systems are widespread in bacteria, but also found in fungi. In this study, we have characterized TcsC, the only Group III two-component sensor kinase of *Aspergillus fumigatus*. TcsC is required for growth under hyperosmotic stress, but dispensable for normal growth, sporulation and conidial viability. A characteristic feature of the Δ*tcsC* mutant is its resistance to certain fungicides, like fludioxonil. Both hyperosmotic stress and treatment with fludioxonil result in a TcsC-dependent phosphorylation of SakA, the final MAP kinase in the high osmolarity glycerol (HOG) pathway, confirming a role for TcsC in this signaling pathway. In wild type cells fludioxonil induces a TcsC-dependent swelling and a complete, but reversible block of growth and cytokinesis. Several types of stress, such as hypoxia, exposure to farnesol or elevated concentrations of certain divalent cations, trigger a differentiation in *A. fumigatus* toward a "fluffy" growth phenotype resulting in white, dome-shaped colonies. The Δ*tcsC* mutant is clearly more susceptible to these morphogenetic changes suggesting that TcsC normally antagonizes this process. Although TcsC plays a role in the adaptation of *A. fumigatus* to hypoxia, it seems to be dispensable for virulence.

Editor: Robert A. Cramer, Montana State University, United States of America

Funding: This work was supported by the LMUexcellent program (grant number: ZUK22; www.unimuenchen.de/excellent/), a grant of the Wilhelm-Sander-Foundation to F.E. (grant number: 2007.102.2; www.sanst.de/) and a grant of the Center of Integrated Protein Science (Munich) to J.H. (grant: Forschungssäule B; www.cipsm.de/). The funders had no role in study design, data collection and analysis, decision to publish, or preparation of the manuscript.

Competing Interests: The authors have declared that no competing interests exist.

* E-mail: ebel@mvp.uni-meunchen.de

Introduction

Aspergillus fumigatus is a mold causing severe and systemic infections in immunocompromised patients [1]. The high mortality of these infections is largely due to the limited therapeutic options. Since *A. fumigatus* seems to lack sophisticated virulence factors, alternative therapeutic targets must be considered. The ability to respond to a plethora of environmental changes and to cope with different stress situations is vital for growth and survival of all microorganisms. This applies in particular to microbial pathogens that have to adapt to changing environments and a hostile immune response during colonization and invasion of the host. In fungi, sensing and responding to environmental stress is mediated by a set of receptors that are linked to a network of down-stream signaling pathways [2]. Interference with these signal transduction cascades can impede the fungal adaptation to stress and is considered a promising option to identify novel therapeutic targets. However, this approach is hampered by the conservation of many central signaling molecules in fungi and humans.

In bacteria sensing and processing of stress signals relies largely on two-component systems (TCS) that consist of a sensor histidine kinase and a response regulator. In fungi and other eukaryotes, hybrid histidine kinases (HHK) integrate both functions in a single protein. Fungal TCS are multistep phospho-relays composed of a sensor kinase (HHK), a histidine-containing phosphotransfer protein (HPt) and one or two response regulators. HHK are conserved within the fungal kingdom and depending on the species they govern the response to various stress signals, including osmotic stress, oxidative stress, hypoxia, resistance to anti-fungals and sexual development [3,4]. In contrast to other signaling molecules, TCS are attractive candidates for new therapeutic targets since they contribute to the virulence of fungal pathogens and are not found in vertebrates [3,5].

In fungi, eleven families of HHK have been described according to their protein sequence and domain organization [6]. Of several potential HHK present in the genome of *A. fumigatus* only two have been studied so far. Deletion of the Group VI HHK gene *tcsB* (AFUA_2G00660) had no severe impact on growth and stress resistance of *A. fumigatus*, but led to a slightly increased sensitivity to SDS [7]. A mutant in the Group IV HHK *tcsA/fos1* (AFU6G10240) showed normal growth, no increased sensitivity to osmotic stress, but resistance to dicarboximide fungicides, like iprodione, and enzymatic cell wall degradation [8]. This is remarkable, since dicarboximide fungicides commonly target Group III HHK [9]. Several lines of evidence link Group III HHK to the high osmolarity glycerol (HOG) pathway that was initially described as a signaling module enabling yeasts to adapt to high external osmotic pressure [10]. However, recent evidence suggests that in pathogenic fungi the HOG pathway is furthermore involved in the response to diverse kinds of stress [4].

In this study, we have analyzed TcsC, the sole representative of the Group III HHK in *A. fumigatus*. Group III HHK are found in bacteria, plants and fungi. They contain a characteristic cluster of HAMP domains that mediate signaling in histidine kinases, adenylyl cyclases, methyl-accepting chemotaxis proteins and certain phosphatases. Conformational changes in the spatial organization of the amphipathic helices in HAMP domains allow two conformations that either activate or inactivate the kinase activity of the output domain [11]. Single HAMP domains of membrane-bound HHK are found in close proximity to the membrane-spanning segment and transduce signals from the external input to the internal output domain. Group III HHK contain clusters of 4-6 HAMP domains, that according to a model developed recently for the osmo-tolerant yeast *Debaryomyces hansenii*, form a functional unit that is able to sense external signals. Changes in external osmolarity are supposed to alter the pattern of HAMP domain interactions and thereby modulate the inherent kinase activity of the protein [12]. The facts that Group III HHK are exclusively found in fungi and that certain fungicides can activate these sensor kinases in an uncontrolled and harmful manner makes them a potential Achilles heel of fungal pathogens that merits further investigations.

Results

The Group III HHK TcsC of *A. fumigatus*

The genome of *A. fumigatus* contains only one putative Group III HHK (AFU2G03560). The corresponding protein comprises a histidine kinase acceptor domain, a histidine kinase-like ATPase domain, a receiver domain and six HAMP domains. It lacks a transmembrane segment and is presumably localized in the cytoplasm. We designated this protein Two-component system protein C (TcsC) following the nomenclature of the previously studied *Aspergillus* TCS sensor kinases TcsA (Fos-1; AFU6G10240) and TcsB (AFU2G00660) [7,8,13].

Generation and Characterization of a ΔtcsC Mutant

To analyze the function of TcsC, we deleted the gene and complemented the mutant by ectopic insertion of the *tcsC* gene under control of its native promoter. The complementation procedure and the analysis of the genotype of the resulting strain are shown in Figure S1. On AMM, YG or Sabouraud medium the mutant grew well, but the colonies had a distinct appearance characterized by a broader white rim and fewer extending hyhae at the periphery (Figure 1A to D), whereas the complemented strain was indistinguishable from the wild type (data not shown). At 48°C growth of the mutant was comparable to the controls (Figure 1E) demonstrating that it is not particularly sensitive to temperature stress. Radial growth of the ΔtcsC mutant was slightly slower on AMM supplemented with ammonium tartrate (Figure 1E), whereas a remarkable reduction in growth was found on AMM plates supplemented with NaNO₃ instead of ammonium tartrate. This defect was not observed for the complemented strain indicating that TcsC is required for normal growth with nitrate as sole nitrogen source (Figure 1F).

In *A. nidulans* deletion of the homologous *nikA* gene had severe consequences for the production and viability of asexual spores [14,15,16]. We therefore compared sporulation and conidial viability of the ΔtcsC mutant and its parental strain. After four days at 37°C both strains produced a confluent and sporulating mycelial layer. No obvious difference in sporulation was apparent and this was confirmed by determining the conidial yield per cm^2 (mutant: $9.7\pm0.8\times10^7$, parental strain: $9.3\pm0.8\times10^7$). Conidia of the ΔnikA gene lose their viability within a few days when stored in water at 4°C. In contrast, conidia of the ΔtcsC mutant remained fully viable after storage for one month (mutant: $93.7\%\pm3.0\%$, parental strain: $95.5\%\pm3.0\%$). Thus, deletion of the Group III HHK gene in *A. fumigatus* does not affect sporulation or conidial viability, thus disclosing a remarkable difference between the two homologous sensor kinases in *A. fumigatus* and *A. nidulans*.

Conidial viability in *A. nidulans* was recently shown to depend on the presence phosphorylated SakA in resting conidia [17]. Several Group III HHK have been linked to the HOG pathway and shown to influence the phosphorylation state of HOG proteins, like *Aspergillus* SakA. In immunoblot experiments we detected only a slight decrease in the level of SakA phosphorylation in resting conidia of the ΔtcsC mutant when compared to its parental strain (Figure 2A), demonstrating that TcsC is not essentially required for SakA phosporylation in resting conidia.

Group III HHK have been shown to be required for resistance to osmotic stress in several fungi, but not in *A. nidulans*. Our data revealed a strong growth inhibition of the ΔtcsC mutant under hyperosmotic stress, e.g. on plates containing 1.2 M sorbitol (Figure 3B), 1 M KCl (Figure 3C) and 1 M NaCl (data not shown). This demonstrates that TcsC is clearly important for adaptation to high osmolarity. Immunoblot analysis revealed that SakA phosphorylation is much weaker in germlings than in resting conidia (Figure 2A and B). However, both 1.2 M sorbitol and the antifungal agent fludioxonil induced SakA hyper-phosphorylation in a TcsC-dependent manner (Figure 2B). Thus, TcsC is required for activation of the HOG pathway by hyperosmotic stress and the phenylpyrrole antifungal agent fludioxonil.

We found no evidence for an enhanced sensitivity of the ΔtcsC mutant to calcofluor white, several clinically relevant antifungals (amphotericin B, posaconazol and caspofungin), pH (pH 5-9), temperature (20°C–48°C) or oxidative stress (H₂O₂ and t-BOOH) (data not shown). In fact, the mutant turned out to be slightly more resistant to the cell wall stressor congo red and UV light (Figure 3D and data not shown). Thus, TcsC activity is required for adaptation to hyper-osmotic stress, but is not essential for the general stress response.

TcsC is Essential for the Fungicidal Acitivity of Fludioxonil and Related Compounds

An interesting feature of Group III HHK mutants is their resistance to fludioxonil and related fungicides. Accordingly, the ΔtcsC mutant grew normally in liquid medium containing 10 µg/ml fludioxonil, whereas growth of the wild type was completely abrogated at 1 µg/ml fludioxonil (data not shown). This phenotype was also evident in drop dilution assays on plates supplemented with fludioxonil (1 µg/ml; Figure 3E) or the functionally related fungicides quintozene (25 µg/ml) and iprodione (25 µg/ml) (Figure S2 B and C, respectively).

To obtain more information on the impact of fludioxonil at the level of individual cells, germlings were incubated in the presence of 1 µg/ml fludioxonil. No obvious morphological changes were apparent after 2 h (Figure 4A and B), but 4 h and 6 h after addition of fludioxonil growth of the wild type (Figure 4D and F) and the complemented mutant (data not shown) stopped and the cells began to swell, whereas the growth and morphology of the ΔtcsC mutant remained normal (Figure 4C and E). Similar results were obtained with 25 µg/ml iprodione (data not shown). DAPI staining of germlings treated with fludioxonil for 6 h revealed a normal distribution of nuclei in hyphae of the mutant

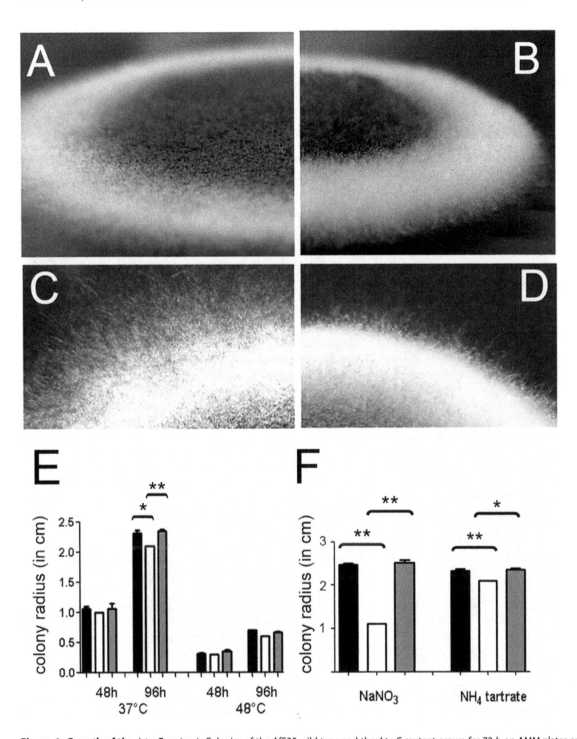

Figure 1. Growth of the Δ*tcsC* mutant. Colonies of the AfS35 wild type and the Δ*tcsC* mutant grown for 72 h on AMM plates are shown in panels A/C and B/D, respectively. Magnifications of the edge of the colonies are depicted in panels C and D. Note the reduced number of extending hyphae in the mutant. Panel E: Quantification of the radial growth of AfS35 (black), Δ*tcsC* mutant (white) and complemented mutant colonies (gray) on AMM plates after 48 h and 96 h at 37°C or 48°C. Panel F: Quantification of the radial growth after 96 h of AfS35 (black), Δ*tcsC* mutant (white) and complemented mutant colonies (gray) on AMM plates supplemented with 1.4 M NaNO3 or 0.2 M ammonium tartrate at 37°C. The experiments shown in panels E and F were done in triplicate. Standard deviations are indicated. Student's *t*-test: *p<0.005; **p<0.001.

(Figure 4G), but an unusually high number of nuclei in the swollen cells of the wild type (Figure 4H) and the complemented mutant (data not shown).

We also analyzed the impact of fludioxonil on the germination of resting conidia. Spores were incubated in medium supplemented with 1 μg/ml fludioxonil. After 28 h, the wild type produced only small germlings (Figure S3 A and B), while abundant hyphae were found in the fludioxonil-treated Δ*tcsC* mutant and an untreated wild type control (data not shown). Thus, germination of wild type spores was impaired, but not completely abolished by fludioxonil. An additional 18 h incubation in fludioxonil yielded cells whose growth was arrested and these exhibited irregular, swollen

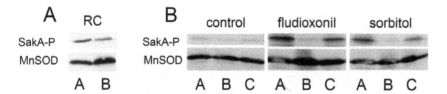

Figure 2. The role of TcsC in the phosphorylation of SakA. Protein extracts of resting conidia (RC)(panel A) and germlings (panel B) were analyzed by immunoblot using specific antibodies to phosphorylated SakA and as a loading control mitochondrial MnSOD. Extracts were prepared from germlings treated with 10 μg/ml fludioxonil and 1.2 M sorbitol for 2 and 20 min, respectively. A: parental strain AfS35, B: ΔtcsC mutant, C: complemented mutant.

morphologies (Figure S3 C-F). As observed for germlings, fludioxonil treatment during germination resulted in unusually high numbers of nuclei that were often clustered in the cytoplasm (Figure S3 A, C and E). Only few fludioxonil-treated cells showed signs of leakage after 46 h (data not shown). We therefore replaced the medium and incubated the cells for another 15 h without fludioxonil to analyze their ability to recover. Although fludioxonil

had induced severe morphological changes the cells were able to restore growth and the resulting hyphae had a normal appearance and a normal distribution and number of nuclei (Figure S3 G and H).

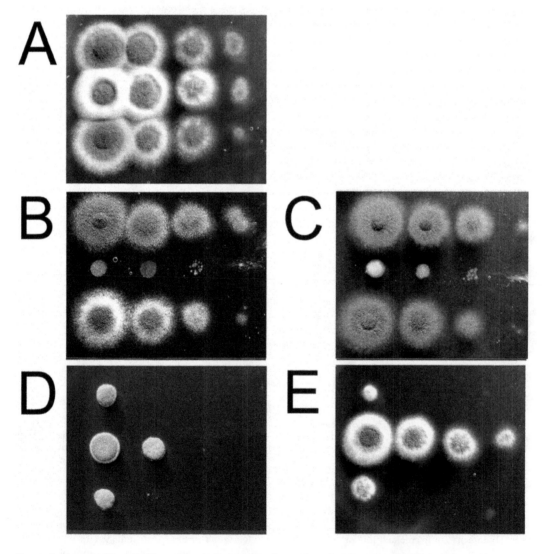

Figure 3. The ΔtcsC mutant is sensitive to hyperosmotic stress and resistant to fludioxonil. Drop dilution assays were performed on AMM plates (supplemented with ammonium). Panel A: control; B: 1.2 M sorbitol; C: 1 M KCl; D: 100 μg/ml congo red; E: 1 μg/ml fludioxonil. The depicted colonies were obtained after 48 h at 37°C. Top: AfS35; middle: ΔtcsC; bottom: complemented strain.

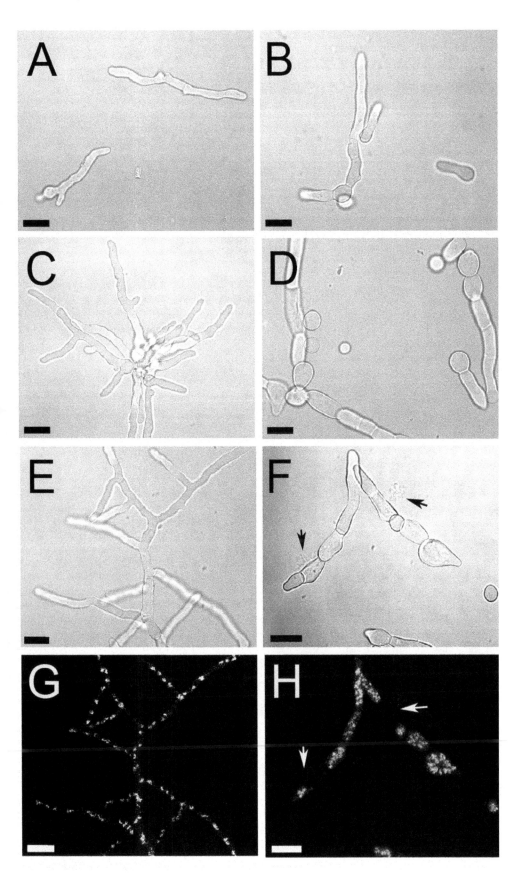

Figure 4. Impact of fludioxonil on *A. fumigatus* germ tubes. Conidia of the Δ*tcsC* mutant (panels A, C, E, G) and its parental strain AfS35 (panels B, D, F, H) were seeded on glass cover slips and incubated overnight in AMM at 30°C. The resulting germ tubes were treated with 1 µg/ml fludioxonil

for 2 h (A, B), 4 h (C, D) and 6 h (E–H) at 37°C. A DAPI staining is shown in panels G and H. Arrows indicate lysed cells that lack intracellular nuclei and are associated with amorphous extracellular material. All bars represent 10 μm.

The Role of TcsC in the *fluffy* Growth Phenotype in *A. fumigatus*

Tco1, the Group III HHK of *Cryptococcus neoformans*, is required for growth under hypoxic conditions [18]. Oxygen limitation is also encountered by *A. fumigatus* during infection and it was recently shown that its ability to grow under hypoxic conditions is a prerequisite for virulence [19]. Adaptation of *A. fumigatus* to 1% oxygen results in colonies that are characterized by a massive production of aerial hyphae, resulting in a dome-shaped morphology, and a complete lack of sporulation (Figure 5A). At 1% oxygen the Δ*tcsC* mutant was indistinguishable from the control strains with respect to growth and colony morphology. At 2% oxygen flat and sporulating colonies were found for the control strains, whereas the mutant colonies remained white and dome-shaped (Fig. 5B). Similar *A. nidulans* colonies, also characterized by the formation of abundant aerial hyphae and the lack of sporulation, were described previously as having a 'fluffy' developmental phenotype [20]. Thus, oxygen limitation seems to activate a specific morphogenetic program and the threshold level of hypoxic stress required to trigger this developmental process is clearly lower in the Δ*tcsC* mutant.

A fluffy phenotype is also apparent in the presence of 2 mM of the acyclic sesquiterpene alcohol farnesol (Figure 5C and E; [21]). Titration of farnesol revealed that at lower concentrations the fluffy growth was restricted to the Δ*tcsC* mutant (Figure 5D). Thus, the absence of TcsC renders *A. fumigatus* more sensitive to oxygen limitation and farnesol. Further experiments revealed a third trigger for fluffy growth in *A. fumigatus*. White, dome-shaped colonies of the mutant, but not of the control strains were obtained on plates containing 100 mM CaCl$_2$ and 100 mM MgCl$_2$ (Figure 5G and data not shown). This phenotypic switch was also induced by 100 mM MgSO$_4$ (Figure 5F and H), but not by 200 mM NaCl (data not shown), indicating that divalent cationic ions, but not the slight increase in osmolarity or elevated chloride concentration induced the fluffy growth. The phenotypic differentiation was already obvious with 50 mM CaCl$_2$ (Figure 5I), and could be enforced by addition of 20 μM farnesol, which *per se* had no impact on the colony morphology (data not shown), suggesting a synergistic mode of action for these stimuli. A further increase of the calcium concentration to 500 mM induced the fluffy growth phenotype in the control strains, but concomitantly abrogated growth of the mutant (Figure 5J). Thus, oxygen limitation, farnesol and divalent cations activate the fluffy developmental program and the lack of *tcsC* renders cells more susceptible to this developmental reprogramming.

The fluffy growth phenotype in *A. nidulans* is regulated by a heterotrimeric G protein that has been functionally linked to the cAMP-dependent protein kinase pathway [22,23]. For *A. fumigatus*, addition of 5 mM cAMP partially rescued the sporulation defect caused by farnesol (Figure 6A), but not that triggered by 100 mM CaCl$_2$ or hypoxia (1% oxygen) (data not shown). We also tested the influence of light that stimulates sporulation in many fungi. Exposure of colonies to white light rescued the sporulation defect induced by 1% oxygen in the parental and the complemented strain, but not in the Δ*tcsC* mutant. Moreover, light also reduced the formation of aerial hyphae and resulted in colonies with a normal appearance (Figure 6B). 100 mM CaCl$_2$ or 2% oxygen are weaker activators of the fluffy program. They only influence the growth of the Δ*tcsC* mutant and this effect can also be prevented by light (Figure 6B and data not shown). The impact of

light on the farnesol-induced sporulation defect could not be analyzed due to the known sensitivity of this agent to light. Thus, light and cAMP can antagonize the development towards a fluffy growth phenotype. In doing so cAMP was only able to neutralize the effect of farnesol, whereas light seems to have a broader impact.

Analysis of the Virulence of the Δ*tcsC* Mutant

The ability to respond to certain kinds of stress is clearly impaired in the Δ*tcsC* mutant. In order to investigate whether this negatively affects its virulence potential, cortisone-acetate treated mice were infected via the intra-nasal route. Survival of mice infected with the Δ*tcsC* mutant was comparable to those infected with the control strains (Figure 7) and the histological analysis of samples from the lungs of mice that succumbed to infection also revealed no apparent differences (data not shown). A normal virulence was furthermore observed in a alternative infection model using embryonated eggs [24](data not shown).

Discussion

In an often hostile environment pathogenic microorganisms rely on the ability to sense and respond to environmental changes. Two-component signaling (TCS) systems are sensing entities that are abundant in bacteria, but also found in fungi and plants. Because they are absent in mammals, TCS systems and their hybrid histidine kinases (HHK) are potential targets for novel anti-microbial strategies. Group III HHK are predicted to localize in the cytoplasm, but are nevertheless supposed to sense changes in the environment. The resulting signals are then transferred via a phospho-relay system to two response regulators that directly or indirectly trigger an appropriate transcriptional response [4]. In this study we have analyzed TcsC, the only Group III HHK of the pathogenic mold *A. fumigatus*. Deletion of the homologous *nikA* gene in *A. nidulans* has been reported to cause a significantly reduced growth on solid medium [14,15], whereas the Δ*tcsC* mutant grows normally on complex media and on minimal medium (AMM) supplemented with ammonium. Growth was however impaired on AMM supplemented with nitrate, suggesting that TcsC is required for efficient nitrogen assimilation. In this context it is noteworthy that the growth defect of the Δ*nikA* mutant was observed using minimal medium with nitrate as the sole nitrogen source [15] and it would be interesting to test the growth of this mutant on a medium containing ammonium.

Although Group III HHK are often linked to the high osmolarity glycerol (HOG) pathway, their relevance for adaptation to hyperosmotic stress seems to vary in different fungi. While the Δ*nikA* mutant showed a normal ability to adapt to hyperosmotic stress [15], the Δ*tcsC* mutant turned out to be highly sensitive. Another striking difference between both mutants exists with respect to their conidial viability. Conidia of the Δ*nikA* mutant showed a dramatic loss of viability when stored in water for several days [14,15], whereas conidia of the Δ*tcsC* mutant remained fully viable upon storage for several weeks. Thus, TcsC and NikA although closely related, appear to differ in their biological activities.

A characteristic feature of mutants lacking Group III HHK is their resistance to fungicides, like fludioxonil. These compounds are currently used in agriculture, but are also of potential interest for the development of novel therapeutic anti-fungals. Their mode

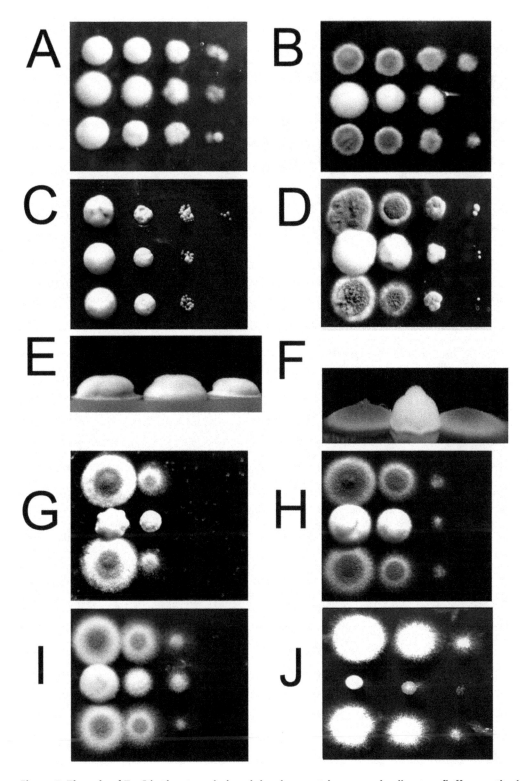

Figure 5. The role of TcsC in the stress-induced developmental program leading to a fluffy growth phenotype. Drop dilution assays were performed on AMM plates (supplemented with ammonium). Panel A: 1% oxygen; B: 2% oxygen; C: 2 mM farnesol; D: 200 µM farnesol; E: 2 mM farnesol; F: 100 mM MgSO$_4$; G: 100 mM CaCl$_2$; H: 100 mM MgSO$_4$; I: 50 mM CaCl$_2$; J: 500 mM CaCl$_2$. Side views of colonies from C and D are shown in panels E and F. The depicted colonies were photographed after 48 h at 37°C. AfS35 (top/left); Δ*tcsC* (middle); complemented strain (bottom/right).

of action is unique in that they activate a fungal signaling process, the HOG pathway. The hallmark of this activation is the phosphorylation and subsequent translocation of SakA/Hog1 to the nucleus [10]. In *A. fumigatus* fludioxonil induces a rapid, transient phosphorylation and translocation of the MAP kinase SakA that leads to a tremendous cellular swelling. Fludioxonil

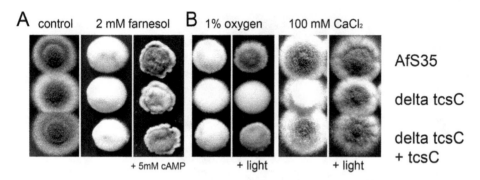

Figure 6. The impact of cAMP and light on the fluffy growth phenotype. Drop dilution assays were performed on AMM plates (supplemented with ammonium). The plates were supplemented or treated as indicated and incubated in incubator. When indicated plates were incubated under white light produced by an LED light source. Pictures were taken after 48 h at 37°C.

blocks growth of germ tubes and hyphae, but it is unable to completely prevent germination of resting conidia. Prolonged incubation in the presence of fludioxonil results in rather odd cellular morphologies. These phenotypic changes are stable as long as the agent is present, but normal growth can be restored after removal of the agent. Apart from their swelling, fludioxonil-treated *A. fumigatus* cells are remarkable because of their large number of nuclei. A block in nuclear division, as recently suggested for fludioxonil-treated *A. nidulans* [17], was not detectable; instead cytokinesis and mitosis seem to be transiently uncoupled, resulting in the accumulation of many more nuclei per cell than normal. These fludioxonil-induced phenotypic changes are dependent on TcsC, since they do not occur in the Δ*tcsC* mutant.

The complete resistance of the Δ*tcsC* mutant to fludioxonil and related fungicides correlates with its high sensitivity to hyperosmotic stress. It has been shown for several plant-pathogenic fungi that fludioxonil mediates its anti-fungal effect by activating the HOG pathway via a Group III HHK [9]. It is therefore conceivable that the characteristic swelling of fludioxonil-treated *A. fumigatus* cells results from a hyperactivation of SakA. This is already detectable after 2 minutes and seems to trigger an uncontrolled increase in the intracellular osmotic pressure. In *A. fumigatus*, TscC is clearly required for the activation of the HOG

pathway by both, fludioxonil and hyperosmotic stress. Thus, the inability of the Δ*tcsC* mutant to adapt to hyperosmotic stress and its resistance to fludioxonil both reflect the important role of the TcsC-SakA signaling axis in the control of the internal osmotic pressure of *A. fumigatus*.

The life cycle of *A. fumigatus* is tightly controlled by environmental cues. In contact with air hyphae initiate the formation of conidiophores and the production of conidia. The 'fluffy' developmental program impedes sporulation and leads to the massive formation of aerial hyphae and the appearance of white, dome-shaped colonies. Fluffy *A. nidulans* colonies were initially described after treatment with 5-azacytidine [20]. The phenotypical stability of these mutants indicates that a developmental program is permanently activated in these cells. We have recently identified the sesquiterpene alcohol farnesol as a trigger for transient fluffy growth in *A. fumigatus* [21]. In the current study, we observed similar phenotypic switches in response to hypoxia and elevated concentrations of certain divalent cations. The fluffy growth type likely provides an advantage enabling the fungus to survive under certain kinds of stress. The Δ*tcsC* mutant shifts earlier towards this phenotype than the wild type. White, dome-shaped colonies appeared at lower concentrations of farnesol and divalent cations and at less pronounced hypoxia. The earlier adaptation of the mutant does not result in a higher robustness,

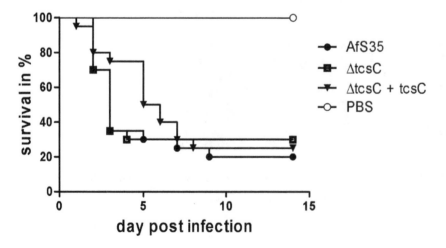

Figure 7. Infection of immuno-compromized mice. Intranasal infection of cortisone-acetate treated mice infected with 1×10^6 conidia of the Δ*tcsC* mutant (n = 20), the parental strain AfS35 (n = 20) and the complemented strain (n = 20). Controls received PBS only. Survival of mice is shown over time.

but seems to be the consequence of a reduced stress resistance. The limited compensatory potential of the fluffy growth was in particular evident at elevated calcium concentrations. The mutant shifts already at 50 mM calcium chloride, but its growth is abolished at 500 mM calcium chloride, when the wild type is still growing well.

So far, little is known about the mechanisms that underlie the fluffy growth phenotype and cause its peculiar morphological changes. In *A. nidulans* the fluffy growth seems to be controlled by a heterotrimeric G protein that is linked to the cAMP-dependent protein kinase pathway [22,23]. This and the recent finding that farnesol blocks adenylyl cyclase activity in *Candida albicans* [25] prompted us to study the relevance of the intracellular cAMP level. Addition of cAMP abrogated the farnesol-induced block in sporulation in the wild type, but cAMP was unable to rescue the sporulation defect caused by hypoxia or elevated calcium concentrations. Light is an environmental signal that stimulates sporulation in many fungi. Exposure to light restored normal growth and sporulation under hypoxic conditions and in the presence of elevated concentrations of divalent cations. Thus, light, cAMP and the TcsC protein are factors that impede an activation of the fluffy growth program caused by hypoxia, farnesol or divalent cations (Figure 8).

A stable fluffy *A. fumigatus* mutant secretes more proteases and has an increased angioinvasive growth capacity [26]. This suggests that fluffy hyphae may be well adapted to the specific requirements during infection. In line with this hypothesis, we identified oxygen limitation as another trigger for a fluffy growth. It will be interesting to analyze to what extent the fluffy growth program observed *in vitro* resembles the morphogenetic program that is active during infection.

The ΔtcsC mutant shows a normal sensitivity to oxidative, temperature and pH stress as well as clinically relevant anti-fungal agents. On the other hand, TcsC activity is important for the response to a limited array of stress signals including hypoxia (Figure 8). The ability to adapt to oxygen limitation is an essential

characteristic of many pathogenic microorganisms. Tco1, the homologous group III HHK in *Cryptococcus neoformans* regulates growth under hypoxic conditions and is also required for virulence [27]. In *A. fumigatus* the situation seems to be different, since we observed no significant attenuation in virulence for the ΔtcsC mutant. However, TcsC is required for the anti-fungal activity of fludioxonil and related compounds and may therefore be an attractive target for new therapeutic anti-fungals. Further studies are underway to define the precise mode of action of the TcsC stress sensing pathway and the impact of fludioxonil on growth and survival of *A. fumigatus*.

Materials and Methods

Strains Media and Growth Conditions

The *A. fumigatus* strain AfS35, a derivative of strain D141, has been described in [28]. AMM and YG medium were prepared as described [29]. AMM was either supplemented with 1.4 M $NaNO_3$ [30] or 0.2 M ammonium tartrate. For hypoxic growth plates were incubated at 37°C in a HERAcell 150i incubator (Thermo Fisher Scientific) adjusted to 5% CO_2 and the desired oxygen concentration.

Sequence Analysis and Data Base Searches

Domains were predicted using SMART (http://smart.embl-heidelberg.de/) and alignments were performed using CLUSTAL (http://www.ebi.ac.uk/Tools/msa/clustalw2/).

Construction of the ΔtcsC Mutant Strain

All oligonucleotides used in this study are listed in Table S1. To construct a suitable replacement cassette a 3.5 kb hygromycin resistance cassette was excised from pSK346 using the SfiI-restriction enzyme. The flanking regions of the *tcsC* gene (approx. 900 bp each) were amplified by PCR from chromosomal DNA using the oligonucleotide pairs tcsC-upstream and tcsC-downstream. These oligonucleotides harbor ClaI and Sfi sites. After

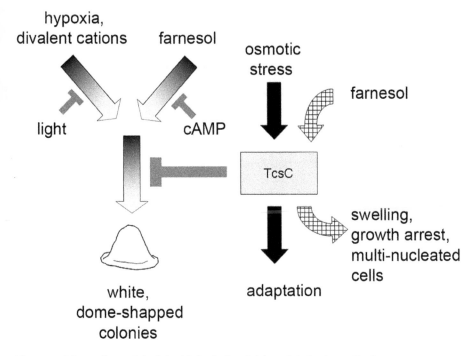

Figure 8. Schematic model of the biological activities of *A. fumigatus* TcsC.

digestion with ClaI and SfiI, ligation of the three fragments (resistance cassette and flanking regions) yielded a 5.3 kb deletion cassette that was purified using the Wizard SV Gel and PCR Clean-Up System (Promega). The fragment was cloned into the pCR2.1 vector (Invitrogen) using oligonucleotide-derived ClaI sites. A 9.2 kb fragment from the resulting plasmid was linearized with SpeI and used for transformation of *Aspergillus*. The construct used for complementation of Δ*tcsC* was generated by amplifying *tcsC* and its native promoter (1.5 kb region upstream of the gene) from chromosomal DNA using the oligonucleotides tcsC + native promoter-forward and tcsC-reverse. The *gpdA* promoter was excised from the pSK379 vector using EcoRV and NsiI; the latter enzyme generates sticky ends compatible with those generated by PstI. The amplified *tcsC* + native promoter fragment was cloned into this modified version of pSK379 using oligonucleotide-derived NsiI sites. The resulting plasmid was purified as above and used for transformation of the Δ*tcsC* mutant.

A. fumigatus protoplasts were generated and fungal transformation was performed essentially as described previously [29]. The resulting protoplasts were transferred to AMM plates containing 1.2 M sorbitol and either 200 μg/ml hygromycin (Roche, Applied Science) or 0.1 μg/ml pyrithiamine (Sigma-Aldrich).

Genomic DNA Analysis

A. fumigatus clones which showed the expected resistance on selective plates were further analyzed by PCR. The correct integration of the deletion cassette was analyzed at the 5′ end using oligonucleotides tcsC-upstream-forward and hph-3-SmaI (PCR1) and at the 3′ end using oligonucleotides trpCt-forward and tcsC-downstream-reverse (PCR2) (Figure S2). To detect the presence of *tcsC* in the complementation mutant, the entire *tcsC* gene was amplified using primers at the 5′ and 3′ ends of the gene (tcsC-forward and tcsC-reverse, PCR3). Primer sequences are listed in Table S1.

Quantification of Sporulation Efficiency

For each strain tested, three small tissue culture flasks (25 cm^2; Sarstedt, Nürnbrecht, Germany) with YG agar were inoculated with 4×10^6 conidia per flask. After incubation for 4 days at 37°C conidia were harvested and counted using a Neubauer chamber.

Spore Viability Assay

To determine their viability, 2×10^4 resting conidia were transferred to 1 ml YG medium in a 24 well plate. After overnight incubation at 37°C samples were fixed by addition of 100 μl 37% formaldehyde. The percentage of germinated cells was determined microscopically. These experiments were done in triplicate.

Protein Extraction and Western Blot

For protein extractions from resting conidia, 75 cm^2 flasks containing YG agar were inoculated with AfS35 or Δ*tcsC* conidia (in triplicate) and grown at 37°C for 3 days. Conidia were harvested in sterile water and the pellet frozen overnight at –20°C. Frozen conidia pellet was lyophilized overnight at 6°C. The dry pellet was ground with a mortar and pestle in liquid nitrogen. The ground conidia powder was added to 300 μl Laemmlie buffer (2% [w/v] SDS, 5% [v/v] mercaptoethanol, 60 mM Tris/Cl pH 6.8, 10% [v/v] glycerol, 0.02 [w/v] bromophenol blue), heated at 95°C and immediately extracted twice using a Fast Prep 24 (M.P. Biomedical, Irvine, CA) with a speed of 5.5 m/s for 20 s, followed by a final heat denaturation at 95 C for five minutes. 20 μl protein extract was used for SDS-PAGE on 12% SDS gel. Proteins were blotted onto 0.45 μm nitrocellulose membranes and labelled with

an α-phospho-p38 MAP kinase antibody (Cell Signaling Technology [#9211], MA, USA). A monoclonal antibody directed against mitochondrial MnSOD (P118-H3) kindly provide by Bettina Bauer was used as a loading control. For protein extractions from germ tubes, 4×10^7 resting conidia were inoculated in 10 ml AMM and incubated 9 h at 37°C. The germ tubes were treated with 10 μg/ml fludioxonil or 1.2 M sorbitol for 2 min or 20 min, respectively, at 37°C. Protein was then extracted from the cell pellet as above and used for SDS-PAGE and immunoblot in the same manner.

Phenotypic Plate Assays

Isolated conidia were counted using a Neubauer chamber. For drop dilution assays, a series of tenfold dilutions derived from a starting solution of 1×10^8 conidia per ml were spotted in aliquots of 1 μl onto plates. These plates were supplemented with the indicated agents and incubated at the indicated temperatures. For quantification of the radial growth, 3 μl containing 3×10^4 conidia were spotted in the centre of a 9 cm Petri dish. The radius of the colonies was determined over time.

E-test strips of voriconazole, amphotericin B and caspofungin were obtained from Inverness Medical (Cologne, Germany). Each E-test strip was placed onto an AMM agar plate spread with 8×10^5 conidia. Plates were incubated 36–48 h at 37°C.

Paper disk assays were performed by spreading 8×10^5 conidia on AMM, Sabouraud, or YG agar plates and placing a sterile paper disk containing fludioxonil, iprodione, or quintozene (Sigma-Aldrich; 46102, 36132 and P8556, respectively), or H_2O_2 or tert-butyl hydroperoxide (t-BOOH; Sigma-Aldrich). Plates were incubated 36–48 h at 37°C. Fludioxonil and iprodione were dissolved at 100 mg/ml stock concentrations in DMSO and quintozene was dissolved at 10 mg/ml stock concentration in chloroform. The influence of light was analyzed using an LED light (Osram DOT-it, Osram, Munich, Germany) that was affixed 15 cm above the Petri dish.

Microscopic Analysis

To visualize the effects of fludioxonil on germ tubes and resting conidia, AfS35 or Δ*tcsC* resting conidia were inoculated in 24-well plates containing 1 ml AMM and glass cover slips. Germ tubes were generated by incubating at 30°C overnight before adding 1 μg/ml fludioxonil. After incubation at 37°C for the indicated times, cells were fixed in 3.7% formaldehyde for five minutes at room temperature. Cover slips were mounted to glass slides in Vecta Shield containing DAPI (Vector Laboratories, Burlingame, California, USA). Cells were then visualized using a Leica SP-5 microscope (Leica Microsystems).

Infection Experiments

To analyze the impact of TcsC we used an intranasal infection model using immunocompromised female outbred CD-1 mice. Mice were immunosuppressed by intraperitoneal injection of cortisone acetate (25 mg/mouse, Sigma-Aldrich) on days –3 and 0. On day 0 the mice were anesthetized with fentanyl (0.06 mg/kg, Janssen-Cilag, Germany), midazolam (1.2 mg/kg, Roche, Germany) and medetomidin (0.5 mg/kg, Pfizer, Germany) and infected intranasally with 1×10^6 conidia in 20 μl PBS. Controls received PBS only. Survival was monitored for 14 days. During this period, mice were examined clinically at least twice daily and weighed individually every day. Kaplan-Meier survival curves were compared using the log rank test (SPSS 15.0 software). Mice were cared for in accordance with the principles outlined by the European Convention for the Protection of Vertebrate Animals Used for Experimental and Other Scientific Purposes (European

Treaty Series, no. 123; http://conventions.coe.int/Treaty/en/Treaties/Html/123). All animal experiments were in compliance with the German animal protection law and were approved (permit no. 03-001/08) by the responsible Federal State authority and ethics committee.

Supporting Information

Figure S1 (A) Schematic drawing of the genomic *tcsC* gene and the deleted *tcsC*::*hph/tk* locus. Approximately 1 kb of the 5′ and 3′ regions of *tcsC* gene were used for construction of the deletion cassette. The positions of the primers employed for the PCR amplifications and the resulting PCR products (PCR 1-3) are indicated. (B) Equal amounts of genomic DNA of AfS35, Δ*tcsC* and Δ*tcsC*+*tcsC* were used as template for PCR amplification of the regions indicated in panel A (PCR 1-3).

Figure S2 Resistance of the Δ*tcsC* mutant to iprodione and quintozene. The sensitivity to iprodione and quintozene was analyzed in drop dilution assays. AfS35 (top) and its Δ*tcsC* mutant (bottom) were spotted on plates without fungicides (panel A) or plates containing either 25 μg/ml quintozene (panel B) or 25 μg/ml iprodione (panel C). Pictures were taken after 48 h at 37°C.

Figure S3 Impact of fludioxonil during germination of *A. fumigatus* conidia. Conidia of *A. fumigatus* strain AfS35 were seeded on glass cover slips and incubated at 37°C in the presence of 1 μg/ml fludioxonil for 28 h (A, B) and 46 h (C to F). After 46 h the medium was replaced by fresh medium. Fungal cells fixed after another 15 h in the absence of fludioxonil are shown in G and H. DAPI stainings are shown in panels A, C, E and G. All bars represent 10 μm.

Table S1 Oligonucleotides used in this study.

Acknowledgments

We thank Kirsten Niebuhr for critical reading of the manuscript and Sven Krappmann for providing strain AfS35 and plasmids.

Author Contributions

Conceived and designed the experiments: FE AM. Performed the experiments: AM IDJ MB JB. Analyzed the data: FE AM JH. Wrote the paper: FE AM.

References

1. McCormick A, Loeffler J, Ebel F (2010) *Aspergillus fumigatus*: contours of an opportunistic human pathogen. Cell Microbiol 12: 1535–1543.
2. Bahn YS, Xue C, Idnurm A, Rutherford JC, Heitman J, Cardenas ME (2007) Sensing the environment: lessons from fungi. Nat Rev Microbiol 5: 57–69.
3. Santos JL, Shiozaki K (2001) Fungal histidine kinases. Sci STKE 98: re1.
4. Bahn YS (2008) Master and commander in fungal pathogens: the two-component system and the HOG signaling pathway. Eukaryot Cell 7: 2017–2036.
5. Li D, Agrellos OA, Calderone R (2010) Histidine kinases keep fungi safe and vigorous. Curr Opin Microbiol 13: 424–430.
6. Catlett NL, Yoder OC, Turgeon BG (2003) Whole-genome analysis of two-component signal transduction genes in fungal pathogens. Eukaryot Cell 2: 1151–1161.
7. Du C, Sarfati J, Latge JP, Calderone R (2006) The role of the sakA (Hog1) and tcsB (sln1) genes in the oxidant adaptation of *Aspergillus fumigatus*. Med Mycol 44: 211–218.
8. Pott GB, Miller TK, Bartlett JA, Palas JS, Selitrennikoff CP (2000) The isolation of FOS-1, a gene encoding a putative two-component histidine kinase from *Aspergillus fumigatus*. Fungal Genet Biol 31: 55–67.
9. Kojima K, Takano Y, Yoshimi A, Tanaka C, Kikuchi T, et al. (2004) Okuno T. Fungicide activity through activation of a fungal signalling pathway. Mol Microbiol 53: 1785–1796.
10. Hohmann S (2002) Osmotic stress signaling and osmoadaptation in yeasts. Microbiol Mol Biol Rev 66: 300–372.
11. Parkinson JS (2010) Signaling mechanisms of HAMP domains in chemorecep-tors and sensor kinases. Annu Rev Microbiol 13: 101–122.
12. Meena N, Kaur H, Mondal AK (2010) Interactions among HAMP domain repeats act as an osmosensing molecular switch in group III hybrid histidine kinases from fungi. J Biol Chem 285: 12121–1232.
13. Miskei M, Karányi Z, Pócsi I (2009) Annotation of stress-response proteins in the aspergilli. Fungal Genet Biol 46: S105–20.
14. Hagiwara D, Matsubayashi Y, Marui J, Furukawa K, Yamashino T, et al. (2007) Characterization of the NikA histidine kinase implicated in the phosphorelay signal transduction of *Aspergillus nidulans*, with special reference to fungicide responses. Biosci Biotechnol Biochem 71: 844–847.
15. Vargas-Pérez I, Sánchez O, Kawasaki L, Georgellis D, Aguirre J (2007) Response regulators SrrA and SskA are central components of a phosphorelay system involved in stress signal transduction and asexual sporulation in *Aspergillus nidulans*. Eukaryot Cell 6: 1570–1583.
16. Hagiwara D, Mizuno T, Abe K (2009) Characterization of NikA histidine kinase and two response regulators with special reference to osmotic adaptation and asexual development in *Aspergillus nidulans*. Biosci Biotechnol Biochem 73: 1566–1571.
17. Lara-Rojas F, Sánchez O, Kawasaki L, Aguirre J (2011) *Aspergillus nidulans* transcription factor AtfA interacts with the MAPK SakA to regulate general stress responses, development and spore functions. Mol Microbiol 80: 436–454.
18. Chun CD, Liu OW, Madhani HD (2007) A link between virulence and homeostatic responses to hypoxia during infection by the human fungal pathogen *Cryptococcus neoformans*. PLoS Pathog 3: e22.
19. Willger SD, Puttikamonkul S, Kim KH, Burritt JB, Grahl N, et al. (2008) A sterol-regulatory element binding protein is required for cell polarity, hypoxia adaptation, azole drug resistance, and virulence in *Aspergillus fumigatus*. PLoS Pathog 4: e1000200.
20. Tamame M, Antequera F, Villanueva JR, Santos T (1983) High-frequency conversion to a "fluffy" developmental phenotype in *Aspergillus* spp. by 5-azacytidine treatment: evidence for involvement of a single nuclear gene. Mol Cell Biol 3: 2287–2297.
21. Dichtl K, Ebel F, Dirr F, Routier FH, Heesemann J, et al. (2010) Farnesol misplaces tip-localized Rho proteins and inhibits cell wall integrity signalling in *Aspergillus fumigatus*. Mol Microbiol 76: 1191–1204.
22. Yu JH, Wieser J, Adams TH (1996) The Aspergillus FlbA RGS domain protein antagonizes G protein signaling to block proliferation and allow development. EMBO J 15: 5184–5190.
23. Shimizu K, Keller NP (2001) Genetic involvement of a cAMP-dependent protein kinase in a G protein signaling pathway regulating morphological and chemical transitions in *Aspergillus nidulans*. Genetics 157: 591–600.
24. Jacobsen ID, Grosse K, Slesiona S, Hube B, Berndt A, Brock M (2010) Embryonated eggs as an alternative infection model to investigate *Aspergillus fumigatus* virulence. Infect Immun 78: 2995–3006.
25. Hall RA, Turner KJ, Chaloupka J, Cottier F, De Sordi L, et al. (2011) The quorum-sensing molecules farnesol/homoserine lactone and dodecanol operate via distinct modes of action in *Candida albicans*. Eukaryot Cell 10: 1034–1042.
26. Ben-Ami R, Varga V, Lewis RE, May GS, Nierman WC, et al. (2010) Characterization of a 5-azacytidine-induced developmental *Aspergillus fumigatus* variant. Virulence 1: 164–173.
27. Bahn YS, Kojima K, Cox GM, Heitman J (2006) A unique fungal two-component system regulates stress responses, drug sensitivity, sexual de-velopment, and virulence of *Cryptococcus neoformans*. Mol Biol Cell 17: 3122–3135.
28. Krappmann S, Sasse C, Braus GH (2006) Gene targeting in *Aspergillus fumigatus* by homologous recombination is facilitated in a nonhomologous end-joining-deficient genetic background. Eukaryot Cell 5: 212–215.
29. Kotz A, Wagener J, Engel J, Routier FH, Echtenacher B, et al. (2010) Approaching the secrets of N-glycosylation in *Aspergillus fumigatus*: Characteriza-tion of the AfOch1 protein. PLoS One 5: e15729.
30. Hill T, Käfer E (2001) Improved protocols for *Aspergillus* minimal medium: trace elements and minimal medium stock solution. Fungal Genet Newslett 48: 20–21.

5

Isolation and Characterization of Carbendazim-degrading *Rhodococcus erythropolis* djl-11

Xinjian Zhang[1⊙], **Yujie Huang**[1⊙], **Paul R. Harvey**[1,2], **Hongmei Li**[1], **Yan Ren**[1], **Jishun Li**[1], **Jianing Wang**[1], **Hetong Yang**[1]*

1 Shandong Provincial Key Laboratory of Applied Microbiology, Biotechnology Center of Shandong Academy of Sciences, Jinan, Shandong Province, People's Republic of China, 2 CSIRO Sustainable Agriculture National Research Flagship and CSIRO Ecosystem Sciences, Glen Osmond, South Australia, Australia

Abstract

Carbendazim (methyl 1*H*-benzimidazol-2-yl carbamate) is one of the most widely used fungicides in agriculture worldwide, but has been reported to have adverse effects on animal health and ecosystem function. A highly efficient carbendazim-degrading bacterium (strain djl-11) was isolated from carbendazim-contaminated soil samples via enrichment culture. Strain djl-11 was identified as *Rhodococcus erythropolis* based on morphological, physiological and biochemical characters, including sequence analysis of the 16S rRNA gene. *In vitro* degradation of carbendazim (1000 mg·L^{-1}) by djl-11 in minimal salts medium (MSM) was highly efficient, and with an average degradation rate of 333.33 mg·L^{-1}·d^{-1} at 28°C. The optimal temperature range for carbendazim degradation by djl-11 in MSM was 25–30°C. Whilst strain djl-11 was capable of metabolizing cabendazim as the sole source of carbon and nitrogen, degradation was significantly ($P<0.05$) increased by addition of 12.5 mM NH_4NO_3. Changes in MSM pH (4–9), substitution of NH_4NO_3 with organic substrates as N and C sources or replacing Mg^{2+} with Mn^{2+}, Zn^{2+} or Fe^{2+} did not significantly affect carbendazim degradation by djl-11. During the degradation process, liquid chromatography-mass spectrometry (LC-MS) detected the metabolites 2-aminobenzimidazole and 2-hydroxybenzimidazole. A putative carbendazim-hydrolyzing esterase gene was cloned from chromosomal DNA of djl-11 and showed 99% sequence homology to the *mheI* carbendazim-hydrolyzing esterase gene from *Nocardioides* sp. SG-4G.

Editor: Stephen J. Johnson, University of Kansas, United States of America

Funding: This work was supported by the grant 2010DFA32330 from the Ministry of Science and Technology of China (http://www.most.gov.cn), the grant 2012GNC11004 and Taishan Research Fellowship from Department of Science and Technology of Shandong province (http://www.sdstc.gov.cn/). The funders had no role in study design, data collection and analysis, decision to publish, or preparation of the manuscript.

Competing Interests: The authors have declared that no competing interests exist.

* E-mail: yanght@sdas.org

⊙ These authors contributed equally to this work.

Introduction

Carbendazim (methyl 1*H*-benzimidazol-2-yl carbamate, MBC) is a systemic benzimidazole fungicide widely used in many countries to control a broad range of fungal diseases of agricultural crops [1]. MBC is the hydrolytic product and active component of some other widely used benzimidzaole fungicides such as benomyl and thiophanate methyl [2,3]. MBC is relatively stable in soil and water and is reported to have an environmental half-life of up to 12 months [4]. The soil persistence and the plant systemic nature of MBC can in turn, lead to the contamination of water and plant products [5]. This causes serious concerns because MBC is a suspected mutagen, teratogen and carcinogen and is reported to be toxic to mammalian liver, endocrine and reproductive tissues [6,7]. Residual MBC in soil has also been reported to alter the taxonomic structure of soil bacterial communities and may therefore adversely affect microbial-mediated ecosystem functions [8].

There is an increasing demand to remediate soils contaminated with MBC because of the prolonged use of the fungicide in agriculture, its environmental persistence and adverse impacts on animal health. Degradation rates of MBC by physical and abiotic chemical processes are reported to be slow, with microbial metabolism thought to be the principal degradative process in natural soils [4,9,10]. Only a limited number of MBC-degrading bacterial strains have been previously reported [4,5,11] and highly efficacious, ecologically competitive microbes are required to remediate a range of MBC contaminated environments. A gene-enzyme system for MBC degradation has been previously reported in *Nocardioides* sp. [4], but mechanisms utilized by other MBC-degrading microbes are yet to be elucidated. In this study, we describe the isolation of a highly efficacious MBC-degrading *Rhodococcus erythropolis* strain djl-11, conditions affecting MBC biodegradation by this strain and sequence characterization of the djl-11 MBC-hydrolyzing esterase gene.

Materials and Methods

Chemicals and growth media

Analytical-grade carbendazim (MBC), 2-aminobenzimidazole (2-AB) and 2-hydroxybenzimidazole (2-HB) were purchased from Sigma-Aldrich Inc. All other chemicals and solvents were of highest analytical-reagent grade.

Liquid minimal salts medium (MSM) consisted of 1.0 g NH_4NO_3, 1.0 g NaCl, 1.5 g K_2HPO_4, 0.5 g KH_2PO_4, 0.2 g $MgSO_4·7H_2O$ per liter. Unless otherwise stated, MSM was adjusted to pH 7.0 and MBC was added at a final concentration of

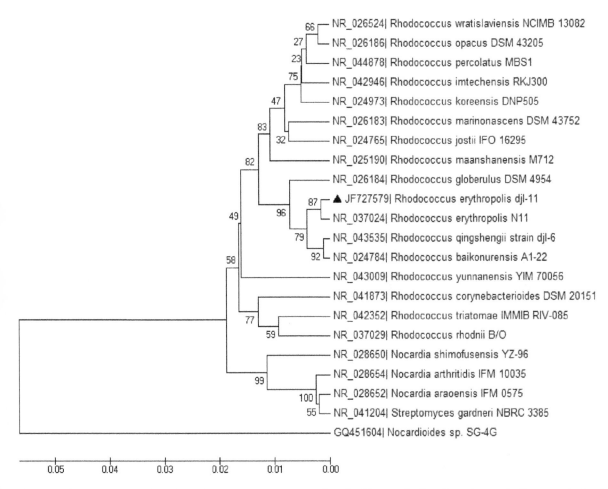

Figure 1. Phylogenetic tree based on the 16S rDNA sequence of strain djl-11 and related species. Strain djl-11 is marked with "▲". The percentages of replicate trees in which the associated species clustered together in the bootstrap test (1000 replicates) are shown next to the branches. Genbank accession numbers are shown.

1000 mg·L^{-1} in powdered form. Solid carbendazim- amended MSM contained 15 g agar L^{-1}, the carbendazim solution added to the cooled medium after autoclaving. Luria-Bertani (LB) medium was used for general bacterial growth.

Isolation of MBC degrading microorganisms

Soil samples were taken from vineyards in Rizhao (Shandong Province, China) with a 10-year history of repeated MBC applications. To select for MBC-degrading microbes, 10 g of soil was placed in a 500 mL Erlenmeyer flask containing 100 mL of MSM supplemented with 1000 mg·L^{-1} MBC (*i.e.* MSM-C$_{1000}$) as the sole carbon source and incubated at 28°C on a rotary shaker (150 rpm). After 7 days, 5 mL of culture was inoculated to 100 mL fresh MSM-C$_{1000}$ and incubated under the same conditions for another 7 days. After 5 sequential rounds of enrichment (*i.e.* 35 days exposure to MSM-C$_{1000}$), 100 μL of culture was plated onto MSM-C$_{1000}$ agar and incubated at 28°C for 5 days. Colonies showing transparent halos indicative of MBC-degradation were streaked onto fresh MSM-C$_{1000}$ agar plates to confirm MBC degradation [4] and single cell colonies were purified for further analyses.

Identification of MBC-degrading bacteria

Identification of MBC-degrading bacteria was based on morphological, physiological and biochemical characterization

according to Bergey's Manual of Systematic Bacteriology [12]. Molecular taxonomy was based on PCR amplification and DNA sequencing of the 16S rRNA gene with the universal primers 8F (5'-AGAGTTTGATCCTGGCTCAG-3') and 1541R (5'-AAG-GAGGTGATCCAGCCGCA-3') according to established protocols [13]. PCR primer synthesis and DNA sequencing (Applied Biosystems) were conducted at Sangon Biotech Co. Ltd., (Shanghai, China). The resulting nucleotide sequences were compared to those in GenBank using a BLAST search.

Microbial MBC degradation

MBC biodegradation was quantified by monitoring decreasing concentrations of the fungicide in liquid culture over time. Strain djl-11 was grown in LB broth at 28°C on a rotary shaker (150 rpm) for 24 h. Cells were collected by centrifugation (6000 g for 5 min.), washed twice and re-suspended to an OD$_{600}$ = 0.8 (Lambda Bio Spectrophotometer, Perkin Elmer, USA) in sterile water. The cell suspension (approx. 1×10^8 cells·mL^{-1}) was used to inoculate (1% v/v) 100 mL flasks of MSM-C$_{1000}$ and incubated at 28°C on a rotary shaker (150 rpm). Uninoculated MSM-C$_{1000}$ served as the negative control and each treatment was replicated 3 times. Culture samples were collected at 6 h intervals over a 72 h growth period, and 4 volumes of acetone were added to each sample. Samples were mixed well and stored at 4°C until analyzed by

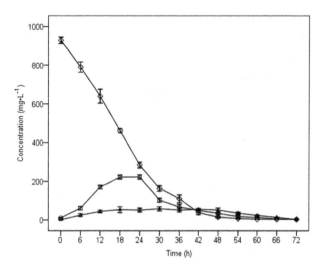

Figure 2. Chemical kinetics of MBC, 2-AB and 2-HB in liquid culture of strain djl-11. ◇, MBC; ○, 2-AB; △, 2-HB. Values are the means of three replicates with standard deviation.

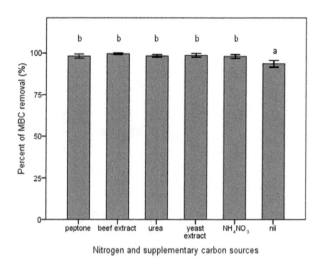

Figure 3. Effects of omitting and substituting NH₄NO₃ with organic substrates (N and C sources) on MBC biodegradation. The means of three independent experiments were plotted with error bars indicating standard deviations. Different letters above each column indicate significant differences among treatments (P<0.05).

liquid chromatography (LC) and LC-mass spectrometry (LC-MS).

To study the effects of MBC concentration on degradation by djl-11, cell suspensions were prepared as described above, inoculated (1% v/v) to 100 mL of MSM containing 200, 400, 600, 800 and 1000 mg·L^{-1} MBC and incubated at 28°C on a rotary shaker (150 rpm) for 48 h. Uninoculated flasks of MSM comprising the 5 MBC concentrations served as the negative controls. Each treatment was replicated 3 times. All samples were collected and prepared as described above for subsequent LC and LC-MS analyses.

Identification and quantification of MBC and its metabolites

Quantitative analysis of MBC, 2-AB, and 2-HB was conducted with an Agilent series LC system (Agilent Technologies, USA). Chromatographic separation was achieved on an Eclipse XDB-C18 column (150 mm×4.6 mm, 5-μm particle size) at 25°C. MBC and its metabolites were monitored at 270 nm using an acetonitrile-water mixture (16:84 [v/v] containing 0.1% [v/v] formic acid) as a mobile phase, at a flow rate of 1 mL·min^{-1}. Metabolites were qualitatively analyzed by a LC-MS mass spectrometer (Agilent Technologies, USA). The separated substrate and metabolites were ionized with positive polarity and scanned within a mass range of 29 to 500 m/z.

Effects of organic substrates, bivalent cations, pH and temperature on MBC degradation

Four experiments were established to determine the individual effects of organic substrates (N and C sources), cations, pH and temperature on MBC degradation by djl-11. Unless otherwise indicated, bacterial inoculum was prepared, cultured in MSM-C₁₀₀₀ (48 h) and MBC degradation analyzed using the methods described above. Each treatment of the 4 experiments was replicated 3 times.

To examine the effects of organic substrates as alternative sources of nitrogen and supplementary sources of carbon on MBC biodegradation, 0.1% peptone, 0.1% beef extract, 0.1% urea, and 0.1% yeast extract (w/v) were respectively added to MSM-C₁₀₀₀ in the absence of NH₄NO₃. MSM with no nitrogen source was

included as the negative control. Similarly, substitutions of the 810.8 μM MgSO₄ in MSM-C₁₀₀₀ with equimolar amounts of ZnSO₄, MnSO₄, CuSO₄, CaSO₄ or FeSO₄ were used to determine the effects of alternative bivalent cations on MBC biodegradation.

The effect of initial MSM-C₁₀₀₀ pH on MBC biodegradation was observed on a scale of pH 4 to pH 9 at increments of 1 pH unit. Optimal culture incubation temperatures for biodegradation of MBC were examined using 5°C increments on a scale of 20°C to 40°C.

Cloning of the MBC-hydrolyzing esterase gene

A putative MBC-hydrolyzing esterase gene was amplified by PCR from chromosomal DNA of djl-11. The primers MheI-F (5′-gcatggccaacttcgtcctcg-3′) and MheI-R (5′-gcgcccagcgccgccagc-3′) were designed according to the sequence of a MBC-hydrolyzing esterase encoding gene *mheI* (GenBank accession GQ454794) [4]. For PCR amplification, approximately 270 ng of djl-11 gDNA,

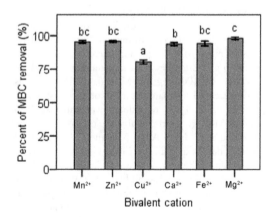

Figure 4. Effects of metal ions on MBC biodegradation. The means of three independent experiments were plotted with error bars indicating standard deviations. Different letters above each column indicate significant differences among treatments (P<0.05).

5 µL of 10× PCR buffer (Mg^{2+} free), 3 µL of 25 mM MgCl$_2$, 4 µL of 2.5 mM dNTPs, 20 pmol of each primer, and 1.25 U of TaKaRa Taq polymerase (TaKaRa Bio Inc., Dalian) were added to a final volume of 50 µL. PCR amplification was carried out as follows: 4 min at 95°C, 30 cycles of 40 sec at 95°C, 30 sec at 58°C and 40 sec at 72°C, plus a final extension step of 8 min at 72°C. The amplified product was purified using an Agarose gel DNA purification kit (TaKaRa Bio Inc., Dalian), inserted into the T-A cloning vector pMD18-T (TaKaRa Bio Inc., Dalian) and sequenced on an Applied Biosystems DNA analyzer. PCR primer synthesis and DNA sequencing were conducted at Sangon Biotech Co. Ltd. (Shanghai, China). The nucleotide sequence was compared to those in GenBank using a BLAST search.

Statistical analysis

Biodegradation of MBC by strain djl-11 was assessed by comparing differences in MBC concentration between treatments, each consisting of 3 replicates. All data were analyzed by analyses of variance (ANOVA), using SPSS 16.0 statistical software (SPSS Inc., USA). Pairwise comparisons of means were used to compute Fisher's least significant difference values (LSD, P = 0.05).

Ethics statement

We confirm that the owner of vineyards gave permissions to take soil samples from the fields. We confirm that no endangered or protected species were involved in field studies.

Results

Isolation and identification of MBC-degrading strain djl-11

Enrichment cultures established from MBC-contaminated soils were plated onto MSM-C$_{1000}$ agar to select for putative MBC degrading microbes. Bacteria representing different colony morphologies were purified and confirmed to have MBC degradative function via plate-clearing assays. Strain djl-11, qualitatively assessed as the most effective MBC degrader, was selected for further study.

Strain djl-11 was a gram-positive, non-motile, rod-shaped bacterium that formed orange colonies on LB agar after 72 h at 28°C. In physiological and biochemical tests, djl-11 tested positive for catalase, urease and acetoin produciton, but negative for oxidase, starch hydrolysis and nitrate reductase. Strain djl-11 was

able to utilise citrate, mannose, sodium benzoate and maltose as sole carbon sources and acetylamine and asparagine as sole sources of carbon and nitrogen.

A 1.5 kb 16S rRNA fragment was amplified from strain djl-11, sequenced and showed 99% homology to *Rhodococcus erythropolis* 16S rRNA. Phylogenetic analysis (Figure 1) based on 16S rDNA sequences revealed that strain djl-11 clustered with *Rhodococcus* species and was most closely related to *R. erythropolis* N11. Molecular taxonomy, cellular and colony morphologies and physiological and biochemical characteristics identified djl-11 as *R. erythropolis*. Strain djl-11 and its 16s rRNA sequence were deposited in the China General Microbiological Culture Collection Center (Accession No. CGMCC4554) and GenBank (Accession No. JF727579), respectively.

MBC biodegradation and metabolite identification in liquid culture

Seed cultures of djl-11 cells (1% v/v) provided with MBC (1000 mg·L^{-1}) as a sole carbon-source degraded approximately 95% of the fungicide in 48 h, with the remaining MBC completely degraded by 72 h (Figure 2). The average degradation rate of MBC by djl-11 was 333.33 mg·L^{-1}·d^{-1} in MSM-C$_{1000}$ at 28°C. Varying the concentration of MBC (200–1000 mg·L^{-1}) at time of djl-11 inoculation had no significant effect on overall degradation, with 95% of the fungicide removed after 48 h growth in all treatments (data not shown). No significant MBC degradation was observed in any of the non-inoculated controls, regardless of MBC concentration at time of inoculation.

Two major metabolite peaks were detected during the growth of djl-11 on MBC, these degradation intermediates being identified as 2-aminobenzimidazole (2-AB) and 2-hydroxybenzimidazole (2-HB) by LC and LC-MS using authentic standards.

Effects of organic substrates, bivalent cations, pH and temperature on MBC degradation

Strain djl-11 degraded approximately 90% of MBC from MSM-C$_{1000}$ in the absence of NH$_4$NO$_3$, indicating its capability to utilize MBC as a sole source of carbon and nitrogen (Figure 3). Omission of NH$_4$NO$_3$ however, resulted in significantly less (P<0.05) MBC degradation compared to that degraded in the presence of this nitrogen-source (Figure 3). Substitution of NH$_4$NO$_3$ in MSM-C$_{1000}$ with equivalent amounts of organic nitrogen and supplementary carbon sources (*i.e.* peptone, urea, beef extract or yeast extract) had no significant effect on MBC degradation by djl-11.

Substitution of Mg^{2+} in MSM-C$_{1000}$ with equimolar amounts of Mn^{2+}, Zn^{2+} or Fe^{2+} had no significant effect on MBC degradation by djl-11 (Figure 4). In contrast, MBC degradation was significantly decreased (P<0.05) when Mg^{2+} in MSM-C$_{1000}$ was substituted with equimolar Ca^{2+} or Cu^{2+} (Figure 4). Substitution with Cu^{2+} resulted in the lowest overall MBC bio-degradation, significantly less (P<0.05) than that observed in the MSM-C$_{1000}$ containing Mg^{2+} (Figure 4).

Temperature had a significant (P<0.05) effect on MBC degradation by djl-11 (Figure 5), with optimal degradation detected in the 25°C to 30°C range. MBC degradation was significantly (P<0.05) lower at 20°C and inhibited further at elevated temperatures of 35°C to 40°C (Figure 5). In contrast, MSM-C$_{1000}$ pH (range pH 4–9) at time of inoculation has no significant effect on biodegradation of MBC by djl-11 (data not shown).

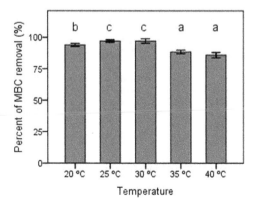

Figure 5. Effect of temperature on MBC biodegradation. The means of three independent experiments were plotted with error bars indicating standard deviations. Different letters above each column indicate significant differences among treatments (P<0.05).

MBC-hydrolyzing esterase gene of strain dj1-11

Primers designed from the previously reported MBC-hydrolyzing esterase gene *mheI* amplified a dj1-11 DNA sequence (*Mhe*) consisting of a 729 bp open reading frame starting with the ATG codon, ending with the stop codon TGA and encoding 242 amino acids residues. The predicted amino acid sequence corresponded to a 26.285 kDa protein with an isoelectric point of 6.27. *R. erythropolis* dj1-11 Mhe exhibited 99% amino acid sequence identity with MBC-hydrolyzing esterase encoded by *mheI* from *Nocardioides* sp. strain SG-4G (GenBank accession number GQ454794). The djl-11 *Mhe* DNA sequence of was deposited in GenBank (accession number HQ874282).

Discussion

At present, only a limited number of bacterial strains capable of degrading MBC have been reported [4,5,11]. Strains from the bacterial genus *Rhodococcus* were most often reported, such as *R. erythropolis* [14], *Rhodococcus qingshengii* [15,16], *Rhodococcus jialingiae* [11]. In this study, strain *R. erythropolis* djl-11 capable of catabolizing and utilizing MBC as the sole carbon and nitrogen sources was isolated. Strain djl-11 showed high MBC-degrading efficacy, with 99% of 1000 mg·L^{-1} MBC being degraded within 72 h. In comparison, *R. qingshengii* djl-6 utilized 100 mg·L^{-1} MBC as the sole carbon source, with an average MBC degradation rate of only 55 mg·L^{-1}·d^{-1} [15].

Varying the concentration of MBC (200–1000 mg·L^{-1}) at time of inoculation had no significant effect on MBC degradation by strain djl-11, with 95% of the fungicide removed after 48h. In contrast, previous researches on MBC degradation by *Bacillus pumilus* NY97-1 [17] and *Pseudomonas* sp. CBW [5] reported enhanced MBC degradation as concentrations of the fungicide increased. However, MBC concentrations (1–300 mg·L^{-1}) in these studies were much lower than those exposed to djl-11, and degradation by *Pseudomonas* sp. CBW was significantly inhibited above MBC concentration of 100 mg·L^{-1} [5].

MBC degradations by *Bacillus pumilus* NY97-1 and *Pseudomonas* sp. CBW were significantly influenced by factors such as pH, temperature and nutrient composition of the culture media [5,17]. In contrast, MBC degradations by strain djl-11 were not significantly affected by varying the initial pH ranging from 4–9, replacing Mg^{2+} with Mn^{2+}, Zn^{2+} or Fe^{2+} or substituting NH$_4$NO$_3$ with organic substrates (peptone, urea, beef and yeast extracts), the latter providing alternative and additional sources of nitrogen and carbon, respectively. MBC degradation by djl-11 was however,

reduced by 5–15% (P<0.05) in the absence of NH$_4$NO$_3$ when using the fungicide as a sole nitrogen and carbon source or at culture temperature ±5–10°C of the optimum for growth (30°C). Whilst significant, these reductions in djl-11 MBC degrading efficacy are relatively small in comparison with other bacterial strains [5,17], indicating the robustness of the process by djl-11 and the potential for MBC bioremediation in different environments.

The metabolites 2-AB and 2-HB were identified during the growth of strain djl-11 on MBC, supporting previous studies of MBC catabolism by *Nocardioides* sp. SG-4G [4] and *Pseudomonas* sp. CBW [5]. Whilst MBC degradation by *R. qingshengii djl-6* [15] and *R. jialingiae* djl-6–2 [11] also produced 2-AB, the intermediate benzimidazole (BI) was also detected either in the presence [11] or absence [15] of 2-HB. Notably, BI was not detected during growth of *R. erythropolis* djl-11, *Nocardioides* SG-4G [4] or *Pseudomonas* CBW [5] on MBC. As 2-AB and 2-HB exhibit relatively benign toxicity [18], no attempt was made to define the downstream metabolites. It was proposed that MBC was first converted to 2-AB, which was then transformed to 2-HB, 1,2-diaminobenzene, catechol, and finally to carbon dioxide by *Pseudomonas* sp. CBW [5].

Till now, only one gene, *mheI* from *Nocardioides* sp. SG-4G, which encodes the first enzyme of the pathway that detoxifies MBC by hydrolyzing it to 2-AB was cloned and reported [4]. In this study, MBC-hydrolyzing esterase (Mhe) gene from *R. erythropolis* djl-11 was cloned, and Mhe exhibited 99% amino acid identity to that of *Nocardioides* sp. SG-4G MheI esterase, suggesting that both strains utilize enzymatic hydrolysis as the first step in catabolism and detoxification of MBC to 2-AB. BLAST searches for *Mhe* in all available *Rhodococcus* genomes, including 3 other *R. erythropolis* strains, did not detect any homologous loci with high similarities. The evolutionary origin of Mhe gene and its frequency among other MBC-degrading bacterial strains need to be further elucidated.

Acknowledgments

The authors would like to thank Dr. Jinpeng Yuan for his excellent assistance in the LC-MS analysis and elucidation.

Author Contributions

Conceived and designed the experiments: XZ YH HY. Performed the experiments: XZ YH HL YR. Analyzed the data: XZ PRH JW JL. Contributed reagents/materials/analysis tools: YH HL YR. Wrote the paper: XZ PRH HY.

References

1. Chen Y, Zhou MG (2009) Characterization of Fusarium graminearum isolates resistant to both carbendazim and a new fungicide JS399–19. Phytopathology 99: 441–446.

2. Sandahl M, Mathiasson L, Jonsson JA (2000) Determination of thiophanate-methyl and its metabolites at thrace level in spiked natural water using the supported liquid membrane extraction and the microporous membrane liquid-liquid extraction techniques combined on-line with highperformance liquid chromatography. J Chromatogr A 893: 123–131.

3. Boudina A, Emmelin C, Baaliouamer A, Grenier-Loustalot MF, Chovelon JM (2003) Photochemical behaviour of carbendazim in aqueous solution. Chemosphere 50: 649–655.

4. Pandey G, Dorrian SJ, Russell RJ, Brearley C, Kotsonis S, et al. (2010) Cloning and biochemical characterization of a novel carbendazim (methyl-1H-benzimidazol-2-ylcarbamate)-hydrolyzing esterase from the newly isolated Nocardioides sp. strain SG-4G and its potential for use in enzymatic bioremediation. Appl Environ Microbiol 76: 2940–2945.

5. Fang H, Wang Y, Gao C, Yan H, Dong B, et al. (2010) Isolation and characterization of Pseudomonas sp. CBW capable of degrading carbendazim. Biodegradation 21: 939–946.

6. Selmanoglu G, Barlas N, Songur S, Kockaya EA (2001) Carbendazim-induced haematological, biochemical and histopathological changes to the liver and kidney of male rats. Hum Exp Toxicol 20: 625–630.

7. Farag A, Ebrahim H, ElMazoudy R, Kadous E (2011) Developmental toxicity of fungicide carbendazim in female mice. Birth Defects Res B Dev Reprod Toxicol 92: 122–130.

8. Wang YS, Huang YJ, Chen WC, Yen JH (2009) Effect of carbendazim and pencycuron on soil bacterial community. J Hazard Mater 172: 84–91.

9. Kiss A, Virag D (2009) Photostability and photodegradation pathways of distinctive pesticides. J Environ Qual 38: 157–163.

10. Yarden O, Salomon R, Katan J, Aharonson N (1990) Involvement of fungi and bacteria in enhanced and nonenhanced biodegradation of carbendazim and other benzimidazole compounds in soil. Can J Microbiol 36: 15–23.

11. Wang Z, Wang Y, Gong F, Zhang J, Hong Q, et al. (2010) Biodegradation of carbendazim by a novel actinobacterium Rhodococcus jialingiae djl-6–2. Chemosphere 81: 639–644.

12. Goodfellow M (1989) Genus Rhodococcus. In: Williams ST, Sharpe ME, Holt JG, editors. Bergey's Manual of Systematic Bacteriology. Baltimore, MD.: Williams & Wilkins. 2362–2371.

13. Baker GC, Smith JJ, Cowan DA (2003) Review and re-analysis of domain-specific 16S primers. J Microbiol Methods 55: 541–555.

14. Holtman MA, Kobayashi DY (1997) Identification of Rhodococcus erythropolis isolates capable of degrading the fungicide carbendazim. Appl Microbiol Biotechnol 47: 578–582.

15. Xu J, Gu X, Shen B, Wang Z, Wang K, et al. (2006) Isolation and characterization of a carbendazim-degrading Rhodococcus sp. djl-6. Curr Microbiol 53: 72–76.

16. Xu JL, He J, Wang ZC, Wang K, Li WJ, et al. (2007) Rhodococcus qingshengii sp. nov., a carbendazim-degrading bacterium. Int J Syst Evol Microbiol 57: 2754–2757.

17. Zhang LZ, Qiao XW, Ma LP (2009) Influence of environmental factors on degradation of carbendazim by *Bacillus pumilus* strain NY97-1. Int J Environ Pollut 38: 309–317.

18. Stringer A, Wright MA (1976) The toxicity of benomyl and some related 2-substituted benzimidazoles to the earthworm *Lumbricus terrestris*. Pestic Sci 7: 459–464.

6

Differential Contributions of Five ABC Transporters to Mutidrug Resistance, Antioxidion and Virulence of *Beauveria bassiana*, an Entomopathogenic Fungus

Ting-Ting Song[1,2◐], **Jing Zhao**[1◐], **Sheng-Hua Ying**[1], **Ming-Guang Feng**[1]*

1 Institute of Microbiology, College of Life Science, Zhejiang University, Hangzhou, Zhejiang, People's Republic of China, **2** Horticulture Institute, Zhejiang Academy of Agricultural Sciences, Hangzhou, Zhejiang, People's Republic of China

Abstract

Multidrug resistance (MDR) confers agrochemical compatibility to fungal cells-based mycoinsecticdes but mechanisms involved in MDR remain poorly understood for entomopathogenic fungi, which have been widely applied as biocontrol agents against arthropod pests. Here we characterized the functions of five ATP-binding cassette (ABC) transporters, which were classified to the subfamilies ABC-B (Mdr1), ABC-C (Mrp1) and ABC-G (Pdr1, Pdr2 and Pdr5) and selected from 54 full-size ABC proteins of *Beauveria bassiana* based on their main domain architecture, membrane topology and transcriptional responses to three antifungal inducers. Disruption of each transporter gene resulted in significant reduction in resistance to four to six of eight fungicides or antifungal drugs tested due to their differences in structure and function. Compared with wild-type and complemented (control) strains, disruption mutants of all the five transporter genes became significantly less tolerant to the oxidants menadione and H_2O_2 based on 22−41% and 10−31% reductions of their effective concentrations required for the suppression of 50% colony growth at 25°C. Under a standardized spray, the killing actions of $\Delta Pdr5$ and $\Delta Mrp1$ mutants against *Spodoptera litura* second-instar larvae were delayed by 59% and 33% respectively. However, no significant virulence change was observed in three other delta mutants. Taken together, the examined five ABC transporters contribute differentially to not only the fungal MDR but antioxidant capability, a phenotype rarely associated with ABC efflux pumps in previous reports; at least some of them are required for the full virulence of *B. bassiana*, thereby affecting the fungal biocontrol potential. Our results indicate that ABC pump-dependent MDR mechanisms exist in entomopathogenic fungi as do in yeasts and human and plant pathogenic fungi.

Editor: Arnold Driessen, University of Groningen, Netherlands

Funding: Funding for this study was provided by the Natural Science Foundation of China (Grant Nos: 30930018 and 30971960) and the Ministry of Science and Technology of China (Grant No: 2011AA10A204). The funders had no role in study design, data collection and analysis, decision to publish, or preparation of the manuscript.

Competing Interests: The authors have declared that no competing interests exist.

* E-mail: mgfeng@zju.edu.cn

◐ These authors contributed equally to this work.

Introduction

Multidrug resistance (MDR) is a major challenge for the control of human, animal and plant pathogenic fungi by antifungal drugs and fungicides [1,2,3] but could be a merit for fungal entomopathogens against arthropod pests [4,5]. This is because fungal cells, such as conidia produced on solid substrates, are the active ingredients of numerous mycoinsecticids and mycoacaricides [6] and MDR may confer their compatibility with chemical fungicides, herbicides and insecticides. Fungal candidate strains with higher MDR are more tolerant to applied chemical pesticides and thus more potential for commercial development and application.

MDR mechanisms in entomopathogenic fungi remain poorly understood although their compatibility with chemical pesticides has been emphasized as one of the determinants to a success of microbial control [7,8]. Previously, some of common β-tubulin point mutations that are attributed to benzimidazole resistance in phytopathogenic fungi [9,10] were found in *Beauveria bassiana* mutants with extraordinarily high carbendazim resistance [11].

However, none of such point mutations was found in *Isaria fumosorosea* mutants that showed not only as high carbendazim resistance as in the *B. bassiana* mutants but also resistance to other compounds different in structure and function [12]. Interestingly, all the *I. fumosorosea* mutants had three common point mutations occurred at the binding sites of the transcription factors Gal4, Abf1 and Raf in the promoter region of an ATP-binding cassette (ABC) transporter gene (*ifT1*) and thus their *ifT1* transcripts were upregulated by 17- to 137-fold. This implies that ABC transporter-dependent MDR mechanism exists in the fungal entomopathogens.

As a large family, ABC transporter proteins can energize the transport of a huge variety of compounds across biological membranes through ATP hydrolysis and confer cellular resistance to a broad spectrum of drug substrates [i.e., MDR or PDR (pleiotropic drug resistance) phenomenon] or a very limited number of substrates [13]. They are structurally featured with essential nucleotide-binding domain(s) (NBD) and one or two hydrophobic transmembrane domains (TMDs) and usually composed of six K-helical transmembrane segments (TMSs), forming the domain architectures of full-size [(TMS$_6$−NBD)$_2$ or (NBD−TMS$_6$)$_2$], half-size (TMS$_6$−NBD) and TMD-lacking (NBD or NBD$_2$) transporters [14]. Those associated with MDR/PDR are all full-size members classified to the subfamilies ABC-B (MDR type), ABC-C (MRP type, i.e., multidrug resistance-associated proteins) and ABC-G (PDR type) [15]. In human and plant pathogens, MDR/PDR results from drug efflux pumped by ABC transporters to reduce intracellular drug accumulation to toxic level at target sites [16,17]. For instance, two PDR-type transporters, Cdr1p and Cdr2p, contribute differentially to azole resistance in *Candida albicans* [18,19] due to their structural differences associated with substrate specificities and transport mechanism [20,21]. ABC transporters also mediate cellular tolerance to natural toxic compounds and xenobiotics and/or virulence in many phytopathogenic fungi [22,23,24,25]. Interestingly, the coding gene of a PDR-type ABC transporter in wheat supports durable resistance to wheat pathogenic fungi [26].

To explore possible MDR mechanisms in *B. bassiana*, we characterized the functions of three types of five representative proteins, which were selected from all full-size ABC transporter proteins by analyzing their phylogenetic and structural features and assessing their expressional responses to three different antifungal drugs. We found that the five transporters made differential contributions to the fungal MDR, antioxidation and virulence by multi-phenotypic comparisons of their single-gene disruption mutants with wild-type and complement strains

Results

Features of ABC transporters in *B. bassiana*

Up to 425 transporter proteins were blasted from the annotated genome of the wild-type strain *B. bassiana* ARSEF 2860 (Bb2860 or wild type herein) [27], including 54 putative ABC pumps coupled with the queries of conserved NBD and TMD regions of budding yeast Ste6p, Pdr5p or Yor1p in the NCBI protein database. The 54 proteins were classified to four ABC subfamilies (Fig. 1A). The largest ABC-C subfamily includes 37 members, of which eight are likely MRP-type transporters based on their membrane topology. Further comparison of domain architecture led to the recognition of seven ABC-B and six ABC-G proteins as potential MDR- and PDR-type transporters respectively. All the 21 recognized transporters of three types are featured with two NBDs and two TMDs

The 21 ABC transporters were assessed for the levels of their gene transcripts in wild-type hyphal cells induced with azoxysyrobin, carbendazim and phosphinothricin for 20 min and 2 h at 25°C respectively. As a result of quantitative real-time PCR (qRT-PCR) with paired primers (Table S1), about half of them were upregulated by the drug inducers within 20 min (Fig. 1B). Longer induction enhanced their transcripts to higher levels (Fig. 1C). Consequently, Pdr1, Pdr2, Pdr5, Mdr1 and Mrp1 were chosen as the representatives of the three types because they were inductively upregulated by all the three drugs. Notably, *Mdr6* and *Mdr7* transcripts were consistently undetectable in the cDNAs from the samples induced or not induced with the drugs (data not shown).

The coding genes of the selected five transporters were disrupted from Bb2860 and complemented into their disruption mutants by integration of the *bar*- and *sur*-inclusive plasmids via *Agrobacterium*-mediated transformation respectively. Putative mutant colonies grown on selective plates were sequentially identified via PCR, reverse transcription PCR (RT-PCR) and Southern blotting with paired primers and amplified probes (Table S2). As a result of the identification, the profiling band or signal for each target gene was consistently present in the wild-type and

complement strains (control strains) but absent in the disruption mutant (Fig. S1). Thus, five single-gene disruption mutants were compared with the control strains to differentiate their phenotypic changes below.

Differentiated MDR responses

All the five disruption mutants showed differential resistance to different types of four fungicides (Fig. 2A) and four antifungal drugs (Fig. 2B) during 6-day growth on 1/4 SDAY at 25°C but their control strains responded equally to each drug (Tukey's HSD, *P*>0.1). Compared with the means of relative growth inhibition (RGI) values observed in the control strains, six, five and four of the tested chemicals were significantly more inhibitory to *ΔPdr1* (12−43%), *ΔPdr2* (11−35%) and three other delta mutants (9−31%), respectively. During the colony growth, dimetachlone exerted inhibitory effect on all the delta mutants while itraconazole, azoxysyrobin and ethirimol were influential only on one or two of them. Null responses were observed in *ΔPdr1* to Congo red and azoxysyrobin and in *ΔPdr2* to carbendazim, itraconazole and 4-nitroquinoline-N-oxide. These data indicated that the spectra and preference of drug substrates were partially different among the five ABC transporters.

Additionally, all the disruption mutants and the control strains grew equally well on drug-free SDAY or 1/4 SDAY at 25°C (*P*>0.15 in *F* tests) and responded equally to hyperosmotic (NaCl) stress ($F_{10,22} = 1.25$, *P*=0.32; data not shown).

Differentiated antioxidation responses

Two oxidants, H$_2$O$_2$ and menadione, were assayed for their effective concentrations (EC$_{50}$s) to suppress 50% colony growth of each strain by modeling analysis of relative growth trend over the gradient concentrations of each oxidant after 6-day incubation at 25°C. Compared with the EC$_{50}$ estimates of menadione [5.6 (±0.09) mM] and H$_2$O$_2$ [41.1 (±0.87) mM] towards the control strains (Fig. 3A), all the five delta mutants were 22−41% less tolerant to menadione and 10−31% less tolerant to H$_2$O$_2$ (Tukey's HSD, *P*<0.01).

Differentiated virulence

Time-mortality trends of all the tested strains against the second-instar larvae of *Spodoptera litura* in the bioassays standardized by a uniform spray of conidial suspension in an automatic spray tower were differentiated by probit analysis. Median lethal time (LT$_{50}$) estimates fell in a narrow range of 4.8−5.3 (average 5.1) days for all the control strains (Tukey's HSD, *P*≥0.15) but significantly increased to 8.1 and 6.8 days for *ΔPdr5* and *ΔMrp1* respectively (Fig. 3D). However, no significant LT$_{50}$ differences were found between other delta mutants and the control strains.

Discussion

B. bassiana harbors 21 full-size ABC transporter genes that may act as potential MDR regulators due to their classification, membrane topology and domain architecture. Of those, however, *Mdr6* and *Mdr7* had no detectable transcriptional signals in the cDNAs from the total RNAs of the wild-type cultures induced with drugs or not induced, suggesting a likelihood of their pseudogene status. Five representatives selected by their transcript levels inductively upregulated by azoxysyrobin, carbendazim and phosphinothricin were confirmed contributing differentially to the fungal MDR, antioxidation and virulence but not involving in osmoregulation, as discussed below.

First of all, up to 425 transporter proteins can be blasted from the genome of *B. bassiana* [27] and the counts of the counterparts

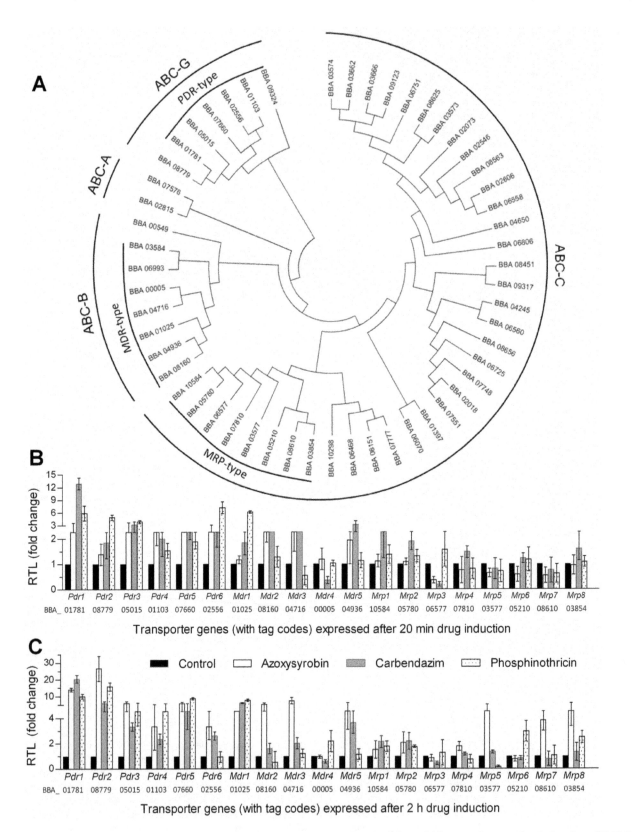

Figure 1. Screening of full-size ABC transporter proteins associated with multidrug resistance in *B. bassiana* (Bb2860). (A) Phylogenetic analysis of 54 full-size ABC proteins. (B), (C) Relative transcript levels (RTL) of 21 ABC transporter genes in the wild-type SDB cultures induced with carbendazim (5 µg/ml), azoxysyrobin (100 µg/ml) and phosphinothricin (100 µg/ml) for 20 min and 2 h at 25°C respectively. Error bars: SD of the mean from three cDNA samples assessed via qRT-PCR with paired primers (Table S1).

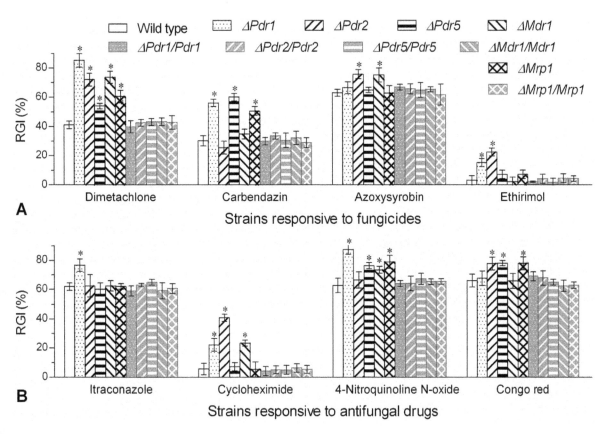

Figure 2. Changes in multidrug resistance of five single-gene disruption mutants of *B. bassiana*. (A) Relative growth inhibition (RGI) of fungal colonies after 6-day incubation at 25°C on 1/4 SDAY supplemented with the fungicides dimetachlone (0.1 mg/ml), carbendazim (0.5 μg/ml), azoxysyrobin (0.1 mg/ml) and ethirimol (1 mg/ml) respectively. (B) RGI values of fungal colonies after 6-day incubation at 25°C on 1/4 SDAY supplemented with the antifungal drugs itraconazole (5 μg/ml), cyclonheximide (20 μg/ml), 4-nitroquinoline-N-oxide (5 μg/ml) and Congo red (0.5 mg/ml) respectively. The bars of each group marked with asterisks differed significantly from those unmarked (Tukey's HSD, $P<0.05$). Error bars: SD of the mean from three repeated assays.

in the genomes of *Metarhizium robertsii* (previously in *M anisopliae* sensu lato) and *M. acridum* [28] are 304 and 307 respectively. Despite a remarkable diversity, only a small proportion of them are full-size ABC pumps of *B. bassiana* (Fig. 1A) and an even smaller proportion are likely associated with the fungal MDR/ PDR in terms of their membrane topology and main domain architecture (14,15,21). The five transporters more responsive to the three inducers (Fig. 1) were all proven to regulate MDR in *B. bassiana* because their single-gene disruption mutants showed less resistance to four to six of the eight antifungal drugs (Fig. 2), which were used in MDR assays due to differences in structure and function. Apparently, ABC pump-dependent MDR mechanisms exist in entomopathogenic fungi as do in yeasts and human and plant pathogenic fungi [13,16].

Despite partially overlapping drug spectra, the examined five transporters showed some degree of substrate preference based on the MDR changes in their delta mutants. The preferred substrate was dimetachlone for Pdr1 and Mdr1, carbendazim for Pdr5 and Mrp1, and cycloheximide for Pdr2. As a broad-spectrum fungicide, dimetachlone was a mere common substrate for the five transporters but itraconazole was specific to only Pdr1 among the tested drugs. Moreover, the drug spectrum was broadest for Pdr1, followed by Pdr2 and three others. Regardless of broader or narrower substrate spectrum, Pdr1, Pdr2, Pdr5, Mdr1 and Mrp1 are functionally very close to other fungal ABC transporters, such as those mediating *Aspergillus nidulans* resistance to all major classes

of fungicides [29], *Botrytis cinerea* sensitivity to phenylpyrrole fungicides [30], and *Mycosphaerella graminicola* responses to azole fungicides [25]. Their drug preferences, substrate spectra and MDR levels altered by single-gene disruption are partially different from one to another. This is in accordance with those of documented fungal ABC pumps between different types [13] or within a type [18,19] and likely due to low primary sequence similarity between their TMDs [21]. Thus, the five transporters of *B. bassiana* regulate differentially the fungal MDR/PDR.

Apart from differential responses to the tested antifungal drugs, all five delta mutants showed significantly less, but differential, resistance to the oxidants menadione and H_2O_2 (Fig. 3A). Fungal antioxidant capability has rarely been associated with ABC transporters in previous studies but is important for the success of *B. bassiana* infection. This capability usually depends on the activities of antioxidative enzymes, such as catalases [31] and superoxide dismutases [32,33], and can be regulated by cellular signaling pathways, such as the mitogen-activated protein kinase cascades of Hog1 [34] and Slt2 [35], P-type calcium ATPase [36] and Ras1/Ras2 GTPases [37]. Particularly, fungal tolerance to oxidation is linearly correlated with *B. bassiana* UV resistance and virulence [33,37], two parameters important for the fungal biocontrol potential. Thus, the antioxidant capability reduced by the disruption of each ABC transporter gene implies that the fungal pathogen is less capable of scavenging harmful superoxide anions often generated from infected host cells. We consider that

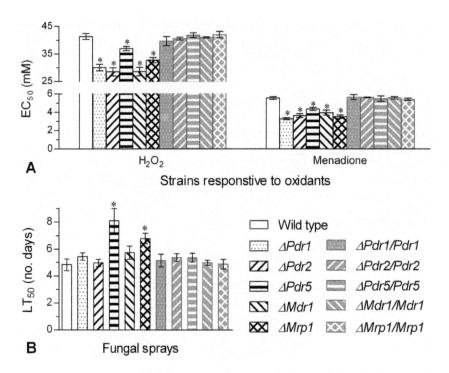

Figure 3. Changes in antioxidant capability and virulence of five single-gene disruption mutants of B. bassiana. (A) Effective concentrations (EC$_{50}$s) estimated for H$_2$O$_2$ and menadione to suppress 50% colony growth by modeling analysis of relative growth trends over the concentrations of 0−80 mM H$_2$O$_2$ or 0−8 mM menadione added to 1/4 SDAY. (B) Median lethal times (LT$_{50}$s) of wild-type and mutant strains against the second-instar larvae of S. litura under a standardized spray. The bars of each group marked with asterisks differed significantly from those unmarked (Tukey's HSD, $P<0.05$). Error bars: SD of the mean from three repeated assays.

the five transporters could pump both oxidants as they usually pump xenobiotic efflux although they are not antioxidant enzymes.

Finally, fungal virulence has been infrequently associated with the effects of ABC transporters but this association has been found in some phytopathogenic fungi. For instance, three ABC pumps, namely ABC1 in *Magnaporthe grisea* [24], NhABC1 in *Nectria haematococca* [22] and BcatrB in *Botrytis cinerea* [38], have proved to influence the fungal virulence due to their pumping action of cytotoxic compound efflux. In this study, only Pdr5 and Mrp1 were found contributing significantly to the virulence of *B. bassiana* to *S. litura* larvae because the killing actions of their delta mutants under a standardized spray were 59% and 33% slower than those of the control strains (Fig. 3B). However, three other transporters we examined showed null effect on the fungal virulence. Taken together with previous reports and our results, not all ABC transporters are contributors to fungal virulence but at least some of them are necessary for the full virulence of a fungal pathogen, thereby affecting the biocontrol potential of *B. bassiana*.

Materials and Methods

Microbial strains and culture conditions

The wild-type strain Bb2860 was cultured on Sabouraud dextrose agar plus 1% yeast extract (SDAY) at 25°C and used as a recipient of gene manipulation and expression. *Escherichia coli* Top10 and *E. coli* DH5α from Invitrogen (Shanghai, China) used for vector propagation were cultured at 37°C in LB medium plus kanamycin (100 µg/ml). For fungal transformation, *A. tumefaciens* AGL-1 was cultured in YEB medium [39] at 28°C.

Phylogenetic, structural and transcriptional analyses of B. bassiana ABC transporters

The conserved NBD and TMD regions of the typical ABC transporters Ste6p, Pdr5p and Yor1p in *Saccharomyces cerevisiae* (NCBI accession codes: NC_001143.9, NC_00147.6 and NC_00113.9 respectively) were used as queries to locate PDR-, MDR- and MRP-type transporters respectively in the sequenced genome of Bb2860 under the NCBI accession ADAH00000000 [27] via BLASTP (http://blast.ncbi.nlm.nih.gov/blast.cgi) and BioEdit analysis (http://www.mbio.ncsu.edu/bioedit/bioedit.html). The resultant protein sequences were screened to remove half-size transporters not associated with MDR in terms of their domain architecture and then classified to ABC subfamilies following a documented system [15] via phylogenetic analysis with MEGA 4.0 software [40]. Their membrane topological features were further analyzed to assess the likelihood of their involvements in fungal MDR/PDR [13] via CD search (http://www.ncbi.nlm.nih.gov/Structure/cdd/wrpsb.cgi), generating 21 full-size ABC transporters (Mdr1−7, Mrp1−8 and Pdr1−6) for study.

To assess the transcriptional expression levels of the selected transporters in response to different antifungal chemicals, Bb2860 was grown in 50 ml aliquots of Sabouraud dextrose broth (SDB) inoculated to 1×10^6 conidia/ml. After 2-day shaking by 120 rpm at 25°C, hyphal cells were harvested from the cultures and transferred to the same volume of fresh SDB supplemented with carbendazim (benzimidazole fungicide, 5 µg/ml), azoxysyrobin (broad-spectrum fungicide, 100 µg/ml) or phosphinothricin (herbicide, 100 µg/ml) for the induction of 20 min and 2 h at 25°C respectively. Total RNAs were extracted from the drug-induced and drug-free (control) cultures. Three samples of 5 µg RNA from each extract was reversely transcribed with PrimeScript^{TM} RT kit

(Takara, Dalian, China). The cDNA samples (diluted to 10 µg/ml) synthesized with the kit were assessed for the transcript levels of all the 21 transporter genes via qRT-PCR with paired primers (Table S1) using *B. bassiana* 18S rRNA as internal standard. The relative transcript level of each gene in a drug-induced sample versus control was calculated as its transcript ratio using the method $2^{-\Delta\Delta Ct}$ [41]. Five transporter genes whose transcript levels were inductively upregulated by all the three drugs were selected for further study, including *Pdr1*, *Pdr2*, *Pdr5*, *Mdr1* and *Mrp1*.

Single-gene disruption and complementation

The plasmid p0380-bar vectoring the *bar* marker and P*trpC* promoter [33] was used as backbone to construct the disruption plasmids of all selected genes except *Pdr1*. The 5′ and 3′ fragments of *Pdr2* (1261 and 1874 bp), *Pdr5* (1560 and 1666 bp), *Mdr1* (1480 and 1147 bp) and *Mrp1* (1531 and 1320 bp) were separately amplified from Bb2860 via PCR with paired primers (Table S2), digested with specific restriction enzymes, and inserted into p0380-bar, generating p0380-5′x-bar-3′x for the disruption of each target gene (x). To delete *Pdr1*, alternatively, its ORF fragment (2500 bp) was amplified from Bb2860 with Pdr1-F/R (Table S2) and inserted into p0380-bar linearized with *Bam*HI/*Hind*III. After digestion with *Xho*I/*Sac*I, the plasmid released a fragment of ~200 bp to separate the ORF into two fragments and a P*trpC-bar* cassette amplified from p0380-bar with the Insert-F/R primers was inserted between the separated fragments, yielding p0380-pdr1D for *Pdr1* disruption.

To rescue each of the disrupted genes, p0380-sur-gateway [33] vectoring the *sur* marker gene was used as backbone. The full-length sequences with flanking regions of *Pdr1* (7899 bp), *Pdr2* (7926 bp), *Pdr5* (7754 bp), *Mdr1* (7850 bp) and *Mrp1* (7389 bp) were separately amplified from Bb2860 with paired primers (Table S1) under the action of LA*Taq* polymerase (TaKaRa) and ligated into the backbone to replace the gateway fragment under the action of Gateway® BP ClonaseTM II Enzyme Mix (Invitrogen), forming p0380-sur-x, where x denotes one of the rescued target genes.

All the disruption and complement plasmids were individually transformed into *A. tumefaciens* AGL-1 for further transformation into Bb2860 or the corresponding disruption mutants using a documented protocol [39] with slight modification. Briefly, the recipient strain (wild type or delta mutant) was co-cultivated with the vector-integrated AGL-1 on induced medium for 48 h at 25°C in dark, followed by washing with ~5 ml of 0.02% Tween 80. The suspension was spread onto M-100 plates [39] supplemented with cefotaxime (300 µg/ml for suppressing AGL-1 growth) and phosphinothricin (200 µg/ml for the selective growth of disruption mutants) or chorimuron ethyl (10 µg/ml for the selective growth of rescued mutants). All the plates were incubated for 6 days at 25°C and 12:12 h (light:dark cycle). Colonies grown on the selective plates were identified via PCR, RT-PCR and Southern blotting with paired primers or amplified probes (Table S2). For Southern blotting, 30 µg genomic DNA extracted from the monoclonal culture of each putative mutant on SDAY was digested with *Xba*I/*Hind*III, separated via electrophoresis in 0.7% agarose gel, and then transferred to Biodyne B nylon membrane (Gelman Laboratory, Shelton, WA, USA) in Trans-Blot SD Electrophoretic Transfer Cell (Bio-Rad, Hercules, CA, USA). Probe preparation, membrane hybridization and visualization were carried out using DIG High Prime DNA Labeling and Detection Starter Kit II (Roche, Mannheim, Germany). Positive disruption and complement mutants of each target gene were assayed together with wild type for their phenotypic changes in triplicate experiments below.

Assays of multidrug responses

The aliquots of 200 µl conidial suspension (2×10^7 conidia/ml) were evenly spread onto cellophane-overlaid SDAY plates. After 3-day incubation at 25°C and 12:12 h, cellophane discs (5 mm diameter) with growing mycelia were cut from the culture of each strain and attached centrally onto the plates (9 cm diameter) of 1/4 SDAY (SDAY nutrients diluted to 1/4) supplemented with the antifungal drugs azoxysyrobin (100 µg/ml), carbendazim (0.5 µg/ml), dimetachlone (pyrrole fungicide, 100 µg/ml), ethirimol (pyrimidine fungicide, 1 mg/ml), itraconazole (triazole agent, 5 µg/ml), cyclonheximide (protein biosynthesis inhibitor, 20 µg/ml), 4-Nitroquinoline N-oxide (potent mutagenic agent, 10 µg/ml) and Congo red (cell wall biosynthesis inhibitor, 500 µg/ml) for MDR assays and NaCl (40 mg/ml) for osmosensitivity assay respectively. All the plates were incubated for 6 days at the same regime, followed by cross-measuring the diameters of their colonies. For each strain stressed with a given drug, relative growth inhibition (RGI) was calculated as $(C-N)/(C-19.6)\times100$, where the constant is the area of inoculated disc, and C and N denote the measurements of colony area (mm^2) from the control (free of drug) and drug treatment respectively.

To quantify antioxidant capability of each strain, the plates of the same medium supplemented with the gradient concentrations of menadione $(0-8$ mM) or H_2O_2 $(0-80$ mM) for varying intensity of oxidative stress were inoculated with the culture discs as above. After 6-day incubation at the same regime, colony diameters were cross-measured and the ratio of a stressed colony size over the size of the control colony was defined as relative growth rate (R_g). The R_g trend of each strain over the concentrations (C) of menadione or H_2O_2 was fitted to the equation $R_g = 1/[1-\exp(a+bC)]$. Solving the fitted equation gave an effective concentration of each oxidant (EC_{50}) to suppress 50% colony growth when $R_g = 0.5$.

Virulence bioassay

All the fungal strains were bioassayed for their changes in virulence to the second-instar larvae of *S. litura* using a standardized method described elsewhere (36,37). Briefly, batches of $30-40$ larvae on cabbage leaf discs (~10 cm diameter) were separately sprayed with 1 ml of conidial suspension (2×10^7 conidia/ml) as treatment or 0.02% Tween 80 (used for suspending conidia) as control in automatic Potter Spray Tower (Burkard Scientific Ltd, Uxbridge, UK). After spray, all larvae were reared on the leaf discs in large Petri dishes (15 cm diameter) for 7 days at 25°C and 12:12 h and fresh leaf discs were supplied daily for their feeding. Mortality in each plate was daily examined during the period. The resultant time-mortality trends were subjected to probit analysis, generating an estimate of medial lethal time (LT_{50}) for each fungal strain against the pest species.

Author Contributions

Conceived and designed the experiments: MGF TTS SHY. Performed the experiments: TTS JZ. Analyzed the data: MGF TTS JZ. Contributed reagents/materials/analysis tools: MGF SHY. Wrote the paper: MGF TTS.

References

1. Bardas GA, Veloukas T, Koutita O, Karaoglanidis GS (2010) Multiple resistance of *Botrytis cinerea* from kiwifruit to SDHIs, QoIs and fungicides of other chemical groups. Pest Manag Sci 66: 967–973

2. Cowen LE, Steinbach WJ (2008) Stress, drugs, and evolution: the role of cellular signaling in fungal drug resistance. Eukaryot Cell 7: 747–764.

3. Del Sorbo G, Schoonbeek HJ, De Waard MA (2000) Fungal transporters involved in efflux of natural toxic compounds and fungicides. Fungal Genet Biol 30: 1–15.

4. Feng MG, Poprawski TJ, Khachatourians GG (1994) Production, formulation and application of the entomopathogenic fungus *Beauveria bassiana* for insect control: current status. Biocontrol Sci Technol 4: 3–34.

5. Roberts DW, St Leger RJ (2004) *Metarhizium* spp., cosmopolitan insect-pathogenic fungi: mycological aspects. Adv Appl Microbiol 54: 1–70.

6. de Faria MR, Wraight SP (2007) Mycoinsecticides and mycoacaricides: a comprehensive list with worldwide coverage and international classification of formulation types. Biol Control 43: 237–256.

7. Majchrowicz I, Poprawski TJ (1993) Effects *in vitro* of nine fungicides on growth of entomopathogenic fungi. Biocontrol Sci Technol 3: 321–336.

8. Shi WB, Jiang Y, Feng MG (2005) Compatibility of ten acaricides with *Beauveria bassiana* and Enhancement of fungal infection to *Tetranychus cinnabarinus* (Acari: Tetranychidae) eggs by sublethal application rates of pyridaben. Appld Entomol Zool 40: 659–666.

9. Ma ZH, Yoshimura MA, Michailides TJ (2003) Identification and character-ization of benzimidazole resistance in *Monilinia fructicola* from stone fruit orchards in California. Appl Environ Microbiol 69: 7145–7152.

10. Ma ZH, Yoshimura MA, Michailides TJ (2005) Characterization and PCR-based detection of benzimidazole-resistant isolates of *Monilinia laxa* in California. Pest Manag Sci 61: 449–457.

11. Zou G, Ying SH, Shen ZC, Feng MG (2006) Multi-sited mutations of beta-tubulin are involved in benzimidazole resistence and thermotolerance of fungal biocontrol agent *Beauveria bassiana*. Environ Microbiol 8: 2096–2105.

12. Song TT, Ying SH, Feng MG (2012) High resistance of *Isaria fumosorosea* to carbendazim arises from the overexpression of an ABC transporter (ifT1) rather than tubulin mutation. J App Microbiol 112: 175–184.

13. Klein C, Kuchler K, Valachovic M (2011) ABC proteins in yeast and fungal pathogens. Essays Biochem 50: 101–119.

14. Lamping E, Baret PV, Holmes AR, Monk BC, Goffeau A, et al. (2010) Fungal PDR transporters: Phylogeny, topology, motifs and function. Fungal Genet Biol 47: 127–142.

15. Kovalchuk A, Driessen AJM (2010) Phylogenetic analysis of fungal ABC transporters. BMC Genomics 11: 177. doi:10.1186/1471-2164-11-177.

16. Ernst R, Kueppers P, Stindt J, Kuchler K, Schmitt L (2010) Multidrug efflux pumps: Substrate selection in ATP-binding cassette multidrug efflux pumps – first come, first served? FEBS Journal 277: 540–549.

17. Morschhäuser J (2010) Regulation of multidrug resistance in pathogenic fungi. Fungal Genet Biol 47: 94–106.

18. Holmes AR, Lin YH, Niimi K, Lamping E, Keniya M, et al. (2008) ABC transporter Cdr1p contributes more than Cdr2p does to fluconazole efflux in fluconazole-resistant *Candida albicans* clinical isolates. Antimicrob Agents Che-mother 52: 3851–3862.

19. Tsao S, Rahkhoodaee F, Raymond M (2009) Relative contributions of the *Candida albicans* ABC transporters Cdr1p and Cdr2p to clinical azole resistance. Antimicrob Agents Chemother 53: 1344–1352.

20. Tanabe K, Lamping E, Nagi M, Okawada A, Holmes AR, et al. (2011) Chimeras of *Candida albicans* Cdr1p and Cdr2p reveal features of pleiotropic drug resistance transporter structure and function. Mol Microbiol 82: 416–433.

21. Zolnerciks JK, Andress EJ, Nicolaou M, Linton KJ (2011) Structure of ABC transporters. Essays Biochem 50: 43–61.

22. Coleman JJ, White GJ, Rodriguez-Carres M, VanEtten HD (2011) An ABC transporter and a cytochrome P450 of *Nectria haematococca* MPVI are virulence factors on pea and are the major tolerance mechanisms to the phytoalexin pisatin. Mol Plant-Microbe Interact 24: 368–376.

23. Schoonbeek HJ, Del Sorbo G., De Waard MA (2001) The ABC transporter BcatrB affects the sensitivity of *Botrytis cinerea* to the phytoalexin resveratrol and the fungicide fenpiclonil. Mol Plant-Microbe Interact 14: 562–571

24. Urban M, Bhargava T, Hamer JE (1999) An ATP-driven efflux pump is a novel pathogenicity factor in rice blast disease. EMBO J 18: 512–521

25. Zwiers LH, De Waard MA (2000) Characterization of the ABC transporter genes MgAtr1 and MgAtr2 from the wheat pathogen *Mycosphaerella graminicola*. Fungal Genet Biol 30: 115–125.

26. Krattinger SG, Lagudah ES, Spielmeyer W, Singh RP, Huerta-Espino J, et al. (2009) A putative ABC transporter confers durable resistance to multiple fungal pathogens in wheat. Science 323: 1360–1363.

27. Xiao GH, Ying SH, Zheng P, Wang ZL, Zhang SW, et al. (2012) Genomic perspectives on the evolution of fungal entomopathogenicity in *Beauveria bassiana*. Sci Rep 2: 483. doi:10.1038.srep00483.

28. Gao Q, Jin K, Ying SH, Zhang YJ, Xiao GH, et al. (2011) Genome sequencing and comparative transcriptomics of the model entomopathogenic fungi *Metarhizium anisopliae* and *M. acridum*. PLoS Genet 7: e1001264.

29. Andrade AC, Del Sorbo G, Van Nistelrooy JGM, De Waard MA (2000) The ABC transporter AtrB from *Aspergillus nidulans* mediates resistance to all major classes of fungicides and some natural toxic compounds. Microbiology-(UK) 146: 1987–1997.

30. Vermeulen T, Schoonbeek H, De Waard MA (2001) The ABC transporter BcatrB from *Botrytis cinerea* is a determinant of the activity of the phenylpyrrole fungicide fludioxonil. Pest Manag Sci 57: 393–402.

31. Wang ZL, Zhang LB, Ying SH, Feng MG (2013) Catalases play differentiated roles in the adaptation of a fungal entomopathogen to environmental stresses. Environ Microbiol 15: 409–418.

32. Xie XQ, Wang J, Ying SH, Feng MG (2010) A new manganese superoxide dismutase identified from *Beauveria bassiana* enhances virulence and stress tolerance when overexpressed in the fungal pathogen. Appl Microbiol Biotechnol 86: 1543–1553.

33. Xie XQ, Li F, Ying SH, Feng MG (2012) Additive contributions of two manganese-cored superoxide dismutases (MnSODs) to antioxidation, UV tolerance and virulence of *Beauveria bassiana*. PLoS One 7: e30298. doi:10.1371/journal.pone.00302988.

34. Zhang YJ, Zhao JH, Fang WG, Zhang JQ, Luo ZB, et al. (2009) Mitogen-acticated protein kinase hog1 in the entomopathogenic fungus *Beauveria bassiana* regulates environmental stress responses and virulence to insects. Appl Environ Microbiol 75: 3787–3795.

35. Luo X, Keyhani NO, Yu X, He Z, Luo Z, et al. (2012) The MAP kinase Bbslt2 controls growth, conidiation, cell wall integrity, and virulence in the insect pathogenic fungus *Beauveria bassiana*. Fungal Genet Biol 49: 544–555.

36. Wang J, Zhou G, Ying SH, Feng MG (2013) P-type calcium ATPase functions as a core regulator of *Beauveria bassiana* growth, conidiation and responses to multiple stressful stimuli through cross-talk with signaling networks. Environ Microbiol 15: 967-979.

37. Xie XQ, Guan Y, Ying SH, Feng MG (2013) Differentiated functions of Ras1 and Ras2 proteins in regulating the germination, growth, conidiation, multi-stress tolerance and virulence of *Beauveria bassiana*. Environ Microbiol 15: 447–462.

38. Stefanato FL, Abou-Mansour E, Buchala A, Kretschmer M, Mosbach A, et al. (2009) The ABC transporter BcatrB from *Botrytis cinerea* exports camalexin and is a virulence factor on *Arabidopsis thaliana*. Plant J 58: 499–510.

39. Fang WG, Zhang YJ, Yang XY, Zheng XL, Duan H, et al. (2004) *Agrobacterium tumefaciens*-mediated transformation of *Beauveria bassiana* using an herbicide resistance gene as a selection marks. J Invertebr Pathol 85: 18–24.

40. Kumar S, Nei M, Dudley J, Tamura K (2008) MEGA: a biologist-centric software for evolutionary analysis of DNA and protein sequences. Brief Bioinform 9: 299–306.

41. Livak KJ, Schmittgen TD (2001) Analysis of relative gene expression data using real-time quantitative PCR and the $2^{-\Delta\Delta CT}$. Methods 25: 402–408.

7

Clonal Expansion and Emergence of Environmental Multiple-Triazole-Resistant *Aspergillus fumigatus* Strains Carrying the TR$_{34}$/L98H Mutations in the *cyp*51A Gene in India

Anuradha Chowdhary[1]*, **Shallu Kathuria**[1], **Jianping Xu**[2], **Cheshta Sharma**[1], **Gandhi Sundar**[1], **Pradeep Kumar Singh**[1], **Shailendra N. Gaur**[3], **Ferry Hagen**[4], **Corné H. Klaassen**[4], **Jacques F. Meis**[4,5]

1 Department of Medical Mycology, Vallabhbhai Patel Chest Institute, University of Delhi, Delhi, India, **2** Department of Biology, McMaster University, Hamilton, Ontario, Canada, **3** Department of Pulmonary Medicine, Vallabhbhai Patel Chest Institute, University of Delhi, Delhi, India, **4** Department of Medical Microbiology and Infectious Diseases, Canisius Wilhelmina Hospital, Nijmegen, The Netherlands, **5** Department of Medical Microbiology, Radboud University Nijmegen Medical Centre, Nijmegen, The Netherlands

Abstract

Azole resistance is an emerging problem in *Aspergillus* which impacts the management of aspergillosis. Here in we report the emergence and clonal spread of resistance to triazoles in environmental *Aspergillus fumigatus* isolates in India. A total of 44 (7%) *A. fumigatus* isolates from 24 environmental samples were found to be triazole resistant. The isolation rate of resistant *A. fumigatus* was highest (33%) from soil of tea gardens followed by soil from flower pots of the hospital garden (20%), soil beneath cotton trees (20%), rice paddy fields (12.3%), air samples of hospital wards (7.6%) and from soil admixed with bird droppings (3.8%). These strains showed cross-resistance to voriconazole, posaconazole, itraconazole and to six triazole fungicides used extensively in agriculture. Our analyses identified that all triazole-resistant strains from India shared the same TR$_{34}$/L98H mutation in the *cyp*51 gene. In contrast to the genetic uniformity of azole-resistant strains the azole-susceptible isolates from patients and environments in India were genetically very diverse. All nine loci were highly polymorphic in populations of azole-susceptible isolates from both clinical and environmental samples. Furthermore, all Indian environmental and clinical azole resistant isolates shared the same multilocus microsatellite genotype not found in any other analyzed samples, either from within India or from the Netherlands, France, Germany or China. Our population genetic analyses suggest that the Indian azole-resistant *A. fumigatus* genotype was likely an extremely adaptive recombinant progeny derived from a cross between an azole-resistant strain migrated from outside of India and a native azole-susceptible strain from within India, followed by mutation and then rapid dispersal through many parts of India. Our results are consistent with the hypothesis that exposure of *A. fumigatus* to azole fungicides in the environment causes cross-resistance to medical triazoles. The study emphasises the need of continued surveillance of resistance in environmental and clinical *A. fumigatus* strains.

Editor: Oscar Zaragoza, Instituto de Salud Carlos III, Spain

Funding: This work was supported with a grant from the Department of Biotechnology (Grant Reference Number: BT/39/NE/TBP/2010), New Delhi, India. J.F. Meis received grants from Astellas, Merck, Schering Plough, Gilead and Janssen Pharmaceuticals. C.H. Klaassen received a grant from Pfizer. J.P. Xu received grants from NSERC of Canada. The funders had no role in study design, data collection and analysis, decision to publish, or preparation of the manuscript.

Competing Interests: J.F. Meis received grants from Astellas, Merck, Schering-Plough, Gilead and Janssen Pharmaceuticals, is a consultant to Basilea and Merck and received speaker's fees from Merck, Schering-Plough, Gilead and Janssen Pharmaceutica. C.H. Klaassen received a grant from Pfizer. J.P. Xu received grants from NSERC of Canada.

* E-mail: dranuradha@hotmail.com

Introduction

Aspergillus fumigatus is the commonest etiologic agent of various clinical forms of bronchopulmonary aspergillosis including allergic, acute invasive and chronic pulmonary aspergillosis (CPA). The disease has a global distribution and it is widespread in India [1]. Invasive aspergillosis is the most severe manifestation with an overall annual incidence varying from 2 to 10% in the immunosuppressed patient population whereas CPA affects primarily immunocompetent individuals with an estimated prevalence of 3 million worldwide [2,3]. Azoles, such as itraconazole, voriconazole, and posaconazole are among the recommended

first-line drugs in the treatment and prophylaxis of aspergillosis [4,5]. Azole resistance is an emerging problem in *A. fumigatus* in Europe and has been shown to be associated with increased probability of treatment failure [6–8]. Azole resistance is commonly due to mutations in the *cyp*51A gene, which encodes 14-α-demethylase in the ergosterol biosynthesis pathway. In azole-resistant clinical *A. fumigatus* isolates a wide variety of mutations in the *cyp*51A gene have been found, such as substitutions at codons G54, G138, P216, F219, M220 and G448 [9–12]. However, in the Netherlands a different resistance mechanism consisting of the L98H substitution, together with a 34-bp tandem repeat (TR$_{34}$) in the promoter region of this gene (TR$_{34}$/L98H) was found to be present in over 90% of azole resistant isolates [13]. The TR$_{34}$/L98H resistance mechanism has been endemic in the Netherlands and subsequently reported from other European countries such as Denmark, France, Germany, Spain and the United Kingdom [12,14–19].

Isolates of *A. fumigatus* with TR$_{34}$/L98H mutations exhibit a pan-azole resistant phenotype and were recovered primarily from azole-naive patients and from environmental sources in the Netherlands and Denmark [15,17,20,21]. These observations suggest that patients acquire azole-resistant *Aspergillus* from environmental sources rather than arising through azole therapy. The consequence of this type of resistance development is that patients at risk can be exposed to and infected by azole-resistant strains in the environment. Furthermore, TR$_{34}$/L98H isolates were cross-resistant to certain azole fungicides employed extensively in agriculture for crop protection against phytopathogenic molds, to prevent post-harvest spoilage [21]. An environmental route of resistance development poses a major challenge because multiplication and spread of resistant strains in the environment can be anticipated. Recently, we reported from India the occurrence of TR$_{34}$/L98H mutations in the *cyp*51A gene in *A. fumigatus* isolates from patients with chronic respiratory disease who had not previously been exposed to azoles [22]. This emergence of resistance in Indian clinical isolates prompted us to undertake a wide environmental survey of azole resistant *A. fumigatus* isolates in India. Herein, we report multi-triazole resistant environmental *A. fumigatus* isolates from India harboring TR$_{34}$/L98H mutations in the *cyp*51A gene, from soil samples of paddy fields, tea gardens, cotton trees, flower pots and indoor air of hospital. Furthermore, we investigated the cross resistance of these environmental and clinical TR$_{34}$/L98H *A. fumigatus* isolates to registered and commonly used azole fungicides in India and determined the genetic relatedness of Indian environmental and clinical *A. fumigatus* isolates harboring the TR$_{34}$/L98H mutations and compared them with isolates from Europe and China.

Results

Isolation of Environmental Strains of *A. fumigatus*

Of the 486 environmental samples tested, 201 (41.4%) showed the presence of *A. fumigatus* in all types of substrates tested except nursery plants soil and decayed wood inside tree trunk hollows. The data of state-wise distribution and prevalence of azole resistant *A. fumigatus* in soil and air samples is presented in Table 1 and Figure 1. Of the 201 *A. fumigatus* positive samples, 630 individual *A. fumigatus* colonies were obtained from Sabourauds dextrose agar (SDA) plates. The count of *A. fumigatus* on primary SDA plate ranged from one colony to confluent growth. Besides *A. niger*, *A. flavus*, *A. terreus*, other molds such as mucorales, and *Penicillium* species were also observed in soil samples. Out of 630 *A. fumigatus* colonies tested, 44 (7%) isolates originating from 24 samples grew on SDA plates containing 4 mg/L itraconazole. Among these 44 itraconazole-resistant (ITC+) isolates, 15 were obtained from different potted plants of the V. P. Chest Institute (VPCI) garden, Delhi, 12 rice paddy fields in Bihar, 9 from tea gardens in Darjeeling, 3 each from soil beneath cotton trees (*Bombax ceiba*) from Kolkata and from aerial sampling of patient rooms of the VPCI hospital, and 2 from soil containing bird droppings in Tamil Nadu (Table 1). Overall, 5% (24/486) of the samples tested harbored itraconazole resistant *A. fumigatus*. Among the positive samples, 11.9% (24/201) showed at least one colony of resistant *A. fumigatus*. The isolation rate of itraconazole resistant *A. fumigatus* was highest 33% (9/27) from the soil of tea gardens followed by soil from flower pots of the hospital garden 20% (15/

75), soil beneath cotton trees 20% (3/15), rice paddy fields 12.3% (12/97), air samples of hospital wards 7.6% (3/39) and from soil admixed with bird droppings 3.8% (2/52). There was no isolation of resistant *A. fumigatus* isolates from soil samples of public parks and gardens inside the hospital premises and red chilly fields in Tamil Nadu.

Evidence for Cross-Resistance to Triazole Antifungal Drugs

All the 44 ITC+ *A. fumigatus* isolates from the environment showed reduced susceptibility to azoles. The geometric mean (GM) MIC of itraconazole (GM, 16 mg/L) was the highest, followed by voriconazole (GM, 8.7 mg/L), and posaconazole (GM, 1.03 mg/L). All the antifungal drugs tested showed reduced efficacy against all the ITC+ *A. fumigatus* isolates (Table 2), consistent with cross-resistance of these isolates to the tested azoles. Among the triazoles, the MIC difference between wild type and TR$_{34}$/L98H isolates were the highest for itraconazole (r = 0.96) followed by voriconazole (r = 0.91) and posaconazole (r = 0.72). Of the10 fungicides, 7 showed dissimilarity between the MICs with greatest differences found for bromuconazole, difenoconazole, tebuconazole (r = 0.96 each) followed by hexaconazole (r = 0.95), epoxiconazole (r = 0.92), metconazole (r = 0.89) and lowest for cyproconazole (r = 0.22) (Table 2).

Evidence for Clonal Spread of a Single Triazole-Resistant *A. fumigatus* Genotype

Our genotype analyses identified that all of the 44 ITC+ *A. fumigatus* isolates from India exhibited the same TR$_{34}$/L98H genotype at the *cyp*51A gene. Furthermore, these strains had the same allele across all nine examined microsatellite loci (Fig. 2). In contrast to the genetic uniformity of azole-resistant strains from India, the azole-susceptible isolates from both patients and environments in India were genetically very diverse. Indeed, all nine loci were highly polymorphic in populations of azole-susceptible isolates from both clinical and environmental samples.

Origin(s) of the Azole-resistant *A. fumigatus* Genotype in India

The widespread occurrence of a single azole-resistant genotype across India contrasts with those found in several other regions outside of India. In our analyses, a diversity of genotypes has been found for clinical TR$_{34}$/L98H azole-resistant *A. fumigatus* strains in China, France, Germany and in both clinical and environmental sources in the Netherlands (Figs. 2 and 3). To examine the origin(s) of the azole - resistant genotype in India, we first attempted to isolate azole - susceptible strains from the 24 soil samples that contained the 44 azole-resistant strains. Among these 24 soil samples, we successfully obtained and analyzed eight azole-susceptible isolates from seven of the 24 samples through dilution plating, single colony purification, and screening using itraconazole-containing and non-containing media. Our genotype analyses using the 9 microsatellite markers revealed that none of the eight strains had a genotype identical to the azole-resistant genotype in India. These eight azole-susceptible strains belonged to four different genotypes. Interestingly, three of the genotypes shared no allele with the azole-resistant genotype at any of the nine microsatellite loci while the remaining genotype shared an allele with the azole-resistant genotype at only one of the nine loci.

To further explore the potential origin(s) of the azole-resistant genotype in India, we further analyzed the genotypes of all the azole-susceptible strains from within India. Among the nine microsatellite loci, we were able to find allele-sharing at only six

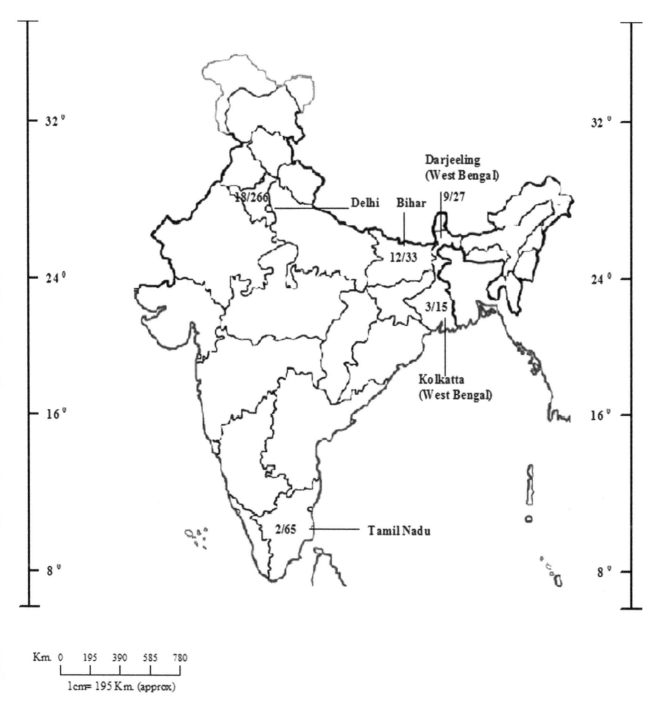

Darjeeling
(West Bengal)
9/27

18/266

Delhi   Bihar

12/33

3/15

Kolkatta
(West Bengal)

2/65 ——— Tamil Nadu

Km. 0   195   390   585   780

1cm= 195 Km. (approx)

Figure 1. An outline map of India showing state-wise isolation of multiple-triazole resistant *Aspergillus fumigatus* isolates from variety of environmental samples.

loci between the Indian azole-resistant genotype and the 35 azole-susceptible clinical and soil/air isolates in India. The highest number of loci with shared alleles between any of the 35 azole susceptible strains and the resistant genotype was at only two of the nine loci. Therefore, even with free recombination among the genotypes represented by the 35 azole susceptible strains in India, the azole-resistant genotype could not be generated due to the lack of corresponding alleles at three of the nine loci (loci 2A, 3A, and 4C, Fig. 2) found only in the azole-resistant strains.

Interestingly, though not identical, several strains from outside of India were found to have genotypes more similar to the Indian azole-resistant strains than the Indian azole-susceptible strains (Fig. 2). For example, ten of the 51 strains from outside of India shared alleles in at least four of the nine loci with the Indian azole–resistant genotype, with four of the 10 strains sharing alleles at five loci. These 10 strains were all similarly resistant to azoles as the Indian azole-resistant genotype and all 10 strains carried the same $TR_{34}/L98H$ mutation. The combined allelic comparisons identified that the azole-resistant strains from outside of India

Table 1. State-wise distribution of environmental *Aspergillus fumigatus* isolates with TR$_{34}$/L98H mutations from India.

	No. of *A. fumigatus* isolates with TR$_{34}$/L98H mutations/No. of isolates tested n = 44/630 (201/486)*									
	Garden soil	Paddy/Rice/Red chilly fields soil	Tea garden soil	Tree trunk hollow wood	Aerial isolations from hospital wards	Nursery flower pots soil	Soil beneath cotton trees	Garden soil of hospitals	Flower pots soil of hospital garden	Soil with bird droppings
UT† of Delhi (n = 266)										
VPCI†, DU†	–	–	–	–	3/7 (14/39)	–	–	0/20 (10/10)	15/120 (30/75)	–
Ashok Vihar Park	0/0(0/10)	–	–	–	–	–	–	–	–	–
Lodhi garden	0/80(20/45)	–	–	–	–	–	–	–	–	–
Central Park, DU	0/27(10/50)	–	–	–	–	–	–	–	–	–
Police Lines, DU	–	–	–	0/0(0/12)	–	–	–	–	–	–
Gulabi Bagh	–	–	–	–	–	0/0(0/25)	–	–	–	–
Tamil Nadu (n = 65) Thorapadi Village	–	0/4(2/13)	–	–	–	–	–	–	–	–
Kanchipuram	–	–	–	–	–	–	–	–	–	2/25 (19/52)
West Bengal (n = 59) Hoogli Dist., Kolkata	–	–	–	–	–	–	3/5(1/15)	–	–	–
Siliguri	–	0/40(13/17)	–	–	–	–	–	–	–	–
Darjeeling	–	–	9/51(16/27)	–	–	–	–	–	–	–
Bihar (n = 33) Munger	–	12/78(26/33)	–	–	–	–	–	–	–	–
Uttrakhand (n = 21) Kedar, Basora	–	0/108(21/21)	–	–	–	–	–	–	–	–
Haryana (n = 21) Jajjhar	–	0/60(15/21)	–	–	–	–	–	–	–	–
Meghalaya (n = 11) Shillong	–	0/0(0/5)	–	–	–	–	–	0/0(0/6)	–	–
Sikkim (n = 6) Gangtok	0/5(4/6)	–	–	–	–	–	–	–	–	–
Himachal Pradesh (n = 4) Dalhousie	0/0(0/4)	–	–	–	–	–	–	–	–	–

*Parenthesis denotes the numerator as number of samples positive for *A. fumigatus*, denominator denotes the number of samples tested; †UT, Union Territory; VPCI, V. P. Chest Institute; DU, Delhi University.

Table 2. In- vitro antifungal susceptibility profile of medical triazoles and triazole fungicides against environmental and clinical Aspergillus fumigatus isolated in India.

Triazole drugs and fungicides	MIC* (mg/L)												Effect size r
	Environment						Clinical						
	TR$_{34}$/L98H (n = 44)			251676672Wild type (n = 22)			TR$_{34}$/L98H (n = 9)			Wild type (n = 13)			
	GM*	MIC$_{50}$*	Range	GM	MIC$_{50}$	Range	GM	MIC$_{50}$	Range	GM	MIC$_{50}$	Range	
Itraconazole	16	16	16->16	0.43	0.5	0.25-1	16	16	16>16	0.11	0.125	0.03-1	0.96
Voriconazole	8.7	8	4-16	0.65	0.5	0.25-1	5.9	8	2-16	0.10	0.125	0.03-0.25	0.91
Posaconazole	1.03	1	0.5-2	0.46	0.5	0.06-1	3.2	2	1->8	0.25	0.25	0.125-1	0.72
Bromuconazole	31.4	32	16->32	2.5	2	1-4	32	32	32->32	2.2	2	1-4	0.96
Cyproconazole	32	32	32->32	30.9	32	16->32	32	32	32->32	29.4	32	16-32	0.22
Difenoconazole	31.4	32	16->32	2.0	2	1-8	32	32	32->32	1.8	2	0.5-4	0.96
Epoxiconazole	32	32	32->32	5.2	4	2-16	32	32	32->32	4.1	4	2-8	0.92
Hexaconazole	31	32	8->32	4.87	4	2-8	32	32	>32	3.39	4	2-8	0.95
Metconazole	3.8	4	1-8	0.3	0.5	0.125-1	4	4	2-16	0.4	0.5	0.25-2	0.89
Penconazole	32	32	32->32	30.9	32	32->32	32	32	32->32	32	32	32->32	0
Tebuconazole	31.4	32	16->32	2.6	2	1-8	32	32	>32	3.0	4	1-8	0.96
Triadimefon	32	32	>32	32	32	>32	32	32	>32	32	32	32->32	0
Tricyclazole	32	32	32->32	32	32	32->32	32	32	>32	32	32	32->32	0

*Minimum inhibitory concentration; GM, geometric mean.

contained alleles at seven of the nine microsatellite loci found in the Indian azole-resistant genotype, one more than all the Indian azole-susceptible strains combined. At locus 2A, only the sample from outside India contained the allele #14 (in 8 of the 51 strains) found in the Indian azole-resistant genotype while the azole-susceptible sample from India did not contain this allele (Fig. 2). However, a reverse situation occurred at locus 2C where allele #9 in the Indian azole-resistant genotype was found in the Indian azole-susceptible population (in one of the 35 strains) but not from outside of India. Finally, different from the other eight loci, locus 4C had a unique allele (#28) found only in the Indian azole-resistant strains and this allele was absent from any other strains in the whole analyzed sample, either from within or outside of India (Fig. 2).

Discussion

The site specific mode of action and intensive use of demethylase inhibitors (DMIs) fungicides for post harvest spoilage crop protection against phytopathogenic molds, has led to the development of resistance in many fungi of agricultural importance. It is anticipated that the excessive use of azoles in agriculture would not only influence the plant pathogenic fungi but also would inevitably influence susceptible species of the saprophytic flora [23]. Many potentially human pathogenic fungi such as *Coccidioides, Histoplasma, Aspergillus,* and *Cryptococcus* have their natural habitats in the environment and in many instances the infecting fungal organisms are acquired from the surrounding environment. Recently the use of azole-based agricultural chemicals has also been implicated as a major factor in the increase in frequency of multiple-triazole-resistant (MTR) isolates of *A. fumigatus* infecting humans by selection of MTR alleles [24,25]. This is supported by a recent report originating from the

Netherlands that showed over 90% of Dutch azole resistant *A. fumigatus* isolates recovered from epidemiologically unrelated patients clustered onto a single lineage [13]. In the present study 7% of the Indian environmental *A. fumigatus* isolates were multi-triazole resistant with a single resistant mechanism carrying the TR$_{34}$/L98H mutation in the *cyp*51A gene (Table 1). The resistant isolates were recovered from soil samples of potted plants, paddy fields and tea gardens where certain triazole fungicides (tebuconazole, hexaconazole, and epoxiconazole) were extensively used. Although, Europe leads the world in usage of agricultural fungicides (40%) followed by Japan and Latin America, in India usage of fungicides is increasing and current fungicide use in India is 19% of the total pesticide use [26]. In the USA the use of azoles in agriculture is insignificant as compared to Europe (http://ec.europa.eu/food/fs/sc/ssc/out278_en.pdf). Consequently, there has been no report of finding the TR$_{34}$/L98H mutation in clinical or environmental isolates in the USA. But this resistance type has been found in the environment in Europe and now also in India. It is noteworthy that so far no environmental survey of TR$_{34}$/L98H *A. fumigatus* isolates outside Europe has been reported. The fungicides belonging to different chemical groups have been registered in India only in the past two decades and these are being used against diverse diseases in fruits, vegetables, plantation crops and some field crops [26]. Triazole fungicides such as hexaconazole, propiconazole, triadimefon, and tricyclazole account for a substantial fungicide market in India [26]. Overall, the highest fungicide usage in India is on pome fruits (12.7%), followed by potatoes (12.2%), rice (12%), tea (9.4%), coffee, chillies, grapevines, other fruits and vegetables [26]. Also, triazole fungicides are characterized by their long persistence in soil. Singh and Dureja demonstrated that hexaconazole persist longer in Indian soil due to its hydrophobic nature [27]. In India, the maximum amounts of fungicide usage are found in southern India, followed by western,

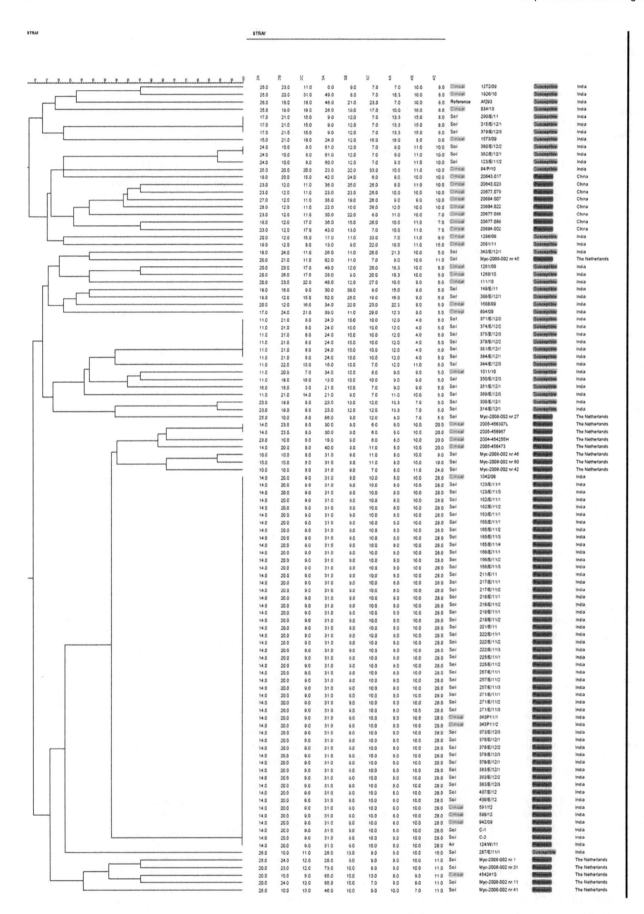

Figure 2. Genotypic relationship between the wild-type and TR$_{34}$/L98H *Aspergillus fumigatus* **(clinical and environmental isolates from India, The Netherlands and France) and TR$_{34}$/L98H** *A. fumigatus* **(clinical isolates from China and Germany).** The dendrogram is based on a categorical analysis of 9 microsatellite markers in combination with UPGMA clustering. The scale bar indicates the percentage identity. Clinical: blue, Environmental: yellow, Resistant: red, Susceptible: green.

eastern and northern Indian states. In this study the multi triazole resistant *A. fumigatus* carrying the TR$_{34}$/L98H genotype was isolated from Union Territory (UT) of Delhi (northern region), West Bengal and Bihar (eastern region of India about 1100 Km from the North) and Tamil Nadu (southern region of India, about 2100 Km from the North) states. The western region of India has yet to be surveyed but considering the high usage of fungicides in this region, isolation of azole resistant *A. fumigatus* may be anticipated.

Previous environmental surveys of azole resistant *A. fumigatus* have only been reported from Europe (the Netherlands and Denmark) and those surveys identified that 12% (6/49) of Dutch soil samples and 8% (4/50) of Danish soil samples were positive for the TR$_{34}$/L98H genotype [15,17]. Only one other mutation in the *cyp*51A gene combined with a different tandem repeat (TR$_{46}$/ Y121F/T289A) that was putatively linked to an environmental origin has been reported from clinical samples [28] and this genotype constituted 36% of resistant isolates in a Dutch referral centre [29]. The present study represents one of the largest environmental surveys of multi-triazole resistant *A. fumigatus* done so far and detected that 7% of the *A. fumigatus* isolates and 5% of soil/aerial samples distributed across large areas of India carried

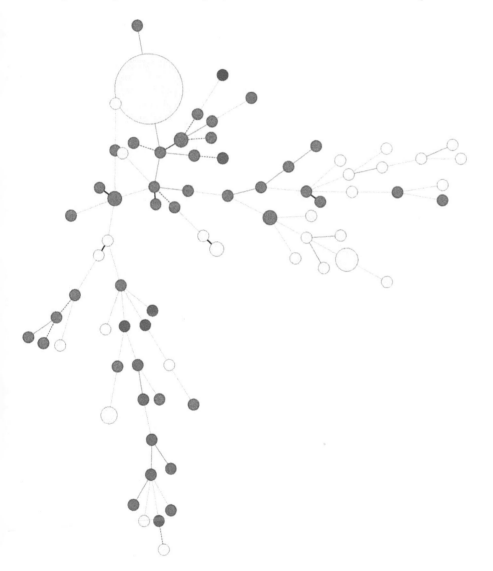

Figure 3. Minimum spanning tree showing wide genotypic diversity in the TR$_{34}$/L98H and wild type *A. fumigatus* **isolates studied.** The figure shows the 74 different genotypes (circles), the number of strains belonging to the same genotype (sizes of the circles), and origin of isolates (circles in yellow indicate Indian isolates; green Dutch isolates; red Chinese isolates; blue French isolates, purple German isolate and white reference strain, AF293). Solid thick and thin branches indicates 1 or 2 microsatellite markers differences, respectively; dashed branches indicates 3 microsatellite markers difference between two genotypes; 4 or more microsatellite markers differences between genotypes are indicated with dotted branches.

one single resistant mechanism. Culture of soil samples taken from potted plants (where commercial compost was used) and kept inside the hospital premises were positive for the same genotype. In contrast, natural soil sampled from the gardens of Delhi and hospitals did not grow the resistant *A. fumigatus* isolates although they were positive for *A. fumigatus*. Our findings corroborate with the findings of a Dutch environmental report where none of the *A. fumigatus* isolates obtained from natural soil was found to be azole resistant [15]. Therefore, environmental surveys for detection of genotype TR_{34}/L98H resistant *A. fumigatus* isolates may focus on sampling of soil from fields and commercial compost where fungicides are invariably used. It is noteworthy that the air samples of patient's wards of VPCI hospital harboured the same genotype of multi-triazole resistant *A. fumigatus*, isolated on two different occasions which raises concern on the exposure of hospitalized patients to this resistant genotype. In this context it is pertinent to mention that previously multi-triazole resistant TR_{34}/L98H *A. fumigatus* isolates have been reported from patients attending the outpatient departments of VPCI who were never exposed to azoles [22]. In addition multi-triazole resistant *A. fumigatus* has also been isolated from admitted patients of VPCI. The presence of *A. fumigatus* resistant to medical triazoles poses a threat to immunocompromised patients as alternative therapy is limited.

Snelders et al. reported that TR_{34}/L98H isolates from clinical and environmental origins were cross resistant to five triazole DMIs fungicides, propiconazole, bromuconazole, tebuconazole, epoxiconazole and difenoconazole and thus supporting the hypothesis that exposure of *A. fumigatus* to azole fungicides in the environment causes cross resistance to medical triazoles. [21]. Furthermore, these investigators also reported that these five triazole DMIs showed very similar molecule structures to the medical triazoles and adopted a similar conformation while docking the target enzyme and exhibit activity against wild type *A. fumigatus* but not against multi-triazole resistant TR_{34}/L98H *A. fumigatus* [21]. Similarly, in the present study four of the five (bromuconazole, tebuconazole, epoxiconazole and difenoconazole) triazole DMIs known to have similar molecule structures as medical triazoles showed significantly higher MICs for multi triazole resistant TR_{34}/L98H *A. fumigatus* from environmental and clinical samples than those of wild type strains (Table 2). In addition, metconazole and hexaconazole also showed high MICs for multi-triazole resistant *A. fumigatus* isolates with the TR_{34}/L98H mutation. Attention is called to the report of Serfling et al., who used the maize anthracnose fungus *Colletotrichum graminicola* model system to study the acquisition of azole resistance and investigated whether isolates that were resistant to an agricultural azole show cross-resistance to azoles and antifungal agents of other chemical classes used in medicine [30]. Their in-vitro data revealed that *C. graminicola* was able to efficiently adapt to medium containing azoles, and strains adapted to tebuconazole were less sensitive to all agricultural and medical azoles tested than the non-adapted control strain. Likewise, azole cross-resistance was observed for yeast isolates from the oropharynx of human immunodeficiency virus-infected patients to agricultural azole drugs and for those from environmental sources to medical azole drugs [31].

It is remarkable that all of the environmental and clinical TR_{34}/L98H *A. fumigatus* isolates in India had the same microsatellite genotype. Although the environmental isolates originated from geographically diverse regions of northern, eastern and southern parts of India were separated from each other by about 2000 Km, they harboured an identical short tandem repeat (STR) pattern. The possibility of contamination during handling of samples was ruled out by processing of the samples by different laboratory

personnel in two different laboratories in India and the Netherlands. Furthermore, we had reported earlier that two clinical TR_{34}/L98H *A. fumigatus* isolates originating from two azole naive patients, who were residents of Bihar and Delhi, shared the same STR pattern [22]. Moreover, azole-resistant strains from the environment of Bihar and Delhi also showed the same STR pattern. Notably, genetic analysis of a collection of MTR isolates showed that all isolates with the TR_{34}/L98H allele were all confined within a single clade and were less variable than susceptible isolates [25], consistent with a single and recent origin of the resistant genotype.

Our results are consistent with the hypothesis that the azole-resistant *A. fumigatus* strains analyzed here from across India were due to the clonal spread of a single genotype. The lack of a single azole-susceptible strain from either clinical origin or the environment in India with the same genotype as the widespread azole-resistant genotype it may be conceivable that the resistant genotype was unlikely the result of a single mutation at the *cyp*51A gene in a common azole-susceptible genotype in India. In addition, our genotype analysis suggest that the azole-resistant genotype in India was likely an extremely adaptive recombinant progeny derived from a cross between an azole-resistant strain migrated from outside of India and a native azole-susceptible strain from within India, followed by mutation. The abundant phylogenetic incompatibility found in each of the sub-samples as well as in the whole sample (where 100% of the loci pairs were phylogenetically incompatible, thus consistent with recombination) supports sexual mating in natural populations of this species in India. Our inferred mechanisms have been similarly suggested for the emergence of many virulent strains of viral, bacterial and protozoan pathogens [32,33]. Once the extremely fit *A. fumigatus* genotype emerged in India, it could spread quickly by producing a large number of airborne asexual spores in the environment. These airborne spores can easily disperse to other geographic areas by air current or anthropogenic means. The widespread application of triazole fungicides in the environment in India in the last two decades could have contributed to its spread by reducing the azole-susceptible genotypes and selecting for this azole-resistant genotype. Whether this resistant genotype has spread to neighbouring countries remain to be determined.

Materials and Methods

Ethics Statement

All necessary permits were obtained for the described field studies.

Collection of Environmental Samples

A total of 486 environmental samples including soil from flowerbeds of nurseries, surrounding parks of hospitals, cotton trees, tea gardens, paddy fields, soil containing bird excreta, decayed wood of tree trunks and aerial samples of the indoor environment of hospital wards from the Union Territory (UT) of Delhi, Haryana, Himachal Pradesh, Uttrakhand, Bihar, West Bengal, Sikkim, Meghalaya and Tamil Nadu States were investigated during July 2011–April 2012. The distribution of the investigated 486 samples was as follows: UT of Delhi (n = 266), Haryana (n = 21), Himachal Pradesh (n = 4), Uttrakhand (n = 21), Bihar (n = 33), West Bengal (n = 59), Sikkim (n = 6), Meghalaya (n = 11) and Tamil Nadu (n = 65).

Soil and Aerial Sampling

About two gram of soil was suspended in 8 ml of 0.85% NaCl, vortexed and allowed to settle for 30 seconds. Subsequently, the suspension was diluted 1:10 and 100 µl was plated in duplicates on Sabouraud dextrose agar plates supplemented with 50 mg/L chloramphenicol and incubated at 37°C for 48 h. One gram of decayed wood was suspended in 10 ml of 0.85% NaCl and allowed to settle after vortexing it for 1 min. Then, 100 µl of suspension was plated in duplicates on SDA and incubated at 37°C for 48 h.

For the indoor aerial sampling of the hospital, duplicate SDA plates were exposed for 1 h in the corners and centre of the general outpatient and wards of the V. P. Chest Institute (VPCI), Delhi, on two different occasions. Plates were incubated for 48 h at 37°C.

Identification

In order to detect overall prevalence of *A. fumigatus* the samples were initially inoculated on SDA plates and maximum of 3 colonies per plate were purified and identified by macro- and microscopic characteristics and growth at 50°C which differentiated *A. fumigatus* from *A. lentulus*. Samples found out to be negative for *A. fumigatus* were again processed without dilution and inoculated directly on SDA plates. All of the *A. fumigatus* isolates were then subcultured on SDA plates supplemented with 4 mg/L itraconazole and incubated at 37°C for 48 h. Identification of all the *A. fumigatus* isolates that grew on 4 mg/L itraconazole containing SDA plates (ITC+ isolates) were confirmed by sequencing of the internal transcribed spacer region. In order to rule out any cryptic species within *Aspergillus* section Fumigati, molecular identification was performed by amplification of parts of the β-tubulin gene and calmodulin gene [34,35].

Antifungal Susceptibility Testing

The in vitro activity of all the standard azole antifungals was investigated using CLSI M38-A2 broth microdilution [36]. A total of 53 itraconazole resistant *A. fumigatus* isolates (44 ITC+ environmental and 9 ITC+ clinical) were subjected to AFST. Nine itraconazole resistant clinical isolates were cultured from patients suspected of bronchopulmonary aspergillosis. Among the 9 ITC+ *A. fumigatus* clinical isolates two have been reported earlier [22]. In addition, 35 itraconazole susceptible *A. fumigatus* isolates comprising 22 randomly selected wild type environmental and 13 azole susceptible clinical *A. fumigatus* isolates cultured from patients of suspected bronchopulmonary aspergillosis were included as controls. The drugs tested included itraconazole (ITC, Lee Pharma, Hyderabad, India, and Janssen Research Foundation, Beerse, Belgium), voriconazole (VRC, Pfizer Central Research, Sandwich, Kent, United Kingdom) and posaconazole (POS, Schering-Plough, Kenilworth, NJ, USA, now Astellas). For the broth microdilution test, RPMI 1640 medium with glutamine without bicarbonate (Sigma-Aldrich, St Louis, MO, USA) buffered to pH 7 with 0.165 M 3-N-morpholinepropanesulfonic acid (Sigma) was used. Isolates were grown on potato dextrose agar for 5 days at 28°C and the inoculum was adjusted to a final density of 0.5–2.5 x 10^4 cfu/ml by measuring 0.09–0.13 OD at 540 nm using spectrophotometer. The final concentrations of the drugs were 0.03 to 16 mg/L for itraconazole and voriconazole and 0.015 to 8 mg/L for posaconazole. Drug-free and mould-free controls were included and microtitre plates were incubated at 35°C for 48 h. CLSI recommended quality control strains, *Candida krusei*, ATCC6258 and *Candida parapsilosis*, ATCC22019 and reference strains *Aspergillus fumigatus*, ATCC204305 and *Aspergillus flavus*, ATCC204304 were included. The MIC end points were read visually which, for azoles were defined as the lowest concentration at which there was 100% inhibition of growth compared with the drug-free control wells. *A. fumigatus* isolates with

high itraconazole MICs were tested twice on different days. Azole resistance was defined for itraconazole, >2 mg/L, voriconazole, >2 mg/L, and posaconazole, >0.5 mg/L as proposed by Verweij et al. [37].

Activity of Azole Fungicides

The commonly used ten azole fungicides registered under the Insecticides Act, 1968 by the Indian Central Insecticide Board and Registration Committee were tested for activity against resistant and wild type environmental and clinical *A. fumigatus* Indian isolates by microdilution method as described above. The azole fungicides tested were bromuconazole, cyproconazole, difenoconazole, epoxiconazole, penconazole, tebuconazole, triadimefon, metconazole (kindly gifted by Dr. P. Verweij, Nijmegen, the Netherlands) hexaconazole (Rallis India, Mumbai, India) and tricyclazole (Cheminova India, Mumbai, India). The fungicides were dissolved in dimethyl sulfoxide and concentration range used was 0.06–32 mg/L.

Statistical Analysis

Point serial correlation was computed between MICs of wild type and TR$_{34}$/L98H *A. fumigatus* isolates of clinical and environmental origin to determine the correlation coefficient which is a measure of the effect size (r), where values of r = 0 indicate no correlation between MICs, r = 1 indicate positive correlation and r = −1 indicate negative correlation. In cases where correlation MICs have similar values for all isolates, correlation effect size was considered r = 0 [21].

Mixed Format Real-time PCR Assay to Detect Mutations

All of the ITC+ *A. fumigatus* isolates were subjected to a mixed-format real-time PCR assay as described previously for detection of TR$_{34}$/L98H, TR$_{46}$/Y121F/T289A, M220, G54 mutations leading to triazole resistance in *A. fumigatus* [38].

Microsatellite Genotypic Analysis

Genotyping was performed with a panel of nine short tandem repeats as described previously [39]. The genetic relatedness between Indian environmental and clinical isolates was determined by using microsatellite typing. A total of 60 ITC+ *A. fumigatus* isolates which included 51 environmental (44 isolated in the Indian laboratory and 7 isolated from Indian soil samples processed in the Netherlands laboratory) and 9 clinical isolates were subjected to microsatellite typing. For phylogenetic analysis, 24 Dutch (15 clinical and 9 environmental), 8 clinical Chinese [40], 3 clinical French [18] and one clinical German [19] isolates of *A. fumigatus* containing the TR$_{34}$/L98H genotype were tested along with the Indian isolates. In addition, 35 (22 environmental and 13 clinical) Indian, 12 environmental Dutch and 2 clinical French *A. fumigatus* isolates without mutations and a reference strain *A. fumigatus* AF293 were included in the analysis.

Genetic Analysis of Microsatellite Genotypes

The composite genotype for each of the 146 strains of *A. fumigatus* was identified based on alleles at all nine microsatellite loci. The genotype information was then used to identify genetic relationships among strains. Gene diversity and genotype diversity within individual samples and the relationships between samples were estimated using the population genetic analyses program GenAlEx 6.1 [41]. The relationships among alleles at different loci were examined for evidence of recombination in natural populations of this fungus, using the computer program Multilocus 2.0 (http://www.agapow.net/software/multilocus/) [42]. Results of these analyses were used to infer the potential source(s) of the triazole-resistant clinical and environmental *A. fumigatus* strains in India.

Acknowledgments

We thank Daniel Diekema (University of Iowa Carver College of Medicine, Iowa City, USA) for Chinese isolates, Andre Paugam (Université Paris Descartes and Hôpital Cochin, AP-HP, Paris, France) for French isolates and Jorg Steinmann and Peter-Michael Rath (Institute of Medical Microbiology, University Hospital Essen, Essen, Germany) for the German isolate which were used as controls. We are grateful to Paul Verweij (Radboud University Nijmegen Medical Centre, Nijmegen, The Netherlands) for providing us several fungicides (bromuconazole, cyproconazole, difenoconazole, epoxiconazole, penconazole, tebuconazole, triadimefon, metconazole). We acknowledge Rallis India, India and Cheminova India, India for kindly providing us hexaconazole and tricyclazole fungicides.

Author Contributions

Conceived and designed the experiments: AC JPX JFM. Performed the experiments: AC SK CS GS PKS FH CHK. Analyzed the data: AC SK JPX FH CHK JFM. Contributed reagents/materials/analysis tools: AC SNG JPX FH. Wrote the paper: AC SK JPX CHK JFM.

References

1. Chakrabarti A, Chatterjee SS, Das A, Shivaprakash MR (2011) Invasive aspergillosis in developing countries. Med Mycol 49 (Suppl 1): S35–47.
2. Verweij PE, Denning DW (1997) The challenge of invasive aspergillosis: increasing numbers in diverse patient groups. Int J Infect Dis 2: 61–63.
3. Denning DW, Perlin DS (2011) Azole resistance in *Aspergillus*: a growing public health menace. Future Microbiol 6: 1229–1232.
4. Herbrecht R, Denning DW, Patterson TF, Bennett JE, Greene RE, et al. (2002) Voriconazole versus amphotericin B for primary therapy of invasive aspergillosis. N Engl J Med 347: 408–415.
5. Cornely OA, Maertens J, Winston DJ, Perfect J, Ullmann AJ, et al. (2007) Posaconazole vs. fluconazole or itraconazole prophylaxis in patients with neutropenia. N Engl J Med 356: 348–359.
6. Howard SJ, Cerar D, Anderson MJ, Albarrag A, Fisher MC, et al. (2009) Frequency and evolution of azole resistance in *Aspergillus fumigatus* associated with treatment failure. Emerg Infect Dis 15: 1068–1076.
7. Arendrup MC, Mavridou E, Mortensen KL, Snelders E, Frimodt-Møller N, et al. (2010) Development of azole resistance in *Aspergillus fumigatus* during azole therapy associated with change in virulence. PLoS One 5:e10080.
8. van der Linden JWM, Snelders E, Kampinga GA, Rijnders BJA, Mattsson E, et al. (2011) Clinical implications of azole resistance in *Aspergillus fumigatus*, the Netherlands, 2007–2009. Emerg Infect Dis 17: 1846–1852.
9. Howard SJ, Arendrup MC (2011) Acquired antifungal drug resistance in *Aspergillus fumigatus*: epidemiology and detection. Med Mycol 49 (Suppl 1): S90–S95.
10. Snelders E, Melchers WJ, Verweij PE (2011) Azole resistance in *Aspergillus fumigatus*: a new challenge in the management of invasive aspergillosis. Future Microbiol 6: 335–347.
11. Camps SM, van der Linden JW, Li Y, Kuijper EJ, van Dissel JT, et al. (2012) Rapid induction of multiple resistance mechanisms in *Aspergillus fumigatus* during azole therapy: a case study and review of the literature. Antimicrob Agents Chemother 56: 10–16.
12. Stensvold CR, Jorgensen LN, Arendrup MC (2012) Azole-resistant invasive *Aspergillus*: relationship to agriculture. Curr Fungal Infect Rep 6: 178–191.
13. Verweij PE, Snelders E, Kema GH, Mellado E, Melchers WJ (2009) Azole resistance in *Aspergillus fumigatus*: a side effect of environmental fungicide use? Lancet Infect Dis 9: 789–795.
14. Rodriguez-Tudela JL, Alcazar-Fuoli L, Mellado E, Alastruey-Izquierdo A, Monzon A, et al. (2008) Epidemiological cutoffs and cross-resistance to azole drugs in *Aspergillus fumigatus*. Antimicrob Agents Chemother 52: 2468–2472.
15. Snelders E, Huis In 't Veld RA, Rijs AJJM, Kema GHJ, Melchers WJ, et al. (2009) Possible environmental origin of resistance of *Aspergillus fumigatus* to medical triazoles. Appl Environ Microbiol 75: 4053–4057.
16. Howard SJ, Pasqualotto AC, Denning DW (2010) Azole resistance in allergic bronchopulmonary aspergillosis and *Aspergillus* bronchitis. Clin Microbiol Infect 16: 683–688.
17. Mortensen KL, Mellado E, Lass-Flörl C, Rodriguez-Tudela JL, Johansen HK, et al. (2010) Environmental study of azole-resistant *Aspergillus fumigatus* and other aspergilli in Austria, Denmark, and Spain. Antimicrob Agents Chemother 54: 4545–4549.
18. Burgel PR, Baixench MT, Amsellem M, Audureau E, Chapron J, et al. (2012) High prevalence of azole-resistant *Aspergillus fumigatus* in adults with cystic fibrosis exposed to itraconazole. Antimicrob Agents Chemother 56: 869–874.
19. Rath PM, Buchheidt D, Spiess B, Arfanis E, Buer J, et al. (2012) First reported case of azole-resistant *Aspergillus fumigatus* due to TR/L98H mutation in Germany. Antimicrob Agents Chemother 56: 6060–6061.
20. Snelders E, van der Lee HA, Kuijpers J, Rijs AJMM, Varga J, et al. (2008) Emergence of azole resistance in *Aspergillus fumigatus* and spread of a single resistance mechanism. PLoS Med 5: e219.
21. Snelders E, Camps SMT, Karawajczyk A, Schaftenaar G, Kema GHJ, et al. (2012) Triazole fungicides can induce cross-resistance to medical triazoles in *Aspergillus fumigatus*. PLoS ONEe 7(3): e31801.
22. Chowdhary A, Kathuria S, Randhawa HS, Gaur SN, Klaassen CH, et al. (2012) Isolation of multiple-triazole-resistant *Aspergillus fumigatus* strains carrying the TR/L98H mutations in the *cyp*51A gene in India. J Antimicrob Chemother 67: 362–366.
23. Hof H (2001) Critical annotations to the use of azole antifungals for plant protection. Antimicrob Agents Chemother 45: 2987–2990.
24. Fisher MC, Henk DA, Briggs CJ, Brownstein JS, Madoff LC, et al. (2012) Emerging fungal threats to animal, plant and ecosystem health. Nature 484: 186–194.
25. Klaassen CH, Gibbons JG, Fedorova ND, Meis JF, Rokas A (2012) Evidence for genetic differentiation and variable recombination rates among Dutch populations of the opportunistic human pathogen *Aspergillus fumigatus*. Mol Ecol 21: 57–70.
26. Thind TS (2007) Changing cover of fungicide umbrella in crop protection. Indian Phytopath 60: 421–433.
27. Singh N, Dureja P (2000) Persistence of hexaconazole, a triazole fungicide in soils. J Environ Sci Health B 35: 549–558.
28. Kuipers S, Brüggemann RJ, de Sévaux RG, Heesakkers JP, Melchers WJ, et al. (2011) Failure of posaconazole therapy in a renal transplant patient with invasive aspergillosis due to *Aspergillus fumigatus* with attenuated susceptibility to posaconazole. Antimicrob Agents Chemother 55: 3564–3566.
29. NETHMAP (2012) Consumption of antimicrobial agents and antimicrobial resistance among medically important bacteria in the Netherlands 59–60. Available: http://www.rivm.nl/dsresource?objectid = rivmp:181194&type = org&disposition = inline. Accessed 2012 Sep 4.
30. Serfling A, Wohlrab J, Deising HB (2007) Treatment of a clinically relevant plant-pathogenic fungus with an agricultural azole causes cross-resistance to medical azoles and potentiates caspofungin efficacy. Antimicrob Agents Chemother 51: 3672–3676.
31. Müller FM, Staudigel A, Salvenmoser S, Tredup A, Miltenberger R, et al. (2007) Cross-resistance to medical and agricultural azole drugs in yeasts from the oropharynx of human immunodeficiency virus patients and from environmental Bavarian vine grapes. Antimicrob Agents Chemother 51: 3014–3016.
32. zur Wiesch PA, Kouyos R, Engelstädter J, Regoes RR, Bonhoeffer S (2011) Population biological principles of drug-resistance evolution in infectious diseases. Lancet Infect Dis 11: 236–247.
33. Fisher MC, Henk DA (2012) Sex, drugs and recombination: the wild life of *Aspergillus*. Mol Ecol 21: 1305–1306.
34. Glass NL, Donaldson GC (1995) Development of primer sets designed for use with the PCR to amplify conserved genes from filamentous ascomycetes. Appl Environ Microbiol 61: 1323–1330.
35. Hong SB, Cho HS, Shin HD, Frisvad JC, Samson RA (2006) Novel *Neosartorya* species isolated from soil in Korea. Int J Syst Evol Microbiol 56: 477–486.
36. Clinical and Laboratory Standards Institute (2008) Reference method for broth dilution antifungal susceptibility testing of filamentous fungi; approved standard, 2nd ed. CLSI document M38-A2. Wayne (PA.
37. Verweij PE, Howard SJ, Melchers WJ, Denning DW (2009) Azole-resistance in *Aspergillus*: proposed nomenclature and breakpoints. Drug Resist Updat 12: 141–147.
38. Klaassen CH, de Valk HA, Curfs-Breuker IM, Meis JF (2010) Novel mixed-format real-time PCR assay to detect mutations conferring resistance to triazoles in *Aspergillus fumigatus* and prevalence of multi-triazole resistance among clinical isolates in the Netherlands. J Antimicrob Chemother 65: 901–905.
39. de Valk HA, Meis JF, Curfs-Breuker IM, Muehlethaler K, Mouton JW et al. (2005) Use of a novel panel of nine short tandem repeats for exact and high-resolution fingerprinting of *Aspergillus fumigatus* isolates. J Clin Microbiol 43: 4112–4120.
40. Lockhart SR, Frade JP, Etienne KA, Pfaller MA, Diekema DJ, et al. (2011) Azole resistance in *Aspergillus fumigatus* isolates from the ARTEMIS global surveillance study is primarily due to the TR/L98H mutation in the *cyp*51A gene. Antimicrob Agents Chemother 55: 4465–4468.
41. Peakall R, Smouse PE (2006) GENALEX 6: genetic analysis in Excel. Population genetic software for teaching and research. Mol Ecol Notes 6: 288–295.
42. Agapow PM, Burt A (2001) Indices of multilocus linkage disequilibrium. Mol Ecol Notes 1: 101–102.

8

Spatial and Temporal Variation of Archaeal, Bacterial and Fungal Communities in Agricultural Soils

Michele C. Pereira e Silva[1]*, Armando Cavalcante Franco Dias[2], Jan Dirk van Elsas[1], Joana Falcão Salles[1]

1 Department of Microbial Ecology, Centre for Life Sciences, University of Groningen, Groningen, The Netherlands, 2 Department of Soil Science, "Luiz de Queiroz" College of Agriculture, University of São Paulo, Piracicaba, São Paulo, Brazil

Abstract

Background: Soil microbial communities are in constant change at many different temporal and spatial scales. However, the importance of these changes to the turnover of the soil microbial communities has been rarely studied simultaneously in space and time.

Methodology/Principal Findings: In this study, we explored the temporal and spatial responses of soil bacterial, archaeal and fungal β-diversities to abiotic parameters. Taking into account data from a 3-year sampling period, we analyzed the abundances and community structures of *Archaea*, *Bacteria* and *Fungi* along with key soil chemical parameters. We questioned how these abiotic variables influence the turnover of bacterial, archaeal and fungal communities and how they impact the long-term patterns of changes of the aforementioned soil communities. Interestingly, we found that the bacterial and fungal β-diversities are quite stable over time, whereas archaeal diversity showed significantly higher fluctuations. These fluctuations were reflected in temporal turnover caused by soil management through addition of N-fertilizers.

Conclusions: Our study showed that management practices applied to agricultural soils might not significantly affect the bacterial and fungal communities, but cause slow and long-term changes in the abundance and structure of the archaeal community. Moreover, the results suggest that, to different extents, abiotic and biotic factors determine the community assembly of archaeal, bacterial and fungal communities.

Editor: A. Mark Ibekwe, U. S. Salinity Lab, United States of America

Funding: This work was supported by the NWO-ERGO Programme and was part of a collaborative project with Utrecht University, Utrecht, The Netherlands. The funders had no role in study design, data collection and analysis, decision to publish, or preparation of the manuscript.

Competing Interests: The authors have declared that no competing interests exist.

* E-mail: m.silva@rug.nl

Introduction

Understanding temporal and spatial patterns in the abundance and distribution of communities has been a fundamental quest in ecology. Such an understanding is crucial to allow an anticipation of responses of ecosystems such as soil to global changes [1]. Because local conditions are never constant, small disturbances that affect the soil microbial communities might occur [2–3] at different temporal and spatial scales. The assessment of microbial communities at a particular locality may result in patterns that vary greatly both within and between years, and these communities may be subjected to changes over longer time scales as a result of processes such as succession and evolutionary change [4]. One approach to investigate temporal (and spatial) variability in complex systems is to explore patterns of β-diversity. Whereas alpha (α-) diversity represents a measure of the total diversity of a given site, β-diversity is the variation of species composition (turnover) across space or time between paired sites. High β-diversity indicates large differences in community composition among different sites. Such high diversity can result from local as well as regional factors, e.g. changes in the local environmental conditions or limitation of dispersal between sites [5].

Temporal variation of conditions is a very common feature of ecosystems. Ecologists have long been interested in how such variation structures natural communities [6,7]. It can presumably affect the rate of microbial turnover, as microorganisms can process resources and adapt to changes in natural environments on a much faster time scale than macroorganisms [8]. Moreover, many functional microbial groups can show dramatic seasonal changes in soils [9].

The number of studies employing the concept of β-diversity to understand how microbial communities respond to biotic and abiotic parameters has increased substantially in soil ecology. Martiny and co-workers [10] studied the mechanisms driving ammonia-oxidizing bacterial (AOB) communities in salt marsh sediments. They found no evolutionary diversification when comparing the AOB community composition between three continents; although a negative relationship was observed between geographic distance and community similarity. Furthermore, in an attempt to determine to which extent a bacterial metacommunity that consisted of 17 rock pools was structured by different assembly mechanisms [11], the authors studied changes in β-diversity across different environmental gradients over time, including phosphorus concentration, temperature and salinity. They found that there were temporal differences in how the communities responded to abiotic factors. β-diversity allows not only the understanding of temporal but of spatial variations as well. For instance, in a survey of bacterial communities across more than 1000 soil cores in Great Britain [12], no spatial patterns were observed, but instead variations in β-diversity according to soil pH were found, which revealed that β-diversity (between sample variance in α-diversity) was higher in acidic soils (pH 4–5) than in more alkaline soils (pH 7–9) [12]. In the former soils, environmental heterogeneity was highest, calculated as the variance in environmental conditions [12]. In another study, different patterns of bacterial β-diversity were observed between different layers in sediment cores, which could be attributed to historical variation and geochemical stratification [13].

Of the soil microbial groups, bacteria have been mostly studied, as they exhibit an estimated species diversity of about 10^3 to up to 10^6 per g soil [14–16]. However, archaea and fungi are also important microorganisms found in soil. Previous studies have shown the ubiquity of archaea in soil, especially the crenarchaeota [17–19]. Fungal abundances in the order of 10^4 fungal propagules per g of dry soil were observed in Antarctic soils [20] and 10^7 per g of soil in soil crusts [21]. Fundamental differences in the physiology and ecology of members of such communities would suggest that their patterns of spatial and temporal variation are controlled by distinct edaphic factors.

In this study, we explored the temporal and spatial fluctuations of soil microbial communities and their relation to local environmental conditions. In order to do so, we investigated the spatiotemporal dynamics of the soil microbiota by analyzing the patterns of α- and β-diversity of archaea, bacteria and fungi in eight agricultural soils across the Netherlands. We sampled the soils eleven times, from 2009 to 2011. Furthermore, to complement the analyses, we applied TLA (time-lag analysis) [22], a distance-based approach to study the temporal dynamics of communities by measuring community dissimilarity over increasing time lags. TLA provides measures of model fit and statistical significance, allowing the quantification of the strength of temporal community change in a numerical framework [23]. We thus interrogated how the relationship between microbial abundance, species composition and the surrounding environment varies in space and time and how this relates to long-term compositional changes.

Materials and Methods

Study area and field sampling

The eight soil sites sampled are located in the Netherlands. Their characteristics and geographical coordinates are found in Table 1 and in Table S1. Sampling points were selected to reflect temporal differences in abiotic parameters. For each soil four replicates were taken. Each replicate consisted of 10 subsamples (15–20 cm deep) collected between plots with a spade, away from the plant roots. Soil samples were collected four times over an annual cycle in 2009 (April, June, September and November), three times in 2010 (April, June and October), and four times in 2011 (February, April, July and September). Each sample was placed in a plastic bag and thoroughly homogenized before analysis. A 100-g subsample was kept at 4°C and used for chemical analyses, whereas the remaining soil was kept at −20°C for subsequent DNA extraction and molecular analysis of community composition and total abundance (see below).

Soil chemical analysis

The environmental variables measured included pH, concentrations of nitrate (N-NO$_3^-$ in mg/kg of soil), ammonium (N-NH$_4^+$ in mg/kg of soil), organic matter (OM in %) and clay content (in %). The pH was measured in CaCl$_2$ suspension 1:4.5 (g/v) (Hanna Instruments BV, IJsselstein, The Netherlands). Organic matter (OM) content is calculated after 4 hours at 550°C. Nitrate (N-NO$_3^-$) and ammonium (N-NH$_4^+$) were determined with a colorimetric method using the commercial kits Nanocolor Nitrat50 (detection limit, 0.3 mg N kg-1 dry weight, Macherey-Nagel, Germany) and Ammonium3 (detection limit, 0.04 mg N kg-1 dry weight; Macherey-Nagel, Germany) according to manufacturer's protocol.

Nucleic acid extraction

DNA was extracted from 0.5 g of soil using Power Soil MoBio kit (Mo Bio Laboratories Inc., NY), according to the manufacturer's instructions, after the addition of glass beads (diameter 0.1 mm; 0.25 g) to the soil slurries. The cells were disrupted by bead beating (mini-bead beater; BioSpec Products, United States) three times for 60 s. Following extraction, the DNA preparations were electrophoresed over agarose gels in order to assess DNA purity, quality (average size) and quantity. The quantity of extracted DNA was estimated on gel by comparison to a 1-kb DNA ladder (Promega, Leiden, Netherlands) and quality was determined based on the degree of DNA shearing (average molecular size) as well as the amounts of coextracted compounds.

Real-time PCR quantification (qPCR)

Absolute quantification was carried out in four replicates on the ABI Prism 7300 Cycler (Applied Biosystems, Germany). The 16S rRNA gene was amplified by qPCR using diluted extracted DNA as template. Specific primers for archaea (group 1 crenarchaeota) 771F/957R [24] and for V5–V6 region of bacteria 16SFP/16SRP [25] were used. We have chosen to focus on Crenarchaeota, as this group is often more common in soil environments than Euryarchaeota [26]. For Fungal community primers 5,8S/ITS1f [27] were chosen. Cycling programs and primer sequences are detailed in Table S2. The specificity of the amplification products was confirmed by melting-curve analysis and on 1.5% agarose gels. Standard curves were obtained using serial dilutions of plasmid DNA containing the cloned 16S rRNA gene obtained from *Burkholderia terrae* BS001 or ITS region of *Rhizoctonia solani* AG3. Dilutions ranged from 10^7 to 10^2 gene copy numbers/μl. The archaeal standard curve was obtained by serial dilution of PCR product generated from *Cenarchaeum symbiosum* with the aforementioned archaeal specific primers [24].

PCR for DGGE analysis

For DGGE analysis, bacterial 16S rRNA genes were PCR amplified using the forward primer F968 [28] with a GC-clamp attached to 5′ and the universal R1401.1b [29]. Archaeal 16S rRNA genes were amplified with the A2F/U1406R primer pair [30], following amplification using the *Archaea*-specific forward primer at position 344 with a 40-bp GC clamp [31] added to the 5′ end, and a universal reverse primer at position 517. The fungal ITS region was amplified with EF4 [32]/ITS4 [33], followed by a second amplification with primers ITS1f-GC [34]/ITS2 [33]. PCR mixtures, primer sequences and cycling conditions are described in Table S3. About 200 ng of amplicons were loaded onto a 6% (w/v) polyacrylamide gel in the Ingeny Phor-U system (Ingeny International, Goes, The Netherlands), with a 20–50%, 45–65% and 40–60% denaturant gradient for the fungal, bacterial

Table 1. List of soils included in this study.

Sampling Site	Soil type	Land use	Crops			North coordinate	East coordinate
			2009	2010	2011		
Buinen (B)	Sandy loam	agriculture	barley	potato	potato	52°55′386″	006°49′217″
Valthermond (V)	Sandy loam	agriculture	barley	potato	potato	52°50′535″	006°55′239″
Droevendaal (D)	Sandy loam	agriculture	triticale	barley	barley	51°59′551″	005°39′608″
Wildekamp (K)	Sandy loam	grassland	grass	grass	grass	51°59′771″	005°40′157″
Kollumerwaard (K)	Clayey	agriculture	potato	grass	sugar beet	53°19′507″	006°16′351″
Steenharst (S)	Silt loam	agriculture	grass	potato	grass	53°15′428″	006°10′189″
Grebedijk (G)	Clayey	agriculture	potato	wheat	wheat	51°57′349″	005°38′086″
Lelystad (L)	Clayey	agriculture	potato	grass	corn	52°32′349″	005°33′601″

and archaeal community, respectively (100% denaturant corresponded to 7 M urea and 40% (v/v) deionized formamide). Electrophoresis was performed at a constant voltage of 100 V for 16 h at 60°C. The gels were stained for 60 min in 0,5× TAE buffer with SYBR Gold (final concentration 0,5 µg/liter; Invitrogen, Breda, The Netherlands). Images of the gels were obtained with Imagemaster VDS (Amersham Biosciences, Buckinghamshire, United Kingdom). Genetic fingerprints were analyzed using GelCompar software (Applied Maths, Sint-Martens Latem, Belgium) [35,36].

Data analyses

The diversity of each of the soil bacterial, archaeal and fungal communities was determined on the basis of the PCR-DGGE profiles. Total diversity (α) of the dominant community members was estimated from these data using the Shannon index, as recommended by Hill et al. [37], as well as the number of DGGE bands (species richness). We calculated the temporal β-diversity of archaeal, bacterial and fungal communities as the mean of all pairwise Bray-Curtis dissimilarities based on the relative abundance of DGGE bands, as previously described [38,39,11]. To support results from the calculated β-diversity and to test the statistical significance and strength of community dynamics we used time-lag analysis (TLA) [22] by plotting Hellinger-transformed [40] distance values against the square root of the time lag for all lags. The time-lag analytical approach can produce a number of general theoretical patterns with time-series data [22]. The square root transformation reduces the probability that a smaller number of points at larger time lags will bias the analysis [41]. The Bray-Curtis matrices as well as Hellinger-transformed distances were determined in PRIMER-E (version 6, PRIMER-E Ltd, Plymouth, UK; [42]).

To test how α-diversity, β-diversity and microbial abundance varied in relation to environmental variables, parametric Pearson correlation coefficients were calculated between α and β diversities, soil pH, organic matter, nitrate, ammonium, clay content and soil moisture, as well as between total abundances and TLA slopes using SPSS v18.0.3 (SPSS Inc., Chicago, IL, USA). All variables except pH were transformed (Log(x+1)) prior to all analyses. Moreover, we applied variance partitioning to evaluate the relative contribution of the drivers of the microbial assemblages. Forward selection was used on CCA (Canonical Correspondence Analysis) to select a combination of environmental variables that explained most of the variation observed in the species matrices. For that, a series of constrained CCA permutations was performed in Canoco

(version 4.0 for Windows, PRI Wageningen, The Netherlands) to determine which variables best explained the assemblage variation, using automatic forward selection and Monte Carlo permutation tests (permutations = 999). The length of the corresponding arrows indicated the relative importance of the chemical factor explaining variation in the microbial communities.

Results

Variability of environmental parameters

Soil pH, nitrate, ammonium and organic matter levels were determined in triplicate across all soil samples. Soil pH was significantly higher (P<0.05) in soils K, G and L (7.32±0.06, n = 57) than in soils B, V, D, W and S (4.88±0.04, n = 99) during the whole experimental period and no significant variation over time was observed. In all soils, significant changes were observed in the levels of nitrate, with lower values at the end of the growing season for most of the soils (September 2009: 32.78 mg/kg±7.77; October 2010: 24.15 mg/kg±3.62; September 2011: 2.45 mg/kg±0.41) and higher at the beginning (April 2009: 75.6 mg/kg±12.5; April 2010: 56.4 mg/kg±5.63; April 2011:100.1 mg/kg±16.5). Levels of ammonium also varied over the whole period, with higher values being observed at the beginning of the season (April 2009: 13.3 mg/kg±1.14; April 2010: 16.0 mg/kg±1.19; April 2011: 12.1 mg/kg±2.72) and lower values at the end (September 2009: 1.93 mg/kg±0.16; October 2010: 8.86 mg/kg±1.22) (Table S1).

Considering each soil individually, they had characteristically different values, with higher levels of nitrate and ammonium found in soils B, V, D and S than in soils W, K, G and L (Table S1). In 2009 and 2010, variations in organic matter (OM) content were observed from September (5.63%±1.20) to November (7.34%±1.45) 2009 and from April (6.28%±0.85) to June (5.04%±0.89) 2010. Small but insignificant variations in OM were observed in 2011. On average, the OM content of all soils was in the range around 4%, except for soil V, which had on average 17% OM.

Temporal variations in the abundance of archaeal, bacterial and fungal communities and their responses to abiotic variables

We studied the variations in the abundances of archaeal, bacterial and fungal communities over time and across all samples in three years. The total bacterial abundance showed significant temporal variation during the whole period, ranging between

8.12±0.23 (mean ± standard error) (September 2011) and 10.93±0.06 (June 2010) log copy numbers per g dry soil and showing comparable copy numbers in sandy (9.65±0.13) and clayey soils (9.64±0.16). The archaeal abundance (crenarchaeota) ranged between 6.96±0.14 (April 2009) and 8.78±0.07 (April 2011) log copy numbers per g dry soil, and showed significant differences between sandy and clayey soils across almost all sampling times, with lower numbers in the sandy soils (7.77±0.13) than in the clayey soils (8.22±0.13). Fungal abundance varied between 8.76±0.16 (February 2011) and 10.00±0.09 (April 2011), and significantly higher abundance was observed in the sandy soils depending on the sampling time (Fig. 1). Overall and on average, the abundance of bacteria was higher than that of the fungi, except in September 2009 and during 2011.

We used Pearson's correlation to examine how soil parameters influenced the abundances of bacterial, archaeal and fungal communities. Whereas the archaeal abundances were positively correlated with soil pH (r = +0.883, P<0.001), they were negatively influenced by nitrate (r = −0.764, P<0.05). A positive relationship was observed between fungal abundance and soil organic matter (r = +0.722, P<0.05), and a negative relationship was observed between fungal abundance and archaeal abundance (r = −0.484, P<0.05). Relationships between the abundance of bacteria and fungi, as well as between bacteria and archaea, were not significant. Interestingly, none of the soil parameters measured influenced bacterial abundance significantly.

Patterns of α-diversity and response to abiotic variables

Understanding how species are distributed in space and time may yield a first avenue towards their assembly rules [43]. We used two ecological measures, i.e. the Shannon index (H′) and species richness, as proxies to study the variations in the α-diversities of the archaeal, bacterial and fungal communities. Differential patterns of archaeal, bacterial and fungal α-diversities were observed, as measured by H′ (Fig. 2). The H′ values of the archaeal communities ranged from 1.68±0.04 in June 2009 to 2.40±0.05 in February 2011, and they were consistently lower than the corresponding bacterial and fungal values. The bacterial H′ values varied from 2.52±0.04 in October 2010 to 3.85±0.04

in April 2009, whereas those of the fungal communities varied from 3.2±0.16 in April 2009 to 4.09±0.04 in April 2010 (Fig. 2). In general, the differences observed between sandy and clayey soils for the bacterial and fungal diversities (Shannon index) were time point-dependent. For archaea, a higher Shannon index was noticed in the sandy soils compared to the clayey ones in 2009 and 2010 but not in 2011.

Concerning correlations with edaphic factors, a positive effect of OM content was observed on the archaeal α-diversity (r = +0.691, P<0.05) (Table 2). When using the number of DGGE bands as a measure of α-diversity (species richness), a significant and strong positive correlation was found between archaeal α-diversity and nitrate levels (r = +0.962 , P<0.001) (Table 2). None of the soil parameters measured correlated significantly with bacterial or fungal α-diversity.

Patterns of temporal β-diversity and responses to abiotic variables

The patterns of temporal β-diversity of the archaeal, bacterial and fungal communities (taking into account the variations in community composition of each microbial group in individual soils over time) showed small but significant variations across soils (Fig. 3A). Bacterial β-diversities were in general higher than fungal ones across soils, except for soil V. There were slight but significant differences (P<0.05) between sandy and clayey soils regarding the temporal β-diversity of archaeal and bacterial but not of fungal communities (Fig. 3B).

Although chemical parameters might show variability over time, significant correlations could still be observed. The patterns in the archaeal temporal β-diversities observed were mainly due to positive correlations with nitrate (r = +0.874, P<0.05) (Table 2). None of the soil parameters measured were correlated with bacterial and fungal temporal β-diversities. Canonical correspondence analysis was used to test the significance of the influence of soil parameters on the community parameters. We used variance partitioning to control for the effect of each individual parameter, while all others are defined as covariables in the constrained analyses [44]. Considering the whole data set, soil parameters explained 45%, 6.6% and 6.9% of the temporal variability in

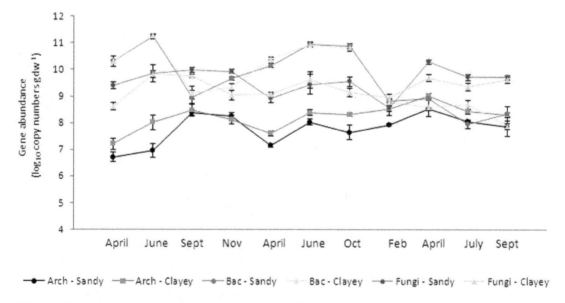

Figure 1. Changes in abundance of archaeal, bacterial and fungal communities. The copy number in each gram of dry soil was estimated by real-time PCR in the eight agricultural soils as an average of sandy and clayey soils at different sampling times. Bars are standard errors (n = 4).

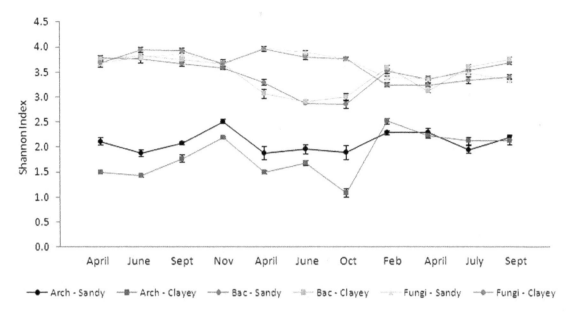

Figure 2. Total α-diversity of archaeal, bacterial and fungal communities. Alpha diversity was calculated as the average of the Shannon Index (H') per soil type (sandy×clayey), from April 2009 to September 2011 (mean ± s.d.).

archaeal, bacterial and fungal community structures, respectively. The archaeal community was mostly affected by changes in OM (11.9%) and nitrogen specimens (nitrate + ammonium; 7.8% each), whereas the bacterial and fungal community variations were mostly related to ammonium (2.1% and 2.2% for bacteria and fungi, respectively) (Fig. S1).

Significant relationships were observed between variation in β-diversity and H' for bacterial (r = +0.602, P<0.05) and fungal communities (r = −0.481, P<0.05) but not for archaeal communities.

Table 2. Pearson's correlation coefficient between soil chemical parameters, biotic parameters (total abundance, alpha diversity, beta diversity and slopes from TLA analysis), calculated from the eight soils over time.

	pH	N-NH$_4^+$ (mg^{-1}kg)	N-NO$_3^-$ (mg^{-1}kg)	OM (%)	Clay (%)
Total abundance					
Total archaeal community	**0.883*****	−0.498^{ns}	**−0.764***	0.030^{ns}	**−0.795***
Total bacterial community	−0.636^{ns}	0.379^{ns}	0.236^{ns}	−0.624^{ns}	0.417^{ns}
Fungi	0.363^{ns}	−0.476^{ns}	−0.356^{ns}	**−0.722***	0.387^{ns}
Alpha Diversity (Shannon)					
Total archaeal community	0.284^{ns}	0.230^{ns}	0.137^{ns}	**0.691***	0.599^{ns}
Total bacterial community	−0.174^{ns}	−0.033^{ns}	0.470^{ns}	−0.158^{ns}	0.442^{ns}
Fungi	0.095^{ns}	−0.149^{ns}	0.175^{ns}	−0.370^{ns}	0.241^{ns}
Alpha Diversity (N° bands)					
Total archaeal community	−0.408^{ns}	−0.056^{ns}	**0.962*****	0.482^{ns}	0.442^{ns}
Total bacterial community	0.441^{ns}	−0.355^{ns}	−0.485^{ns}	−0.497^{ns}	−0.335^{ns}
Fungi	−0.154^{ns}	0.150^{ns}	−0.579^{ns}	−0.416^{ns}	−0.469^{ns}
Temporal Beta diversity					
Total archaeal community	−0.194^{ns}	−0.249^{ns}	**0.874***	0.541^{ns}	−0.415^{ns}
Total bacterial community	0.028^{ns}	−0.313^{ns}	−0.123^{ns}	−0.502^{ns}	0.074^{ns}
Fungi	−0.380^{ns}	0.167^{ns}	−0.456^{ns}	−0.232^{ns}	−0.035^{ns}

Notes: Values in boldface type indicate significant correlations with P values indicated in superscript.
*P<0.05;
**P<0.01;
***P<0.001.
^{ns}not significant at P<0.05.

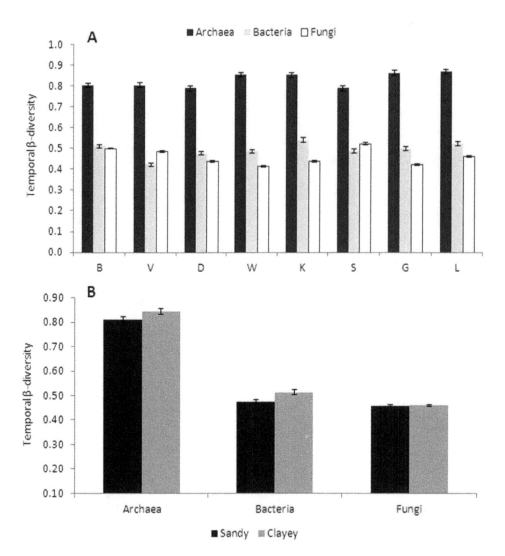

Figure 3. Temporal β-diversity of archaeal, bacterial and fungal communities. Temporal β-diversity, which takes into account temporal changes of each individual soil, was calculated across the different sampling points (A) and separated per soil type (B) (mean ± s.e.).

Quantifying temporal changes of archaeal, bacterial and fungal communities

The temporal changes of the microbial guilds were quantified and statistically tested via TLA. TLA analyses were performed separately per year and also considering all three years. For both analyses, the results and conclusions were similar. Therefore, we decided to include only the whole three year dataset. A statistically significant regression line (P<0.05) was observed for the archaeal community, with an overall slope of 1.835 (Fig. 4A and Table 3). Moreover, all eight soils showed indications of directional changes in community composition, yielding regression lines that were statistically different from zero (P<0.05) with the exception of the G soil (Table 3). Although the slopes were small (Table 3), they were mainly reflected in the positive Pearson correlations with nitrate levels (r = +0.814, P<0.05). The bacterial communities showed a similar trend as observed for the archaeal ones, with significant regression lines and a slope of 0.785 (Fig. 4B and Table 3). Although analyses of the fungal communities in the eight soils showed that only three soils were undergoing directional changes (B, V an L soils), the overall result based on the simultaneous analysis of all soils yielded statistically significant

regression lines (slope of 0.638, P<0.05, Fig. 4C). None of the soil parameters measured had significant effects on the rates of change of bacterial and fungal communities. Significant and contrasting relationships were observed between the TLA slopes and the H′ values of archaeal (r = +0.629, P<0.05) and bacterial (r = −0.523, P<0.05) communities but not of fungal communities.

Discussion

Temporal variation in the abundance of soil microbial communities

In our study, population sizes of archaea, bacteria and fungi, estimated using quantitative PCR, were found to be within the range observed in other soil systems [24,45]. Quantitative PCR of soil DNA, as any PCR based approach, has its inherent limitations, whether it be the biases of soil DNA extraction, PCR, or the core genes targeted. However, the method is highly reproducible and sensitive, enabling the quantification of microbial abundance changes across temporal and spatial scales. Moreover, this study performed multiple qPCR runs in order to ensure results were statistically significant. In our calculations, we also took into account the efficiency and amount of extracted

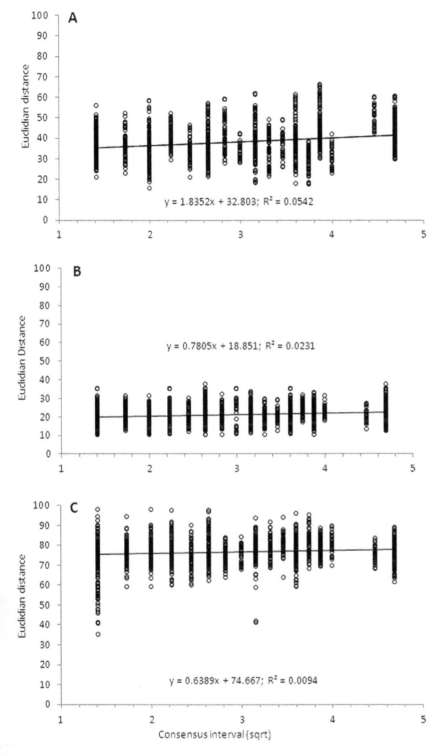

Figure 4. Quantification of archaeal, bacterial and fungal dynamics. Patterns of change (regression of square root of time-lag on Euclidian distance) of archaeal (A, slope 1.835), bacterial (B, slope 0.785) and fungal community (C, slope 0.638) in eight soils. The best-fit line is shown.

DNA from the soil samples. Therefore, we argue that our results are representative of the fluctuations observed between different times rather than pure noise.

A high abundance of crenarchaeota in soils has been previously observed [24,46], possibly indicating a crucial functional role for such organisms in agricultural soils. Furthermore, the bacterial

abundance was often higher across soils than the fungal abundance (except at the end of 2009 and the end of 2011), supporting the finding that bacterial:fungal (B:F) ratios are quite high in agricultural or grassland soils as compared to, say, forest soils [45,47–49]. These comparisons are very important in the context of whether soils are thought of a being fungal (more

Table 3. Results of the time-lag analyses (TLA) performed for bacterial, archaeal and fungal communities for all soils separately, and an overall result considering all soils.

Sampling Site	Archaeal community			Bacterial community			Fungal community		
	Slope	P	R^2	Slope	P	R^2	Slope	P	R^2
Buinen (B)	1.749	0.000	0.116	3.298	0.000	0.227	0.989	0.007	0.031
Valthermond (V)	1.876	0.000	0.064	1.903	0.000	0.139	1.297	0.001	0.059
Droevendaal (D)	2.063	0.000	0.098	1.864	0.000	0.135	0.671	NS	0.013
Wildekamp (W)	1.765	0.000	0.105	1.059	NS	0.036	0.414	NS	0.004
Kollumerwaard (K)	1.378	0.028	0.021	1.860	0.009	0.079	−0.078	NS	0.000
Steenharst (S)	2.038	0.000	0.080	1.697	0.015	0.069	0.817	NS	0.012
Grebedijk (G)	0.853	NS	0.022	2.300	0.000	0.156	−0.029	NS	0.000
Lelystad (L)	1.282	0.016	0.026	0.552	NS	0.009	0.981	0.011	0.034
Overall	1.835	0.000	0.054	0.785	0.000	0.023	0.638	0.000	0.009

"natural") or bacterial (more highly cultivated) dominated. Indeed, such elevated B:F ratios may also reflect anthropogenic disturbances due to agricultural practices.

The variations in microbial abundances could be explained by several parameters, depending on the target group. Soil pH and nitrate explained more than 75% of the variation in archaeal abundance. Previous studies have reported negative effects of pH on group 1.1c Crenarchaeota [50] in acid forest soils and negative relationships between nitrate and archaeal abundance [21]. The positive correlations between archaeal abundance and soil pH observed here suggest that our soils might be dominated by crenarchaeal species that are adapted to conditions of higher soil pH (7.0–7.5) [51], which may be linked to the long agricultural history of the plots studied here.

Interestingly, the bacterial abundances didn't respond to soil pH or any other measured abiotic parameter, although several studies have reported pH as the main determinant of bacterial community composition [52–57]. It has been shown that some specific bacterial taxa abundances decrease or increase with a changing pH, for instance members of the *Acidobacteria* and *Actinobacteria* [56]. Although pH may have driven changes in the relative abundance of some bacterial classes, the abundance of total bacteria remained quite constant in the different pH ranges, indicating that the carrying capacity of the soil was not strongly affected by pH. Fungal abundance was also not affected by pH, but this was expected since the pH range in our soils was within the (wide) pH optimum for this group, often covering 5–9 pH units without significant inhibition of growth [58,59]. We also observed that when conditions apparently favored increases in fungal abundance, archaeal abundance decreased, suggesting that fungi and archaea might compete for similar niches. Nonetheless, fungal abundance was positively affected by OM content, which is consistent with the saprophytic status of most fungi [60].

Temporal variation in α-diversity

None of the soil parameters measured in this study were able to explain the patterns of α-diversity observed for bacteria and fungi. It might be that the taxonomic scale was too broad and a deeper analysis would allow a better understanding of the observed patterns, as observed by Rasche et al. [61]. With PCR-DGGE, only the most abundant taxa, comprising no more than 0.1–1% of the community, can be detected. In other words, only the most abundant organisms are detected with PCR-DGGE. Because of

these caveats, the parameters calculated from PCR-DGGE fingerprints and correlations based thereon should be interpreted as indications and not as absolute conclusions.

Archaeal α-diversity, on the other hand, was shown to respond to nitrate and OM levels. Nitrate had opposite effects on archaeal richness and diversity, depicting a community that responds to increasing nitrate with an increase of richness but a great decrease of evenness. This most likely indicates the outgrowth of previously undetectable OTUs. Thus, in addition to the strong negative effect on archaeal abundance observed by qPCR, nitrate availability seems to be a crucial factor determining archaeal community structure. The positive correlation between archaeal diversity and soil OM content indicate that the OM provides a substantial fraction of carbon to the local archaeal communities. Recently, genomic analyses of *Crenarchaeum symbiosum* and *Nitrosopumilus maritimus* suggested that these organisms are capable of mixotrophy [62,63] and that group 1.1c Crenarchaeota are able to grown on methanol and methane [64]. This suggests that archaea might not be solely sustained by ammonia oxidation [65,66].

Variation in β-diversity over time (species turnover)

To assess how dynamic each soil microbial group was over time, we calculated the temporal β-diversity (average Bray Curtis dissimilarity) for each soil over time and for each microbial group. We observed a higher temporal β-diversity for archaea than for bacteria and fungi across all soils. This indicates that the archaeal communities are much more dynamic than the bacterial or fungal ones along a time gradient. These differences are probably due to the differential physiologies and sensitivities to environmental perturbations of these microorganisms. It has been shown that changes in temperature, moisture [61,67] and resource availability due to seasonal variation [61] can affect soil archaeal as well as bacterial communities. Moreover, a clear pattern was observed for bacterial β-diversity in the metacommunity of 17 rock pools, with higher variations during summer and lower during autumn [11]. The temporal variations of archaeal and bacterial communities were also higher in the clayey soils than in sandy ones, suggesting that the latter harbors more dynamic communities.

One main finding of this study is that, although the β-diversity patterns of the three microbial domains investigated are related with the same set of abiotic factors, the total percentage of variation able to explain those patterns was much higher for archaeal (45.0%) than for bacterial (6.6%) or fungal (6.9%)

communities. This suggests that the archaeal communities might be much more sensitive to environmental changes than the bacterial or fungal ones. Based on these results, we hypothesize that the archaeal communities of agricultural soils with a long history of N-fertilization are more sensitive to disturbances than the corresponding bacterial or fungal communities.

Quantification of β-diversity

To be able to quantify community dynamics, allowing comparisons and providing a general overview of long-term trends in the complex soil system, we used the slopes obtained from TLA. TLA has been intensively used to identify directional changes and to quantify temporal dynamics of macroorganisms [69–73] but very few studies have focused on microorganisms. Although the TLA slopes for archaea and bacteria were small, they were significantly different from neutral. Clearly, even small changes can be part of a long-term trend. On the contrary, changes in the fungal communities were non-significant, suggesting stochastic species dynamics.

Changes in environmental variables within soil sites determine how time affects turnover (β-diversity), as different microbial interactions are favored if prevailing conditions change [74]. Only archaeal communities responded to changes in environmental parameters, being strongly correlated with nitrogen availability and with the degree of temporal variation quantified by TLA. This might suggest that at some level strongly deterministic processes are acting on the archaeal but not on the bacterial and fungal communities in these soils. Another explanation is that archaea are much more limited in their ecoversatility, whereas bacteria and fungi are highly functionally redundant. The observed relation between bacterial richness and TLA slopes, e.g. high turnover at low richness, was also noticed in a study of the distribution of British birds [75]. The authors discuss that low species richness areas tend to have relatively more random mixtures of species than high species richness areas. The relation observed between archaeal turnover and species richness suggests a less random distribution of species, caused mainly by nitrate contents.

In this study we demonstrate that changes in the community composition of bacteria and fungi could be linked to both environmental and biotic factors (e.g. species-species interactions), as their (α-) diversity co-varied significantly with their β-diversity over the time period of the study. Conversely, archaea showed no significant correlation between α- and β-diversity, and the community shifts were mainly driven by the surrounding environment, mostly by the effects of soil pH and nitrate concentrations. This might indicate that changes in archaeal community are mostly driven by environmental factors, as

previously observed by Zinger et al. [68] in a study on the patterns of archaeal, bacterial and fungal communities in an alpine landscape. Furthermore, we propose that different environmental and biological mechanisms act on each microbial niche. A more comprehensive understanding of the rules governing these important soil microorganisms will require additional field work as well as microcosm experiments to identify the key environmental and biotic factors driving the assemblage of these communities.

Ethic statement. No specific permits were required for the described field studies. The locations are not protected. The field studies did not involve endangered or protected species.

Supporting Information

Figure S1 Biplots of canonical correspondence analysis (CCA) of Archaeal, Bacterial and Fungal similarity matrices and vector fitting of the environmental variables. Similarity matrices from DGGE data were obtained from eight soils over three years (2009, 2010 and 2011). Physicochemical data, soil moisture (Humidity), soil nitrate (NO3), soil ammonium (NH4), organic matter (OM), clay content (clay + silt) and soil pH (pH) are presented with black arrows.

Table S1 Soil chemical parameters measured in this study.

Table S2 PCR mixtures for real time quantification of Archaeal 16S rDNA, Bacterial 16S rDNA and Fungal ITS region.

Table S3 PCR mixtures for DGGE analysis of Archaeal 16S rDNA, Bacterial 16S rDNA and Fungal ITS region.

Acknowledgments

We would like to thank our colleagues Alexander V Semenov, Jolanda Brons and our Utrecht partners, Heike Schmitt and Agnieszka Szturc-Koetsier for their help with sampling. We also thank Cyrus A. Mallon for revision and helpful comments on the manuscript.

Author Contributions

Performed the experiments: MCPS. Analyzed the data: MCPS ACFD. Contributed reagents/materials/analysis tools: JDvE JFS. Wrote the paper: MCPS .

References

1. Singh BK, Bardgett RD, Smith P, Reay DS (2010) Microorganisms and climate change: terrestrial feedbacks and mitigation options. Nature Microbiol Rev 8:779–790.
2. Hooper DU, Vitousek PM (1997) The effects of plant composition and diversity on ecosystem processes. Science 277:1302–1305.
3. Tilman D, Knops J, Wedin D, Reich P, Richie M, et al. (1997) The influence of functional diversity and composition on ecosystem processes. Science 277:1300–1302.
4. Bardgett RD, Yeates G, Anderson J (2005) Patterns and determinants of soil biological diversity. In Biological diversity and function in soils. pp. 100–118.
5. Lindström ES, Langenheder S (2011) Local and regional factors influencing bacterial community assembly. Environm Microbiol Rep 4:1–9.
6. Andrewartha HG, Birch LC (1953) The Lotka-Volterra theory of interspecific competition. Aust J Zoology 1:174–177. DOI: 10.1071/ZO9530174.
7. Lewontin R, Cohen D (1969) On population growth in a randomly varying environment. Proc Natl Acad Sci USA 62:1056–1060.
8. Schmidt SK, Costello EK, Nemergut DR, Cleveland CC, Reed SC, et al. (2007) Biogeochemical consequences of rapid microbial turnover and seasonal succession in soil. Ecology 88:1379–1385.

9. Lipson DA, Schadt CW, Schmidt SK (2002) Changes in soil microbial community structure and function in an alpine dry meadow following spring snow melt. Microbial Ecol 43:307–314.
10. Martiny JBH, Eisen JA, Penn K, Allison SD, Horner-Devine MC (2011) Drivers of bacterial β-diversity depend on spatial scale. Proc Natl Acad Sci USA 10: 7850–4.
11. Langenheder S, Berga M, Örjan Ö, Székely AJ (2012) Temporal variation of β-diversity and assembly mechanisms in a bacterial metacommunity. ISME J 6:1107–1114.
12. Griffiths RI, Thomson BC, James P, Bell T, Bailey M, et al. (2011) The bacterial biogeography of British soils. Environm Microbiol 13:1642–1654.
13. Wang J, Wu Y, Jiang H, Li C, Dong H, et al. (2008) High beta diversity of bacteria in the shallow terrestrial subsurface. Environm Microbiol 10:2537–2549.
14. Curtis T P, Sloan WT, Scannell JW (2002) Estimating prokaryotic diversity and its limits. Proc Natl Acad Sci USA 99:10494–10499.
15. Gans J, Wolinsky M, Dunbar J (2005) Computational improvements reveal great bacterial diversity and high metal toxicity in soil. Science 309:1387–1390.

16. Torsvik V, Ovreas L, Thingstad TF (2002) Prokaryotic diversity: magnitude, dynamics, and controlling factors. Science 296:1064–1066.

17. Buckley DH, Graber JR, Schmidt TM (1998) Phylogenetic analysis of nontermophilic members of the kingdom Crenarchaeota and their diversity and abundance in soils. Appl Environ Microbiol 64:4333–4339.

18. Jurgens G, Lindström K, Saano A (1997) Novel group within kingdom Crenarcheota from borest forest soil. Appl Environ Microbiol 63:803–805.

19. Ueda T, Suga Y, Matsuguchi T (1995) Molecular phylogenetic analysis of a soil microbial community in a soybean field. Eur J Soil Sci 46:415–421.

20. Jung J, Yeom J, Kim J, Han J, Lim HS, et al. (2011) Change in gene abundance in the nitrogen biogeochemical cycle with temperature and nitrogen addition in Antarctic soils. Res Microbiol 168:1028–1026.

21. Bates ST, Berg-Lyons D, Caporaso JG, Walters WA, Knight R, et al. (2011) Examining the global distribution of dominant archaeal populations in soil. ISME J 5:511–517.

22. Collins SL, Micheli F, Hartt L (2000) A method to determine rates and patterns of variability in ecological communities. Oikos 91:285–293.

23. Angeler DG, Viedma O, Moreno JM (2009) Statistical performance and information content of time lag analysis and redundancy analysis in time series modeling. Ecology 90:3245–3257.

24. Ochsenreiter T, Selezi D, Quaiser A, Bonch-Osmolovskaya L, Schleper C (2003) Diversity and abundance of Crenarchaeota in terrestrial habitats studied by 16S RNA surveys and real time PCR. Environm Microbiol 5:787–797.

25. Bach H-J, Tomanova J, Schloter M, Munch JC (2002) Enumeration of total bacteria and bacteria with genes for proteolytic activity in pure cultures and in environmental samples by quantitative PCR mediated amplification. J Microbiol Met 49:235–45.

26. Nicol GW, Webster G, Glover LA, Prosser JI (2004) Differential response of archaeal and bacterial communities to nitrogen inputs and pH changes in upland pasture rhizosphere soil. Environm Microbiol 6:861–867.

27. Fierer N, Jackson JA, Vilgalys R, Jackson RB (2005) Assessment of soil microbial community structure by use of taxon-specific quantitative PCR assays. Appl Environ Microbiol 71: 4117–4120.

28. Gomes NCM, Heuer H, Schonfeld J, Costa R, Hagler-Mendonça L, et al. (2001) Bacterial diversity of the rhizosphere of maize (Zea mays) grown in tropical soil studied by temperature gradient gel electrophoresis. Plant and Soil 232:167–180.

29. Brons JK, van Elsas JD (2008) Analysis of bacterial communities in soil by use of denaturing gradient gel electrophoresis and clone libraries, as influenced by different reverse primers. Appl Environm Microbiol 74:2717–2727.

30. Bano N, Ruffin S, Ransom B, Hollibaugh T (2004) Phylogenetic composition of arctic ocean archaeal assemblages and comparison with Antarctic assemblages Appl. Environm Microbiol 70:781–789.

31. Myers RM, Fischer SG, Lerman LS, Maniatis T (1985). Nearly all single base substitutions in DNA fragments joined to a GC-clamp can be detected by denaturing gradient gel electrophoresis. Nucleic Acids Res 13:3131–3145.

32. Smit E, Leeflang P, Glandorf B, van Elsas JD, Wernars K (1999) Analysis of fungal diversity in the wheat rhizosphere by sequencing of cloned PCR-Amplified genes encoding 18S rRNA and temperature gradient gel rlectrophoresis. Appl Environ Microbiol 65:2614–2621.

33. White TJ, Bruns T, Lee S, Taylor JW (1990) Amplification and direct sequencing of fungal ribosomal RNA genes for phylogenetics. In: Innis MA, Gelfand DH, Sninsky JJ and White TJ, editors. PCR Protocols: A Guide to Methods and Applications. Academic Press, New York , pp. 315–322.

34. Gardes M, Bruns TD (1993) ITS primers with enhanced specificity for basidiomycetes - application to the identification of mycorrhizae and rusts. Mol Ecol 2: 113–118.

35. Kropf S (2004) Nonparametric multiple test procedures with data-driven order of hypotheses and with weighted hypotheses. J Statis Plann Inf 125:31–47.

36. Rademaker J L, de Bruijn AF (1999) Molecular microbial ecology manual. In van Elsas, JD, Akkermans ADL and de Bruijn AF, Eds. Molecular Microbial Ecology Manual. Dordrecht, The Netherlands: Kluwer Academic Publishers, pp. 1–33.

37. Hill TCJ, Walsh KA, Harris JA, Moffett BF (2003) Using ecological diversity measures with bacterial communities. FEMS Microbiol Ecol 43:1–11.

38. Legendre P, Borcard D, Peres-Neto PR (2005) Analyzing β-diversity: partitioning the spatial variation of community composition data. Ecological Monographs 75:435–450.

39. Peres-Neto PR, Legendre P, Dray S, Borcard D (2006) Variation partitioning of species data matrices: estimation and comparison of fractions. Ecology 87:2614–25.

40. Legendre P, Gallagher ED (2001) Ecologically meaningful transformations for ordination of species data. Oecologia 129:271–280.

41. Kampichler C, Geissen V (2005) Temporal predictability of soil microarthropod communities in temperate forests. Pedobiologia 49:41–50.

42. Clarke KR, Gorley RN (2006) PRIMER v6: User manual/tutorial, PRIMER-E, Plymouth UK, (192pp).

43. Magurran AE, Dornelas M (2010) Biological diversity in a changing world. Phil Trans R Soc B 365:3593–3597.

44. Leps J, Smilauer P (2003) Multivariate analysis of ecological data using CANOCO. Cambridge:CambridgeUniversityPress.

45. Bailey VL, Smith JL, Bolton H (2002) Fungal-to-bacterial ratios in soils investigated for enhanced C sequestration. Soil Biol Biochem 34:997–1007.

46. Kemnitz D, Kolb S, Conrad R (2007) High abundance of Crenarchaeota in a temperate acidic forest soil. FEMS Microbiol Ecol 60:442–448.

47. Bossuyt H, Denef K, Six J, Frey SD, Merckx R, et al. (2001) Influence of microbial populations and residue quality on aggregate stability. Appl Soil Ecol 16:195–208.

48. Högberg MN, Högberg P, Myrold DD (2007) Is microbial community composition in boreal forest soils determined by pH, C-to-N ratio, the trees, or all three? Oecologia 150:590–601.

49. Treseder KK (2004) A meta-analysis of mycorrhizal responses to nitrogen, phosphorus, and atmospheric CO2 in field studies. New Phytologist 164:347–355.

50. Lethovirta LE, Prosser JI, Nicol GW (2009) Soil pH regulates the abundance and diversity of groups 1.1c Crenarchaeota. FEMS Microbiol Ecol 70:367–376.

51. Bengtson P, Sterngren AE, Rousk J (2012) Archaeal abundance across a pH gradient in an arable soil and its relationship with bacterial and fungal growth rates. Appl Environ Microbiol doi:10.1128/AEM.01476-12.

52. Fierer N and Jackson RB (2006) The diversity and biogeography of soil bacterial communities. Procl Natl Acad Sci USA 103:623–631.

53. Lauber CL, Strickland MS, Bradford MA, Fierer N (2008) The influence of soil properties on the structure of bacterial and fungal communities across land-use types. Soil Biol Biochem 40:2407–2415.

54. Baker BJ, Comolli LR, Dick GJ, Hauser LJ, Hyatt D, et al. (2010) Enigmatic, ultrasmall, uncultivated Archaea. Proc Natl Acad Sci USA 107:8806–8811.

55. King AJ, Freeman KR, McCormick KF, Lynch RC, Lozupone C, et al. (2010) Biogeography and habitat modeling of high-alpine bacteria. Nature Comm 1:53. DOI: 10.1038/ncomms1055.

56. Lauber CL, Hamady M, Knight R, Fierer N (2009) Pyrosequencing-based assessment of soil pH as a predictor of soil bacterial community structure at the continental scale. Appl Environm Microbiol 75:5111–5120.

57. Rousk J, Baath E, Brookes PC, Lauver CL, Lozupone C, et al. (2010) Soil bacterial and fungal communities across a pH gradient in an arable soil. ISME J 4:1340–1351.

58. Wheeler KA, Hurdman BF, Pitt JI. (1991) Influence of pH on the growth of some oxigenic species of Aspergillus, Penicillium and Fusarium. Int J Food Microbiol 12: 141–150.

59. Nevarez L, Vasseur V, Le Madec L, Le Bras L, Coroller L, et al. (2009) Physiological traits of Penicillium glabrum strain LCP 08.5568, a filamentous fungus isolated from bottled aromatised mineral water. Int J Food Microbiol 130: 166–171.

60. de Boer W, Folman LB, Summerbell RC, Boddy L (2005) Living in a fungal world: impact of fungi on soil bacterial niche development. FEMS Microbiol Rev 29:795–811.

61. Rasche F, Knapp D, Kaiser C, Koranda M, Kitzler B, et al. (2011) Seasonality and resource availability control bacterial and archaeal communities in soils of a temperate beech forest. ISME J 5:389–402.

62. Hallam SJ, Konstantinidis KT, Putnam N, Schleper C, Watanabe Y-i, et al. (2006) Genomic analysis of the uncultivated marine crenarchaeote Cenarchaeum symbiosum. Proc Natl Acad Sci USA 103: 18296–18301.

63. Walker CB, de la Torre JR, Klotz MG, Urakawa H, Pinel N, et al. (2010) Nitrosopumilus maritimus genome reveals unique mechanisms for nitrification and autotrophy in globally distributed marine crenarchaea. Proc Natl Acad Sci USA 107:8818–8823.

64. Bomberg M, Timonen S (2007) Distribution of cren- and euryarchaeota in scots pine mycorrhizosphere and boreal forest humus. Microb Ecol 54:406–416.

65. Ouverney CC, Fuhrman JA (2000) Marine planktonic archaea take up amino acids. Appl Environ Microbiol 66:4829–4833.

66. Jia Z, Conrad R (2009) Bacteria rather than Archaea dominate microbial ammonia oxidation in an agricultural soil. Environm Microbiol 11:1658–1671.

67. Tourna M, Freitag TE, Nicol GW, Prosser JI (2008) Growth, activity and temperature responses of ammonia-oxidizing archaea and bacteria in soil microcosms. Environ Microbiol 10:1357–1364.

68. Zinger L, Lejon DPH, Baptist F, Bouasria A, Aubert S, et al. (2011) Contrasting diversity patterns of crenarchaeal, bacterial and fungal communities in an alpine landscape. Plos One 6(5): e19950. doi:10.1371/journal.pone.0019950.

69. Thibault KM, White EP, Ernest KM (2004) Temporal dynamics in the structure and composition of a desert rodent community. Ecology 85:2649–2655.

70. Baez S, Collins ST, Lightfoot D, Koontz TL (2006) Ecology 87:2746–2754.

71. Collins SL, Smith MD (2006) Scale-dependent interaction of fire and grazing on community heterogeneity in tallgrass prairie. Ecology 87:2058–2067.

72. Feeley KJ, Davies SJ, Perez R, Hubbell SP, Foster RB (2011) Directional changes in the spceies composition of a tropical forest. Ecology 92:871–882.

73. Flohre A, Fischer C, Aavik T, Bengtsson J, Berendse F, et al. (2011) Agricultural intensification and biodiversity partitioning in European landscapes comparing plants, carabids and birds. Ecol Appl 21:1772–1781.

74. Chesson P, Huntly N (1997) The roles of harsh fluctuation conditions in the dynamics of ecological communities. Amer Natur 150:519–553.

75. Lennon JJ, Koleff P, Grenwood JJD, Gaston KJ (2001) The geographical structure of british birds distributions: diversity, spatial turnover and scale. J Animal Ecol 70:966–979.

9

Injury Profile SIMulator, a Qualitative Aggregative Modelling Framework to Predict Injury Profile as a Function of Cropping Practices, and Abiotic and Biotic Environment. II. Proof of Concept: Design of IPSIM-Wheat-Eyespot

Marie-Hélène Robin[1,2], Nathalie Colbach[3], Philippe Lucas[4], Françoise Montfort[4], Célia Cholez[1,2], Philippe Debaeke[1,5], Jean-Noël Aubertot[1,5]*

1 Institut National de la Recherche Agronomique, Unité Mixte de Recherche 1248 Agrosystèmes et agricultures, Gestion des ressources, Innovations et Ruralités, Castanet-Tolosan, France, 2 Université de Toulouse, Institut National Polytechnique de Toulouse, Ecole d'Ingénieurs de Purpan, Toulouse, France, 3 Institut National de la Recherche Agronomique, Unité Mixte de Recherche 1347 Agroécologie, Dijon, France, 4 Institut National de la Recherche Agronomique, Unité Mixte de Recherche 1099 Biologie des Organismes et des Populations appliquée à la Protection des Plantes. Le Rheu, France, 5 Université Toulouse, Institut National Polytechnique de Toulouse, Unité Mixte de Recherche 1248 Agrosystèmes et agricultures, Gestion des Ressources, Innovations et Ruralités, Castanet-Tolosan, France

Abstract

IPSIM (Injury Profile SIMulator) is a generic modelling framework presented in a companion paper. It aims at predicting a crop injury profile as a function of cropping practices and abiotic and biotic environment. IPSIM's modelling approach consists of designing a model with an aggregative hierarchical tree of attributes. In order to provide a proof of concept, a model, named IPSIM-Wheat-Eyespot, has been developed with the software DEXi according to the conceptual framework of IPSIM to represent final incidence of eyespot on wheat. This paper briefly presents the pathosystem, the method used to develop IPSIM-Wheat-Eyespot using IPSIM's modelling framework, simulation examples, an evaluation of the predictive quality of the model with a large dataset (526 observed site-years) and a discussion on the benefits and limitations of the approach. IPSIM-Wheat-Eyespot proved to successfully represent the annual variability of the disease, as well as the effects of cropping practices (Efficiency = 0.51, Root Mean Square Error of Prediction = 24%; bias = 5.0%). IPSIM-Wheat-Eyespot does not aim to precisely predict the incidence of eyespot on wheat. It rather aims to rank cropping systems with regard to the risk of eyespot on wheat in a given production situation through *ex ante* evaluations. IPSIM-Wheat-Eyespot can also help perform diagnoses of commercial fields. Its structure is simple and permits to combine available knowledge in the scientific literature (data, models) and expertise. IPSIM-Wheat-Eyespot is now available to help design cropping systems with a low risk of eyespot on wheat in a wide range of production situations, and can help perform diagnoses of commercial fields. In addition, it provides a proof of concept with regard to the modelling approach of IPSIM. IPSIM-Wheat-Eyespot will be a sub-model of IPSIM-Wheat, a model that will predict injury profile on wheat as a function of cropping practices and the production situation.

Editor: Matteo Convertino, University of Florida, United States of America

Funding: This study was carried out within a PhD project co-funded by INRA and INPT EI Purpan, by the project MICMAC design (ANR-09-STRA-06) supported by the French National Agency for Research, and by the Programme "Assessing and reducing environmental risks from plant protection products (pesticides)", funded by the French Ministry in charge of Ecology and Sustainable Development (project "ASPIB"). The funders had no role in study design, data collection and analysis, decision to publish, or preparation of the manuscript.

Competing Interests: The authors have declared that no competing interests exist.

* E-mail: Jean-Noel.Aubertot@toulouse.inra.fr

Introduction

Stem base diseases on cereals and grasses are widespread in many eco-regions of the world and cause important production and economic losses. The most detrimental foot and root pathogens on cereals in temperate areas are *Pseudocercosporella herpotrichoides*; *Fusarium* spp, *Rhizoctonia cerealis* and *Gaeumannomyces graminis* [1]. Eyespot caused by the necrotrophic and soil-borne fungi *Oculimacula yallundae* and *O. acuformis*, anamorph *Pseudocercosporella herpotrichoides* [2–4] is considered to be the most important

stem base disease of cereals in temperate countries [5]. Under cool and wet conditions in autumn and spring, both species sporulate and infect the stem bases of their hosts. Without any host crops (cereals, ryegrass), the pathogen survives on previously infected stubble, on which splash-dispersed conidia and air-dispersed ascospores are produced [6]. Injuries interfere with the circulation of nutrients and water through the base of the stem [7] leading to a weakening and possibly to a breakage of the stem base, causing lodging before harvest [5,8]. Relative yield losses of up to 50%

have been reported for the most severe attacks on winter wheat with lodging [2,7,9–11].

In the past, the control of eyespot has relied largely on chemical protection [12]. However, due to the development of resistance to the main available fungicides in *O. yallundae* and *O. acuformis* populations, adaptation of the entire cropping system to control eyespot on wheat is a sound alternative [13,14]. Furthermore, growing concerns about the impact of pesticides on the environment and human health has led to attempts to limit pesticide use [15,16]. Most governments of developed countries have launched national action plans to reduce pesticide use. For instance, the French government has set as a goal to reduce pesticide use by 50% by 2018 if possible [17]. The European Union has proposed to encourage the use of low-pesticide farming as one of its priorities by the Sustainable Use Directive (SUD) (http://eurlex.europa.eu/LexUriServ/LexUriServ.do?uri = OJ:L:2009:309:0071:0086:FR:PDF, accessed November 2012).

In addition, the USA decided to support and develop Integrated Pest Management (IPM) nationwide in order to reduce pesticide use [18]. It appears necessary therefore to combine various methods (cultural, genetic and chemical) in IPM strategies [19] to control eyespot on wheat. The main cultural practices that can partly control eyespot through a specific adaptation are: a low host frequency in the crop sequence, infected stubble management through adapted tillage, a late sowing date and low sowing rate [10,20,29] The genetic control of eyespot consists of using resistant cultivars. There are several known sources of resistance to eyespot, but only three resistance genes have been described so far [21–23].

IPM strategies, based on these control methods, have to be developed, adapted and applied to a wide range of physical, chemical, biological and socio-economic contexts. However, it is extremely difficult to describe the entirety of the cropping practices*environment*crop*pest system because of the tremendous number of interactions [24]. Modelling is certainly the best way to handle such a level of complexity and to help design sustainable innovative cropping systems less reliant on pesticides.

However, crop models do not deal with injuries caused by pests [25] and few pest models integrate the effects of cultural practices because of the difficulty of describing their numerous consequences on the agroecosystem [26] Thus, different models have been developed to represent eyespot injuries on wheat [27–29] or the associated damage [30] Among these, only one model takes into account the effect of the cropping system (crop succession, tillage, sowing date, sowing rate, total nitrogen fertiliser and its form) on injuries caused by eyespot [29]. However, this model does not take into account soil and climate, along with some cultural practices that can greatly influence the disease development (e.g. cultivar choice). There is therefore a need for a model that predicts as exhaustively as possible the effect of cropping practices on eyespot on wheat in a given production situation.

In this article, we will define the production situation as the physical, chemical and biological components, except for the crop, of a given field (or agroecosystem), its environment, as well as socio-economic drivers that affect farmers' decisions (adapted from [31,32], [33]). In this definition, "environment" refers to climate and the fraction of the territory that can influence pest dynamics through dispersal of harmful or beneficial organisms. In a given production situation, a farmer can design several cropping systems according to his goals, his perception of the socio-economic context and his environment, farm features, his knowledge and cognition. However, it is assumed that a given cropping system in a given production situation, such as defined above, should lead to a unique injury profile. In IPSIM, production situations are partly described by three components: soil, climate, and the biological environment of the field [33]. In the approach used here, the farmer's decision-making process and socio-economic drivers are not taken into account.

The conceptual bases of IPSIM have been described in detail by Aubertot and Robin [33]. The generic hierarchical aggregative modelling framework of IPSIM aims at predicting an injury profile as a function of cropping practices, soil, and climate and the biological field environment for any mono-specific crop production (arable crop, perennial or protected crops). In order to test whether this modelling approach could be successfully applied to represent injuries caused by a single pest, a model, named IPSIM-Wheat-Eyespot, has been developed according to the conceptual framework of IPSIM. It aims at predicting the final incidence of eyespot on wheat as a function of the production situation and cropping practices. IPSIM-Wheat-Eyespot gathers available knowledge in the scientific literature (models, experimental results) and expertise and will help design cropping systems with low risk of eyespot on wheat and perform diagnoses of commercial wheat fields. IPSIM-Wheat-Eyespot will be used as a sub-model for IPSIM-Wheat, a model that will predict the injury profile on winter wheat (i.e. the distribution of injuries caused by the most important detrimental pests on wheat [34]). This paper presents the method used to develop IPSIM-Wheat-Eyespot using the conceptual modelling framework of IPSIM [33], an evaluation of its predictive quality and a discussion on the limitations and benefits of the model.

Materials and Methods

Design of IPSIM-Wheat-Eyespot

1. General Approach. IPSIM-Wheat-Eyespot is based on the DEX method, and is implemented with the software DEXi [35]. DEX is a method for qualitative hierarchical multi-attribute decision modelling and support, based on a breakdown of a complex decision problem into smaller and less complex sub-problems, characterised by indicators (or attributes) that are organised hierarchically into a decision tree. These attributes are characterised by their name, a description and a scale. DEXi is generally used to evaluate and analyse decision problems, e.g. [36]. However, the DEX method has been used here in an original way to model complex agroecosystems. IPSIM-Wheat-Eyespot is therefore a hierarchical and qualitative multi-criteria model, allowing the prediction of eyespot injury according to various factors with sometimes opposite effects. IPSIM-Wheat-Eyespot has the following features (derived from [37]):

i) Processes are hierarchically organised into a tree of attributes that constitutes the structure of the model;

ii) Terminal attributes of the tree (i.e. leaves or *basic attributes*) are input variables of the model and must be specified by users; the "trunk" of the tree (i.e. the final aggregated attribute) is the main model output variable (final eyespot incidence on wheat); internal nodes are called *aggregated attributes*;

iii) All model attributes are qualitative variables (nominal or ordinal) rather than quantitative variables. They take only discrete symbolic values, usually represented by words rather than numbers: e.g. "ploughing, stubble disking, rotary harrowing" for nominal variables, "low, medium, high" for ordinal variables;

iv) The aggregation of values up the tree is defined by aggregating tables for each aggregated attribute based on "if-then" decision rules. These aggregating tables can be seen as equivalents of parameters for quantitative numerical models, whereas the tree of attributes can be viewed as the equivalent of their mathematical structure.

IPSIM-Wheat-Eyespot was designed in 3 steps [37]: (1) identification and organisation of the attributes, (2) definition of attribute scales, and (3) definition of aggregating tables.

2. Identification and Organisation of Attributes. IPSIM-Wheat-Eyespot aims at predicting the incidence of eyespot on wheat in a given field according to a set of input variables. The spatial scale addressed is the field and the temporal scale is the wheat growing season, although some input variables encompass the crop sequence (up to the pre-preceding crop). IPSIM-Wheat-Eyespot is a static deterministic model.

The hierarchical structure presented in Figure 1 represents the breakdown of factors affecting eyespot final incidence into specific explanatory variables, represented by lower-level attributes. This figure represents the adaptation to eyespot of the model structure presented in Figure 2 by Aubertot and Robin [33].

In all, IPSIM-Wheat-Eyespot has 21 attributes, of which 14 are basic (i.e. input variables) and 7 aggregated. The 14 basic attributes are presented as the terminal leaves of the tree and their levels are aggregated into higher levels according to aggregating tables. They represent input variables of the model. Some of them (e.g. those representing the interactions at the territory level) could be omitted since they do not influence the final output. However, they were kept because these basic attributes will be necessary for the modelling of the whole injury profile on wheat. The aggregated attributes are internal nodes. They represent state variables or the output variable of IPSIM-Wheat-Eyespot. They are determined by lower-level basic attributes [38]. The output of IPSIM-Wheat-Eyespot is represented by the attribute "Final eyespot incidence" (eyespot incidence at the "milky grain", stage 7: development of fruit on BBCH scale [39]) which is determined by three main factors: cropping practices, soil and climate and the biological environment of the considered field. This is reflected by the hierarchical structure of the model, which consists of three sub-trees of attributes (Figure 1) split into one main part and two smaller ones. The main sub-tree, "Effect of cropping practices", illustrates the complexity of the effects of cropping practices and the need to consider a

combination of practices in order to evaluate the final eyespot incidence. It uses indicators based on tactical (with a short time-frame) or strategic decisions (with a longer time-frame [40]). These decisions can affect the agroecosystem at several stages.

i) Eyespot is considered as a highly endocyclic disease (as defined in [33]). Upstream, some cropping practices affect the quantity of the endo-inoculum (initial pathogen population present in the field). Crop sequence and tillage determine the vertical distribution of infected stubble and have proven to be of major importance for eyespot control [41–45]. Nevertheless, the effects of tillage on the disease are controversial in the literature. According to several authors [1,41,46–50], minimum tillage is highly favourable to eyespot development in the presence of preceding host-crop residues in the top layer, whereas ploughing significantly reduces its incidence by burying host-crop residues. These results conflict with those that show that eyespot was more severe after soil inversion than after non-inversion under moist, cool conditions [44–47,51–55]. The possible explanation of this apparent contradiction is that non-inversion is more favourable to antagonistic micro-organisms than ploughing (the microbiological activity is higher at the soil surface than in the top 20 cm soil layer and the weather in some experiments, such as those in Italy, was probably too dry for antagonistic biota to flourish on crop debris and thus to control eyespot [1].

ii) Action by escape consists of shifting periods of highest crop susceptibility away from the main periods of pathogen contamination. This is achieved by altering the wheat sowing date. In the case of eyespot, "escape strategies" cannot really be considered. However, early sowing increases the probability of autumn contamination through primary infection, due to the longer time available for eyespot to develop and to affect stems [42].

iii) During the crop cycle, some cropping practices can mitigate infection through crop status by increasing crop competitiveness and/or by creating less favourable conditions for pest development. Low plant density can limit pathogen development through several mechanisms, such

Attribute

```
Final incidence of eyespot
├─Effects of cropping practices
│  ├─Primary inoculum management: interaction between crop sequence and tillage
│  │  ├─Preceding crop
│  │  ├─Pre-preceding crop
│  │  ├─Tillage after harvest of the previous crop
│  │  └─Tillage after harvest of the pre-previous crop
│  ├─Escape: effects of the sowing date
│  ├─Mitigation through crop status
│  │  ├─Cultivar choice
│  │  ├─Level of N fertilisation
│  │  └─Sowing rate
│  └─Chemical control: use of fungicide
├─Effects of soil and climate
│  ├─Soil
│  └─Climate
│     ├─Autumn/winter
│     └─Spring
└─Interactions with the territory
   ├─Beneficial sources
   └─Primary inoculum sources
```

Figure 1. Hierarchical structure of IPSIM-Wheat-Eyespot (screenshot of the DEXi software). Bolded and non-bold terms represent aggregated and basic attributes, respectively.

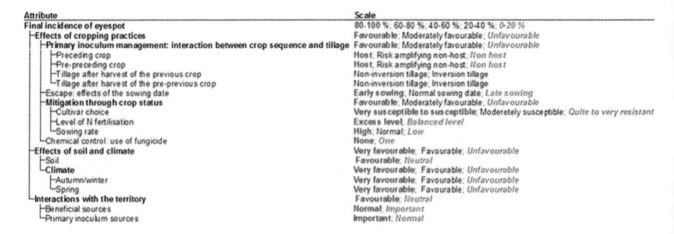

Attribute	Scale
Final incidence of eyespot	80-100 %; 60-80 %; 40-60 %; 20-40 %; 0-20 %
├─Effects of cropping practices	Favourable; Moderately favourable; Unfavourable
│ ├─Primary inoculum management: interaction between crop sequence and tillage	Favourable; Moderately favourable; Unfavourable
│ │ ├─Preceding crop	Host; Risk amplifying non-host; Non host
│ │ ├─Pre-preceding crop	Host; Risk amplifying non-host; Non host
│ │ ├─Tillage after harvest of the previous crop	Non-inversion tillage; Inversion tillage
│ │ └─Tillage after harvest of the pre-previous crop	Non-inversion tillage; Inversion tillage
│ ├─Escape: effects of the sowing date	Early sowing; Normal sowing date; Late sowing
│ ├─Mitigation through crop status	Favourable; Moderately favourable; Unfavourable
│ │ ├─Cultivar choice	Very susceptible to susceptible; Moderetely susceptible; Quite to very resistant
│ │ ├─Level of N fertilisation	Excess level; Balanced level
│ │ └─Sowing rate	High; Normal; Low
│ └─Chemical control: use of fungicide	None; One
├─Effects of soil and climate	Very favourable; Favourable; Unfavourable
│ ├─Soil	Favourable; Neutral
│ └─Climate	Very favourable; Favourable; Unfavourable
│ ├─Autumn/winter	Very favourable; Favourable; Unfavourable
│ └─Spring	Very favourable; Favourable; Unfavourable
└─Interactions with the territory	Favourable; Neutral
│ ├─Beneficial sources	Normal; Important
│ └─Primary inoculum sources	Important; Normal

Figure 2. Attribute scales of IPSIM-Wheat-Eyespot (screenshot of the DEXi software). All the scales are ordered from values detrimental to the crop (i.e. favourable to eyespot) on the left-hand side to values beneficial to the crop on the right-hand side (i.e. unfavourable to eyespot). In the DEXi software, this difference is clearly visible because, by convention, values beneficial to the user are coloured in green, detrimental in red, and neutral in black.

as restricting the contact between plant organs and infectious propagules and lowering the humidity within the canopy. This results in a control of soil-borne diseases like eyespot by low plant density and/or a high shoot number per plant [20]. In addition, low densities increase distances between plants, which limits secondary pathogen cycles, and leads to a drier microclimate. Excessive use of nitrogen fertilisers produces lush crops and favours eyespot through direct and indirect effects [56,57]. However, in the case of eyespot, nitrogen availability in the soil seems to be a minor factor for the development of the disease [10,20,29].

Use of disease-resistant cultivars provides an economic, environmentally friendly and effective strategy to control disease. However, not all resistant cultivars have been assessed in integrated cropping systems [58] and cultivars do not share the same susceptibilities to different diseases [59]. Eyespot resistance is generally not complete and its expression depends widely on environmental factors [22].

iv)   Lastly, a fall-back solution (use of fungicide) can be used when alternative practices are not sufficient. However, several studies have provided evidence for reduced susceptibility to fungicides in populations of O. yallundae and O. acuformis [60]. For the sake of simplicity, resistance to fungicide in pathogen populations was not taken into account in IPSIM-Wheat-Eyespot.

The two other sub-trees describe the biological environment of the considered field, as well as soil and climate. These sub-trees are not affected by cropping practices. Among these factors, climate is the main factor affecting eyespot development [9,43].

3. Definition of the Attribute Scales. The second step in the design of a DEXi model is the choice of ordinal or nominal scales for basic and aggregated indicators. Sets of discrete values were defined for all attributes of the model and described by symbolic value scales defined by words. These values were defined according to the knowledge available in the international literature and some expertise when needed. IPSIM-Wheat-Eyespot uses at most a three-grade value scale (i.e. "Unfavourable", "Favourable", "Very favourable") for the aggregated and basic attributes.

This scale refers to the disease. The value "Favourable" means that the attribute is favourable to the development of the disease and therefore potentially detrimental to the crop.

Some values for basic indicators can be specified using quantitative values that are then translated into qualitative values. For instance, the translation into qualitative values of the sowing date, sowing density or N rate is performed using experimental references or expertise. This translation takes into account the regional context. For example, a sowing date classified as "Early" in the south of France might be classified as "Normal" in northern France. This classification actually depends on the sowing date distribution in the considered region.

Other attributes are directly qualitatively estimated. For instance, the indicators "inversion tillage or non-inversion tillage" or "preceding and pre-preceding crop" are nominal variables and directly monitored as such in experiments [61,62]. The level of cultivar resistance has been described using the official list provided by the French National Seed Station (Groupe d'Etude et de contrôle des Variétés et des Semences; http://cat.geves.info/Page/ListeNationale; accessed November 2012) and published by Arvalis-Institut du végétal (http://www.arvalisinfos.fr/_plugins/WMS_BO_Gallery/page/getElementStream.html?id = 13504&prop = file; accessed November 2012). In this list, cultivars are rated for their susceptibility to eyespot on a 0–9 scale, from very susceptible to resistant.

For the climate attribute, a three-value scale ("Unfavourable"; "Favourable"; "Very favourable") was defined using climatic models [27,43] and data from the INRA Climatik database.

All the scales in Figure 2 are ordered from values detrimental to the crop (i.e. favourable to the disease) on the left-hand side to values beneficial to the crop on the right-hand side (i.e. unfavourable to the disease). In the DEXi software, this difference is clearly visible because, by convention, values beneficial to the user are coloured in green, detrimental in red, and neutral in black. The scales for the "tillage after preceding crop" and "tillage after pre-preceding crop" attributes appear in black since their effects on the disease cannot be defined independently from the crop sequence.

Initial input attribute values (either quantitative or qualitative) are translated into qualitative appreciation, according to two to three scales defined on the basis of available information in the literature, models or expertise. Sometimes, a two-value scale is enough to represent the value of an indicator (e.g. chemical control was applied or not; or the soil has either been ploughed or not after the preceding harvest). However, other attributes usually need a three-value scale to describe the diversity of cropping practices or environment (e.g. the sowing rate attribute requires three grades to describe farmers' practices: the sowing rate can be low, normal or high).

4. Definition of Aggregating Tables. The third step in the design of a DEXi model is the choice of aggregating tables determining the aggregation of attributes in the tree and their interactions. For each aggregated attribute in the model, a set of "*if-then*" rules define the value of the considered attribute as a function of the values of its immediate descendants in the model. The rules that correspond to a single aggregated attribute are gathered together and conveniently represented in tabular form. In this way, each table defines a mapping of all value combinations of lower-level attributes into the values of the aggregate attribute. Figure 3 shows decision rules that correspond to the "mitigation through crop status" aggregated attribute and define the value of this attribute for the 18 possible combinations of the three cultivar choices, the 2 levels of fertilisation and the 3 sowing densities. For example, if the cultivar is quite resistant, the level of N fertilisation balanced and the sowing rate low, then the "mitigation through crop status" attribute will be unfavourable to eyespot (the final incidence will decrease). However, even if the sowing rate and the N application rate are both high, the "mitigation through crop status" attribute during wheat growth will control eyespot significantly because the "cultivar choice" attribute is much more influential than the two other attributes (Figure 3).

The aggregating tables of IPSIM-Wheat-Eyespot have been established using knowledge available in the international literature and summarised in Table 1, and expert knowledge when needed. All aggregating tables of the model are presented in figures S1, S2, S3, S4, S5, S6.

5. Attribute Weights. The influence of each basic and aggregated attribute on the value of the output variable can be characterised with weights. The higher the weight, the more important the attribute. Table 2 summarises the weights of each of the 19 attributes of the model, providing an overview of the model's structure. IPSIM-Wheat Eyespot has 3 levels of aggregation (Figure 1), the third one being the leaves (i.e. the model input basic attributes). The "local" and "global" weights are normalised in two different ways. "Local" weights are given to each aggregated attribute separately so that the sum of weights of its immediate descendants in the hierarchy equals 100%. The "global" weights are calculated at a given level of aggregation and express the influence of each attribute at that aggregation level. They are obtained by multiplying the local weight of a given attribute at a given level of aggregation, by local weighting of its ascendants. For instance, the value of the "soil and climate" attribute is completely defined by the "Climate" attribute (100%, local weight), but this attribute only contributes 53% to the definition of the value of "Eyespot incidence" (global weight at the second level of aggregation). Local and global weights are identical at the first level of aggregation, since in this case there is only one level of aggregation. Global weights of basic attributes are shown in bold in Table 2 in order to ease their identification, since they are distributed among the second and third levels of aggregation of IPSIM-Wheat-Eyespot. The sum of global weights at the third level is only 76%. This is because some basic attributes are directly embedded in the model at the second level of aggregation. The sum of global basic attribute weights is logically equal to 100%. Table 2 can be seen as an equivalent of a sensitivity analysis that would aim at identifying the most influential input (and state) variables of a quantitative model.

6. Simulations with DEXi. The qualitative final attribute value (final incidence of eyespot) is calculated by DEXi. The

	Cultivar choice	Level of N fertilisation	Sowing rate	Mitigation through crop status
1	Very susceptible to susceptible	Excess level	High	**Favourable**
2	Very susceptible to susceptible	Excess level	Normal	**Favourable**
3	Very susceptible to susceptible	Excess level	Low	**Favourable**
4	Very susceptible to susceptible	Balanced level	High	**Favourable**
5	Very susceptible to susceptible	Balanced level	Normal	**Favourable**
6	Very susceptible to susceptible	Balanced level	Low	**Favourable**
7	Moderetely susceptible	Excess level	High	**Moderately favourable**
8	Moderetely susceptible	Excess level	Normal	**Moderately favourable**
9	Moderetely susceptible	Excess level	Low	**Moderately favourable**
10	Moderetely susceptible	Balanced level	High	**Moderately favourable**
11	Moderetely susceptible	Balanced level	Normal	**Moderately favourable**
12	Moderetely susceptible	Balanced level	Low	**Moderately favourable**
13	Quite to very resistant	Excess level	High	**Unfavourable**
14	Quite to very resistant	Excess level	Normal	**Unfavourable**
15	Quite to very resistant	Excess level	Low	**Unfavourable**
16	Quite to very resistant	Balanced level	High	**Unfavourable**
17	Quite to very resistant	Balanced level	Normal	**Unfavourable**
18	Quite to very resistant	Balanced level	Low	**Unfavourable**

Figure 3. Aggregating table for the "Mitigation through crop status" aggregated attribute (screenshot of the DEXi software).
Aggregation rules for the 18 possible combinations of the 3 cultivar choices, the 2 levels of fertilisation and the 3 sowing rates.

Table 1. Available knowledge in the scientific literature describing the effects of cropping practices and the production situation on the incidence of eyespot on wheat.

Factor	Direction of the effect	Intensity of the effect	Impact on eyespot development	References
Tillage	+/−	++	Contradictory results. For some authors, reduced soil tillage decreased eyespot infection. For others, eyespot was often more severe after ploughing than after non-inversion tillage.	[1,11,41,42,44–55]
Preceding and pre-preceding crop	+	++	Preceding and pre-preceding host crops are known to favour eyespot. However, the interaction between tillage and the crop sequence has to be taken into account.	[9,10,29,41–42,55,61,62]
Sowing date	+	++	Eyespot has always been reported to be more severe in early sown crops.	[10,20,27,29,42,55]
N fertilisation rate	+	+	High nitrogen availability generally favoured the disease. However these results were questioned.	[9,10,20,29,56,57]
Sowing rate	+	+	Prevalence was increased by high plant density and/or low shoot number per plant.	[20,29]
Cultivar choice	+	+++	The use of varieties with resistance could obviate the need for fungicide.	[10,21,22,42,58,59]
Cultivar mixture	0	0	No significant difference was found between the disease level in mixtures and the mean of disease level of the mixture components in pure stands.	[70–72]
Climate	+	++	Eyespot strongly depends on climate. Infections require periods of at least 15 h with T° between 4°C and 13°C and HR>80% (from October to April).	[9,27,28,29,43]

Cropping practices and climate can be favourable (+), unfavourable (−) or neutral (0) to the development of eyespot. The intensity of the considered factor is summarised with 4 classes: 0, no effect; +, slight; ++, significant; +++, crucial.

calculation consists in computing all aggregated attribute values according to: (i) the structure of the tree; (ii) a set of input variables (basic attribute values) defining a simulation unit; and (iii) the aggregating tables for the aggregation of attributes. An example of output results obtained for two simulation units is provided in Figure 4 (input basic attributes and calculated aggregated attribute values for the simulation of two systems: an organic and a high-input one).

Table 2. Respective weights of the attributes of IPSIM-Wheat-eyespot.

Attributes defining the final incidence of eyespot	Local level 1	Local level 2	Local level 3	Global level 1	Global level 2	Global level 3
1 Effects of cropping practices	47			47		
1.1 Primary inoculum management		21			10	
1.1.1 Preceding crop			40			4
1.1.2 Pre-preceding crop			12			1
1.1.3 Tillage after the preceding crop			40			4
1.1.4 Tillage after pre-preceding crop			8			1
1.2 Escape: effects of sowing date		9			4	
1.3 Mitigation through crop status		26			12	
1.3.1 Cultivar choice			100			12
1.3.2 Level of N fertilisation			0			0
1.3.3 Sowing rate			0			0
1.4 Chemical control		44			21	
2 Effects of soil and climate	53			53		
2.1 Soil		0			0	
2.2 Climate		100			53	
2.2.1 Autumn/winter			29			15
2.2.2 Spring			71			38
3 Interactions with the rest of the territory	0			0		

The "local" and "global" weights are calculated for each aggregated attribute separately and are distributed in 3 levels of aggregation. Bold and non-bold terms represent basic attributes and aggregated terms, respectively.

Option	Organic system	High input system
. Final incidence of eyespot	20-40 %	60-80 %
.. Effects of cropping practices	Unfavourable	**Moderately favourable**
... Primary inoculum management: interaction between crop sequence and tillage	Unfavourable	Favourable
.... Preceding crop	Non host	Host
.... Pre-preceding crop	Non host	Host
.... Tillage after harvest of the previous crop	Inversion tillage	Non-inversion tillage
.... Tillage after harvest of the pre-previous crop	Inversion tillage	Non-inversion tillage
... Escape: effects of the sowing date	Late sowing	Early sowing
... Mitigation through crop status	Unfavourable	Favourable
.... Cultivar choice	Quite to very resistant	Very susceptible to susceptible
.... Level of N fertilisation	Balanced level	Balanced level
.... Sowing rate	High	Normal
... Chemical control: use of fungicide	None	One
.. Effects of soil and climate	**Very favourable**	**Very favourable**
... Soil	Favourable	Favourable
... Climate	**Very favourable**	**Very favourable**
.... Autumn/winter	Very favourable	Very favourable
.... Spring	Very favourable	Very favourable
.. Interactions with the territory	Neutral	Neutral
... Beneficial sources	Normal	Normal
... Primary inoculum sources	Normal	Normal

Figure 4. Example of 2 simulations carried out with IPSIM-Wheat-Eyespot (screenshot of the DEXi software).

Evaluation of the Predictive Quality of IPSIM-Wheat-Eyespot

1. Description of the Dataset Used. Data representative of a wide range of climate patterns, soils and cropping practices are needed to assess the predictive quality of the model. A large dataset was therefore developed to assess the predictive quality of IPSIM-Wheat-Eyespot. A national survey was conducted to identify relevant data from various research and development institutes. The required datasets had to provide information for input attributes of IPSIM-Wheat-Eyespot (description of cropping practices, soil and climate) and its output (eyespot incidence at the "milky grain", stage 7: development of fruit on BBCH scale [39]). The dataset obtained is summarised in Table 3. It comprises results from multifactorial trials from 1980 to 1994 in 7 contrasting regions in France, which were set up to analyse the effects of various cropping practices on foot and root winter wheat diseases on different soils and with differing climate patterns. Various cultivars were combined with different crop sequences, conventional and reduced tillage, low or high plant densities, early or late sowing dates, low or high N fertilisation, in various areas of production where eyespot epidemics are observed. Most of these trials were specific studies on foot diseases [20,41,61,62], so the experimental conditions were suited to ensure the presence of eyespot (i.e. infected wheat present in the crop sequence and only susceptible cultivars). Other data originated from a regional agronomic diagnosis [63] performed in cereal fields from 1987 to 1994 in 19 French regions to analyse the effects of cultural practices on the incidence and severity of foot and root disease complexes [64]. In this survey, data were collected on 894 cereal fields in a wide range of production situations.

For some situations, the pre-preceding crop (3 possible types of crop in the model: "host", "non-host" and "risk amplifying") and the associated tillage after the harvest of this crop (2 possible values

in the model) were not observed. Instead of ignoring these precious data, simulations were performed for the 3*2 possibilities and only cases for which the 6 simulations led to similar output values were kept for evaluating the model. In all, 526 site-years were used for the evaluation of the model and they represented a large number of combinations of cropping practices and production situations (19 French regions over 9 years).

The data presented in Table 3 were transformed into qualitative values and used as input basic attributes to feed IPSIM-Wheat-Eyespot.

2. Evaluation of the Predictive Quality of IPSIM-Wheat-Eyespot. The evaluation consisted in comparing simulated and observed values. Since the model predicts classes of incidence, observed incidences at wheat stage 7 were transformed into observed incidence classes using the same discretisation as the model (i.e. 0–20%, 20–40%, 40–60%, 60–80%, 80–100%). However, one might want to predict incidences rather than classes of incidence. In order to test the predictive quality of IPSIM-Wheat-Eyespot for incidences, its output main variable was transformed into a numerical value by replacing the predicted incidence class by the centre of the class. The model was therefore evaluated in two ways: first, on its ability to predict incidence classes, and second on its ability to predict eyespot incidences.

For incidence classes, the deviation of the model was characterised by calculating the number of classes of difference between observed and simulated classes. The distribution of simulated classes was displayed according to observed incidence classes. This information was summarised by a multinomial distribution in 9 difference classes (from −4 to +4) since the model has 5 incidence classes. The proportion of situations for which the model correctly predicted the observed incidence class was taken as an indicator of the quality of prediction of the model. In addition, a non-parametric Wilcoxon test was performed to test

Table 3. Main features of the datasets used for the evaluation of IPSIM-Wheat-Eyespot's predictive quality.

Cropping practice	Design	Year	Location	Number of site-years	References
Crop sequence	Multifactorial field trials	1981–1982	Toulouse (Midi-Pyrénées)	11	[61]
Crop sequence including various durations of continuous cereal cropping	Multifactorial field trials	1980–1994	Grignon (Ile-de-France)	29	[62]
Tillage (soil structure)	Multifactorial field trials	1992–1993	Péronne (Picardie)	8	[73]
Tillage (crop residue vertical distribution)	Multifactorial field trials	1992–1993	Chartres (Centre), Grignon (Ile-de-France)	12	[41]
Sowing date, sowing rate, N fertilisation	Multifactorial field trials	1992–1994	Chartres, La Verrière (Ile-de-France), Le Rheu (Bretagne), Nancy (Lorraine), Dijon (Bourgogne)	95	[20]
Tillage, previous crop, fertilisation, sowing rate, sowing date, cultivar choice and use of fungicide	Diagnoses in cereal fields	1987–1994	19 French regions	370	[64]
Crop sequence	Multifactorial field trials	1981–1982	Toulouse (Midi-Pyrénées)	11	[61]
Crop sequence including various durations of continuous cereal cropping	Multifactorial field trials	1980–1994	Grignon (Ile-de-France)	29	[62]
Tillage (soil structure)	Multifactorial field trials	1992–1993	Péronne (Picardie)	8	[73]
Tillage (crop residue vertical distribution)	Multifactorial field trials	1992–1993	Chartres (Centre), Grignon (Ile-de-France)	12	[41]
Sowing date, sowing rate, N fertilisation	Multifactorial field trials	1992–1994	Chartres, La Verrière (Ile-de-France), Le Rheu (Bretagne), Nancy (Lorraine), Dijon (Bourgogne)	95	[20]
Tillage, previous crop, fertilisation, sowing rate, sowing date, cultivar Diagnoses in cereal fields choice and use of fungicide		1987–1994	19 French regions	370	[64]

whether the distribution of errors was zero-centred (in that case, the model can be considered unbiased).

For incidences, the predictive quality of IPSIM-Wheat-Eyespot was characterised using three common statistical criteria [65]: bias (Equation 1), Root Mean Square Error of Prediction (RMSEP, Equation 2), and efficiency (Equation 3).

$$Bias = \frac{1}{n}\sum_{i=1}^{i=n}\left(Y_i^{obs} - Y_i^{sim}\right) \qquad (1)$$

where n is the total number of considered situations, Y_i^{obs} the observed value for situation i, and Y_i^{sim} is the corresponding value simulated by the model. The bias measures the average difference between observed and simulated values. If the model underestimates the considered variable, the bias is positive. Conversely, if the model overestimates the variable, the bias is negative.

$$RMSEP = \sqrt{\frac{1}{n}\sum_{i=1}^{i=n}\left(Y_i^{obs} - Y_i^{sim}\right)^2} \qquad (2)$$

RMSEP quantifies the prediction error when the model parameters have not been estimated using the observations Y_i^{obs} used in the calculation of this criterion.

$$EF = 1 - \frac{\sum_{i=1}^{i=n}\left(Y_i^{obs} - Y_i^{sim}\right)^2}{\sum_{i=1}^{i=n}\left(Y_i^{obs} - \bar{Y}\right)^2} \qquad (3)$$

Where $\bar{Y}$ is the mean of observed data. Nash and Sutcliffe [66] defined the efficiency as a normalised statistic that determines the relative magnitude of the residual variance ("noise") compared with the measured data variance ("information"). The efficiency defines the ability of a model to predict the value of a variable. The efficiency can range from $-\infty$ to 1. If the model perfectly predicts the observations, the efficiency is maximum and is equal to 1. Efficiency values lower than 0 indicate that the mean observed value is a better predictor than the simulated values, which indicate a poor predictive quality of the model. Values between 0 and 1 are generally viewed as acceptable levels of performance. The closer the model efficiency is to 1, the better is the fit between observed and simulated data [65].

Results

Evaluation of the Quality of Prediction for Final Incidence Classes

The high number of observed site-years in the dataset (526) permitted a reliable evaluation of the predictive quality of IPSIM-Wheat-Eyespot. Residuals were distributed around 0 (Figure 5), indicating that the predicted values were close to observations. Nearly half (47.1%) of the simulated classes encompassed the observed values and 80.4% had at most a difference of one class only. In addition, there are nearly as many negative as positive differences of exactly one class. The Wilcoxon test performed over the 9 class differences (from -4 to $+4$) proved that the model was significantly biased (simulated final incidence classes lower than observations, $p < 1.0\ 10^{-10}$). Figure 6 illustrates the distribution of class differences between observed and predicted final eyespot incidences. The overall predictive quality of IPSIM-Wheat-Eyespot was judged fair, even if slightly biased. The predictive

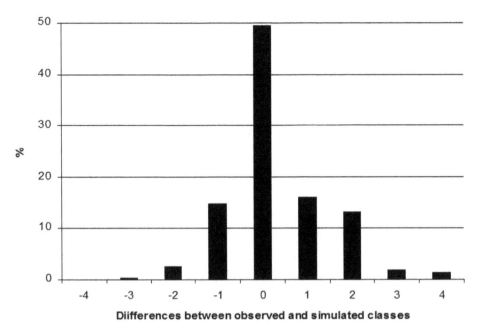

Figure 5. Evaluation of the predictive quality of IPSIM-Wheat-Eyespot. Residuals distribution: number of classes of difference between observed and simulated final eyespot classes (0–20%, 20–40%, 40–60%, 60–80%, 80–100%; 526 fields, over 9 years and 19 French regions).

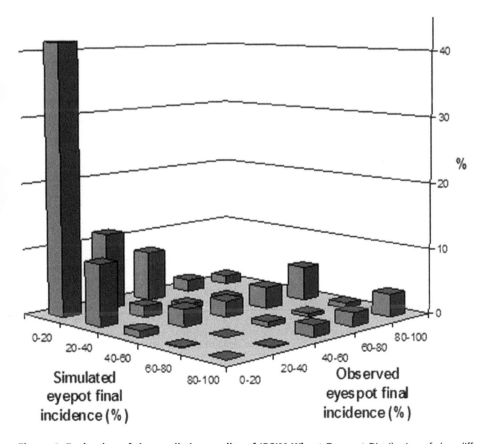

Figure 6. Evaluation of the predictive quality of IPSIM-Wheat-Eyespot Distribution of class differences between observed and predicted final eyespot incidences. (526 fields, over 9 years and 19 French regions).

quality was good for the lowest class (52% of all the observations in the dataset): 80% of the observed values between 0 and 20% were correctly simulated. The model underestimated final incidences for observations higher than 20%.

Evaluation of the Quality of Prediction for Final Incidence Values

For these 526 output values, the overall predictive quality of the model was correct. The model's predictive quality was good as its efficiency value was correct: 0.51. The Root Mean Square Error of Prediction error was quite high, 24%. The bias was positive (5.0%), so the model slightly underestimated final eyespot incidences.

Discussion

Interests and Limitations of IPSIM-Wheat-Eyespot

Several studies have been conducted to analyse the effects of cropping practices on the development of eyespot on wheat [10,29]. However, only one statistical model had been developed in order to predict the incidence of eyespot as a function of few cropping practices [29]. IPSIM-Wheat-Eyespot offers new possibilities for the design of innovative cropping systems since it is the first functional model to encompass simultaneously the effects of soil, climate and the cropping system and to represent the effects of interactions among these many factors.

The development of IPSIM-Wheat-Eyespot was made possible using (1) a schematic representation of the relationships between cropping practices, the production situation and injuries, (2) the translation of this conceptual scheme into a simulation model, and (3) a combination of data from a wide range of production situations (many regions and years) to test its predictive quality.

1. Conceptual Bases of IPSIM-Wheat-Eyespot. The conceptual scheme of IPSIM-Wheat-Eyespot is innovative because i) it encompasses a temporal scale longer than the cropping season (effect of the crop sequence in interaction with tillage over two years); ii) the main cropping practices that can affect the disease are represented; iii) interactions between practices as well as interactions between practices and climate are taken into account. As compared to the conceptual scheme of IPSIM [33], the spatial scale considered was limited to the field because of the lack of interactions at larger scales. In addition, the conceptual scheme of IPSIM-Wheat-Eyespot does not take into account socio-economic drivers, farmer's goals and cognition since it does not aim at simulating decisions. However, this original conceptual model can help design innovative cropping systems less susceptible to eyespot. The information provided by IPSIM-Wheat-Eyespot should be combined with other sources of information (references, other models, or expertise) in order to design new cropping systems, especially since damage (*i.e.* crop loss) caused by the disease are not represented.

2. Hierarchical Tree of Attributes and Aggregating Tables. The qualitative nature of the DEX method is well suited to the modelling of complex systems for which no high level of precision is required. The DEXi software tool [36] offered a suitable environment for the organisation of available knowledge and a rapid development of IPSIM-Wheat-Eyespot. The main breakthrough of the IPSIM platform is to allow the handling of complexity in a simple way [33]. The work presented in this paper provides a proof of concept for this innovative modelling approach in the field of crop protection for a single disease. A major innovation of this modelling approach is to be able to aggregate attributes of different natures (e.g. cultivar choice, a nominal variable and fertilisation rate, a quantitative variable) to describe

the impact of various components of cropping systems and their interactions on eyespot incidence. IPSIM-Wheat-Eyespot is actually the first model which can overcome the lack of data on the relationships between cropping practices and a single pest in a given production situation to help design strategies to control the disease. The qualitative DEXi approach may lead to a loss of precision and sensitivity in the developed model [67]. Increasing the number of attribute scales at the top of the decision tree could be a way to improve the sensitivity [68]. However, it results in more complicated aggregating tables which are consequently more difficult to define. Due to the tremendous complexity of interactions between cropping practices and the production situation, a smaller number of indicator states have been chosen to keep the representation of the complex underlying mechanisms as simple as possible. A correct definition of aggregating tables is of primary importance in DEXi models [68]. The choice of the nature and the number of qualitative scales is also crucial and will partly determine the quality of prediction. The choices of both aggregating tables and qualitative scales of attributes have to be explicit and traceable. Indeed, the scales and the aggregating tables used for the attributes of IPSIM-Wheat-Eyespot could not be determined in a generic way but have been specifically defined according to experimental results, models available in the literature and expert knowledge if need be. Unfortunately, literature to analyse some attributes may not exist, lack certain features, or controversial. For instance, the impact of soil type on eyespot incidence is very poorly described in the international literature and the relationships between tillage and eyespot are subject to much controversy [43]. For these cases, expert knowledge had to be used to complete some aggregating tables. In addition, the model runs using simple "if–then" rules, which are "shallow" in the sense that they only define direct relationships between conditions and consequences, but do not represent any "deeper" (or mechanistic) biological, physical, chemical processes [69]. Since the early stages of development of IPSIM-Wheat-Eyespot, it has been clear that precision was not an objective of the model. It appears more important to focus on accuracy rather than precision when modelling such a complex system.

Table 2 reveals the overall behaviour of IPSIM-Wheat-Eyespot. This is also an additional value of the IPSIM approach: the model is transparent and can be easily discussed. For instance, it is clear that the overall effect of fungicide on the disease is low (21%). This is because fungicide does not always control the disease efficiently [59]. The main factor influencing the disease is the spring weather (38%). This is consistent with Matusinsky et al. [43] who showed that the disease was very dependent on the climatic conditions during spring.

3. Predictive Quality of IPSIM-Wheat-Eyespot. The quality of the analysis of the IPSIM-Wheat-Eyespot predictive quality not only depends on the model itself (hierarchical structure of attributes and aggregating tables) but also on the diversity of the data used, which must reflect a wide range of production situations. These data should represent a variety of soil, climate and cropping practices, but also of final incidences. The dataset used in this study satisfied the three former conditions, but did not fully satisfy the latter. The observed final eyespot incidences were generally quite low, so the predictive quality of IPSIM-Wheat-Eyespot could not be extensively evaluated for high levels of incidence.

The main difference with other models is that IPSIM-Wheat-Eyespot is based on qualitative variables and not quantitative ones. The use of qualitative data requires greater attention to the description of the adopted hypotheses, because qualitative data are more difficult to interpret objectively [68]. This is particularly the case for the transformation of quantitative variables that have to be translated into qualitative input of IPSIM-Wheat-Eyespot (e.g. sowing density expressed in kg/ha or number of seeds m^{-2} and translated into "low", "normal" or "high"). Thus, IPSIM-Wheat-Eyespot can be used in 2 ways. On the one hand, some users can provide directly qualitative basic attributes (i.e. input variables of the model) if they want to test the performance of some technical options in given production situation. On the other hand, other users might want to run the model for real or putative situations where both the production situation and cropping practices are characterised with quantitative, qualitative or nominative data. In this case, an algorithm should be developed in order to rigorously translate these data into appropriate basic attributes based on national or international official references (e.g. a given cultivar will be classified as "very susceptible to susceptible"; "moderately susceptible"; or "resistant" according to official national or international seed classification); regional references (e.g. a given sowing date will be classified as "early"; "normal"; "late" as a function of regional references established by extension services); knowledge available in the literature (e.g. a given crop will be classified as "host", "non-host", or "risk-amplifying crop" according to published scientific articles); references produced by models (e.g. a given weather scenario can be classified as "very favourable"; "favourable"; or "unfavourable" according to a published model).

IPSIM-Wheat-Eyespot proved to fairly represent the variability of the 526 "site-years" used to test its predictive quality. This indicates that the model is already operational and can represent the effects of a wide range of production situations*cropping practices combinations for eyespot epidemics to help design cropping systems less susceptible to the disease. This is remarkable since, unlike most models, no fitting procedure was used.

Prospects

1. Improvements to the Model.   Further refinements could be added in the future. They should keep the balance between: i) modelling of the effects of cropping practices and the production situation on eyespot epidemics as accurately as possible, and ii) keeping the model as simple as possible. In addition to the design of a model, the approach presented in this article allowed us to structure the available knowledge in the literature about the effects of cropping practices and the production situation on eyespot epidemics (Table 1). Aggregating tables derived from Table 1 could be easily adapted according to future advances in the knowledge of underlying mechanisms responsible for the disease. In the same way, the model structure could easily be modified to integrate new knowledge. For instance, the model does not yet take into account the effects of cultivar mixtures, whereas some authors have described a reduction of eyespot by cultivar mixtures [70–72]. However, this cropping practices is not currently widespread, data are sparse and there is no consensus in the literature on this matter.

IPSIM-Wheat-Eyespot requires the provision of qualitative basic attributes. This is a benefit for the *ex-ante* design of innovative cropping systems. However, this requires translating nominative or quantitative variables used to describe cropping practices and the production situation into *ad hoc* qualitative variables. In order to avoid subjectivity when translating these variables, some reference values have to be used. Such values were gathered for several French regions (data not shown) in order to design an algorithm that translates nominative or quantitative variables describing cropping practices and the production situation into relevant basic attributes of the model. This algorithm can be easily adapted to any location where wheat is grown and eyespot is present,

provided that relevant reference values are available. At last, aggregating tables could be adjusted to improve IPSIM-Wheat-Eyespot's predictive quality using statistical procedures, as done for parameter estimation for quantitative models.

2. Future Use of the Model.   The main breakthrough of the IPSIM framework, with a simple hierarchical aggregative structure, is to allow the handling of complexity in a simple way. The input variables of models developed with IPSIM, such as IPSIM-Wheat-Eyespot, are easily obtained [33]. IPSIM-Wheat-Eyespot will help design cropping systems with a lower risk of eyespot on wheat. In order to do so, simulation plans will be defined to assess the performance of cropping practices in a given production situation with regard to the control of the disease. It is obvious that this simulation work will have to be combined with other sources of information such as other models, expert knowledge, diagnoses in commercial fields or experiments to propose innovative sustainable cropping systems.

The model, along with the interface that translates nominative and quantitative variables into relevant qualitative input variables for IPSIM-Wheat-Eyespot (Microsoft® Office Excel 2003), is now available upon request. This model can now be used as a communication, organisation, training and teaching tool for researchers, extension engineers, advisers, teachers or even farmers. Appropriation and adaptation of the model by technicians, advisers or farmers could be useful to exchange knowledge and experience (building up from their technical know-how).

The model presented in this paper only takes into account one pest among the biocenosis of a wheat field. Nevertheless, it is necessary to consider the entirety of the major pests when designing cropping systems because farmers have to manage combinations of pest populations, leading to injury profiles, which can in turn lead to quantitative or qualitative damage and ultimately economic losses. In addition to being a model specific to a given disease, IPSIM-Wheat-Eyespot can also be seen as the first sub-model of IPSIM-Wheat, a model that will predict injury profiles on wheat as a function of cropping practices and the production situation.

Supporting Information

Figure S1   Aggregating table used for the calculation of the value of the aggregative attribute "Final incidence of Eyespot" (screenshot of the DEXi software).

Figure S2   Aggregating table used for the calculation of the value of the aggregative attribute "Effects of cropping practices" (screenshot of the DEXi software).

Figure S3   Aggregating table used for the calculation of the value of the aggregative attribute "Effects of soil and climate" (screenshot of the DEXi software).

Figure S4   Aggregating table used for the calculation of the value of the aggregative attribute "Primary inoculum management: interaction between crop sequence and tillage" (screenshot of the DEXi software).

Figure S5   Aggregating table used for the calculation of the value of the aggregative attribute "Mitigation through crop status" (screenshot of the DEXi software).

Figure S6 Aggregating table used for the calculation of the value of the aggregative attribute "Climate" (screenshot of the DEXi software).

Acknowledgments

We thank Alain Cavelier (INRA, Rennes), Xavier Coquil (INRA, Nancy), Claire Thierry (INRA, Nancy), Michel Bertrand (INRA, Versailles-Grignon) for providing data. We also acknowledge Marc Délos (French Ministry of Agriculture), David Gouache (ARVALIS-Institut du Végétal), Claude Maumené (ARVALIS-Institut du Végétal), Sabrina Gaba (INRA, Dijon), Marie Gosme (INRA, Versailles-Grignon), Robert Faivre (INRA, Toulouse) and Bruno Coulomb (INRA, Toulouse) for their useful advices.

Author Contributions

Conceived and designed the experiments: NC PL FM. Performed the experiments: NC PL FM. Analyzed the data: JNA MHR CC. Contributed reagents/materials/analysis tools: JNA MHR. Wrote the paper: JNA MHR PD NC.

References

1. Montanari M, Innocenti G, Toderi G (2006) Effects of cultural management on the foot and root disease complex of durum wheat. J Plant Pathol 88: 149–156.
2. Lucas JA, Dyer PS, Murray TD (2000) Pathogenicity, host-specificity, and population biology of Tapesia spp., causal agents of eyespot disease of cereals. Advances in Botanical Research Incorporating. Adv Plant Pathol 33: 225–258.
3. Crous PW, Groenewald JZ, Gams W (2003) Eyespot of cereals revisited: ITS phylogeny reveals new species relationships. Eur J Plant Pathol 109: 841–850.
4. Ray RV, Jenkinson P, Edwards SG (2004) Effects of fungicides on eyespot, caused predominantly by Oculimacula acuformis, and yield of early-drilled winter wheat. Crop Prot 23: 1199–1207.
5. Ray RV, Crook MJ, Jenkinson P, Edwards SG (2006) Effect of eyespot caused by Oculimacula yallundae and O. acuformis, assessed visually and by competitive PCR, on stem strength associated with lodging resistance and yield of winter wheat. J Exp Bot 57: 2249–2257.
6. Dyer PS, Nicholson P, Lucas JA, Peberdy JF (1996) Tapesia acuformis as a causal agent of eyespot disease of cereals and evidence for a heterothallic mating system using molecular markers. Mycol Res 100: 1219–1226.
7. Scott PR, Hollins TW (1974) Effects of eyespot on the yield of winter wheat. Ann App Biol 78: 269–279.
8. Clarkson JDS (1981) Relationship between eyespot severity and yield loss in winter-wheat. Plant Pathol 30: 125–131.
9. Fitt BDL, Goulds A, Polley RW (1988) Eyespot (Pseudocercosporella herpotrichoides) epidemiology in relation to prediction of disease severity and yield loss in winter-wheat - a review. Plant Pathol 37: 311–328.
10. Fitt BDL, Goulds A, Hollins TW, Jones DR (1990) Strategies for control of eyespot (Pseudocercosporella herpotrichoides) in UK winter wheat and winter barley. Ann App Biol 117: 473–486.
11. Clarkson JP, Lucas JA (1993) Screening for potential antagonists of Pseudocercosporella herpotrichoides, the causal agent of eyespot disease of cereals 1. Bacteria. Plant Pathol 42: 543–551.
12. Russell PE (2005) A century of fungicide evolution. J Agr Sci 143: 11–25.
13. Leroux P, Gredt M (1997) Evolution of fungicide resistance in the cereal eyespot fungi Tapesia yallundae and Tapesia acuformis in France. Pestic Sci 51: 321–327.
14. Leroux P, Gredt M, Albertini C, Walker AS (2006) Characteristics and distribution of strains resistant to fungicides in the wheat eyespot fungi in France. 8ème Conference Internationale sur les Maladies des Plantes, Tours, France, 5 et 6 Décembre. pp. 574–583.
15. Stoate C, Boatman ND, Borralho RJ, Carvalho CR, Snoo GRd, et al. (2001) Ecological impacts of arable intensification in Europe. J Environ Manage 63: 337–365.
16. Bell EM, Sandler DP, Alavanja MC (2006) High pesticide exposure events among farmers and spouses enrolled in the agricultural health study. J Agri Saf Health 12: 101–116.
17. Paillotin G (2008). Ecophyto2018. Chantier 15 «agriculture écologique et productive» Rapport final du Président du Comité opérationnel. Ministère de l'agriculture et de la pêche. Paris, France. 138 pp. http://agriculture.gouv.fr/IMG/pdf/Rapport_Paillotin_.pdf. Accessed November 2012.
18. Epstein L, Bassein S (2003) Patterns of pesticide use in California and the implications for strategies for reduction of pesticides. Annu Rev Phytopathol 41: 351–375.
19. Birch ANE, Begg GS, Squire GR (2011) How agro-ecological research helps to address food security issues under new IPM and pesticide reduction policies for global crop production systems. J Exp Bot 62: 3251–3261.
20. Colbach N, Saur L (1998) Influence of crop management on eyespot development and infection cycles of winter wheat. Eur J Plant Pathol 104: 37–48.
21. Doussinault G, Delibes A, Sanchez-Monge R, Garcia-Olmedo F (1983) Transfer of a dominant gene for resistance to eyespot disease from a wild grass to hexaploid wheat. Nature 303: 698–700. United Kingdom.
22. Wei L, Muranty H, Zhang H (2011) Advances and prospects in wheat eyespot research: contributions from genetics and molecular tools. J Phytopathol 159: 457–470.
23. Murray TD, Delapena RC, Yildirim A, Jones SS (1994) A new source of resistance to Pseudocercosporella-herpotrichoides, cause of eyespot disease of wheat, located on chromosome-4v of Dasypyrum-villosum. Plant Breeding 113: 281–286.
24. Savary S, Mille B, Rolland B, Lucas P (2006) Patterns and management of crop multiple pathosystems. Eur J Plant Pathol 115: 123–138.
25. Bergez JE, Colbach N, Crespo O, Garcia F, Jeuffroy MH, et al. (2010) Designing crop management systems by simulation. Eur J Agron 32: 3–9.
26. Aubertot JN, Doré T, Ennaifar S, Ferré F, Fourbet JF et al. (2005) Integrated Crop Management requires to better take into account cropping systems in epidemiological models. Proceedings of the 9th International Workshop on Plant Disease Epidemiology. 11–15 April. Landerneau, France.
27. Payen D, Rapilly F, Galliot M (1979) Effect of the sowing date on the climatic potentialities of winter infections of soft winter wheat by eyespot in France. Comptes Rendus des Séances de l'Académie d'Agriculture de France 7 : 473–481.
28. Délos M (1995) Top: a model forecasting eyespot development. Phytoma 474: 26–28.
29. Colbach N, Meynard JM, Duby C, Huet P (1999) A dynamic model of the influence of rotation and crop management on the disease development of eyespot. Proposal of cropping systems with low disease risk. Crop Prot 18: 451–461.
30. Willocquet L, Aubertot JN, Lebard S, Robert C, Lannou C, et al. (2008) Simulating multiple pest damage in varying winter wheat production situations. Field Crop Res 107: 12–28.
31. Savary S, Willocquet L, Elazegui FA, Teng PS, Pham Van D, et al. (2000) Rice pest constraints in tropical Asia: characterization of injury profiles in relation to production situations. Plant Dis 84: 341–337.
32. Breman H, de Wit CT (1983) Rangeland Productivity and Exploitation in the Sahel. Science 221: 1341–1347.
33. Aubertot J-N, Robin M-H (2013) Injury Profile Simulator, a Qualitative Aggregative Modelling Framework to Predict Crop Injury Profile as a Function of Cropping Practices, and the Abiotic and Biotic Environment. I. Conceptual Bases. PLoS ONE 8(9): e73202. doi:10.1371/journal.pone.0073202.
34. Wiese MV (1987) Compendium of wheat diseases. Second edition. American Phytopathological Society. St Paul, Minessota, USA. 112 pp.
35. Bohanec M (2009) DEXi: program for multi-attribute decision making, Version 3.02. Jozef Stefan Institute, Ljubljana. Available: http://www-ai.ijs.si/MarkoBohanec/dexi.html. Accessed November 2012
36. Griffiths BS, Ball BC, Daniell TJ, Hallett PD, Neilson R, et al. (2010) Integrating soil quality changes to arable agricultural systems following organic matter addition, or adoption of a ley-arable rotation. App Soil Ecol 46: 43–53.
37. Bohanec M (2003) Decision support. In: Mladenić D, Lavrač N, Bohanec M, Moyle S (Eds.). Data Mining and Decision Support: Integration and Collaboration, Kluwer Academic Publishers (2003). pp. 23–35.
38. Bohanec M, Cortet J, Griffiths B, Znidarsic M, Debeljak M, et al. (2007) A qualitative multi-attribute model for assessing the impact of cropping systems on soil quality. Pedobiologia 51: 239–250.
39. Lancashire PD, Bleiholder H, Boom Tvd, Langeluddeke P, Stauss R, et al. (1991) A uniform decimal code for growth stages of crops and weeds. Ann Appl Biol 119: 561–601.
40. Kropff MJ, Teng PS, Rabbinge R (1995) The challenge of linking pest and crop models. Agricultural Systems 49: 413–434.
41. Colbach N, Meynard JM (1995) Soil tillage and eyespot - influence of crop residue distribution on disease development and infection cycles. Eur J Plant Pathol 101: 601–611.
42. Meynard JM, Dore T, Lucas P (2003) Agronomic approach: cropping systems and plant diseases. C R Biol 326: 37–46.
43. Matusinsky P, Mikolasova R, Klem K, Spitzer T (2009) Eyespot infection risks on wheat with respect to climatic conditions and soil management. J Plant Pathol 91: 93–101.
44. Vanova M, Matusinsky P, Javurek M, Vach M (2011) Effect of soil tillage practices on severity of selected diseases in winter wheat. Plant Soil Environ 57: 245–250.
45. Jenkyn JF, Gutteridge RJ, Bateman GL, Jalaluddin M (2010) Effects of crop debris and cultivations on the development of eyespot of wheat caused by Oculimacula spp. Ann Appl Biol 156: 387–399.
46. Cox J, Cock IJ (1962) Survival of Cercosporella herpotrichoides on naturally infected straws of wheat and barley. Plant Pathol 11: 65–66.
47. Herrman T, Wiese MV (1985) Influence of cultural practices on incidence of foot rot in winter wheat. Plant Dis 69: 948–950.

48. Smiley RW, Collins HP, Rasmussen PE (1996) Diseases of wheat in long-term agronomic experiments at Pendleton, Oregon. Plant Dis 80: 813–820.

49. Innocenti G, Montanari M, Marenghi A, Toderi G (2000) Influence of cropping systems on eyespot in winter cereals. Ramulispora herpotrichoides in cereali vernini in diverse situazioni colturali. Atti, Giornate fitopatologiche, Perugia, 16–20 aprile, Volume 2: 241–246.

50. Bailey KL, Gossen BD, Lafond GP, Watson PR, Derksen DA (2001) Effect of tillage and crop rotation on root and foliar diseases of wheat and pea in Saskatchewan from 1991 to 1998: univariate and multivariate analyses. Can J Plant Sci 81: 789–803.

51. Jenkyn JF, Christian DG, Bacon ETG, Gutteridge RJ, Todd AD (2001) Effects of incorporating different amounts of straw on growth, diseases and yield of consecutive crops of winter wheat grown on contrasting soil types. J Agr Sci 136: 1–14.

52. Jalaluddin M, Jenkyn JF (1996) Effects of wheat crop debris on the sporulation and survival of Pseudocercosporella herpotrichoides. Plant Pathol 45: 1052–1064.

53. Anken T, Weisskopf P, Zihlmann U, Forrer H, Jansa J, et al. (2004) Long-term tillage system effects under moist cool conditions in Switzerland. Soil Till Res 78: 171–183.

54. Prew RD, Ashby JE, Bacon ETG, Christian DG, Gutteridge RJ, et al. (1995) Effects of incorporating or burning straw, and of different cultivation systems, on winter wheat grown on two soil types, 1985–91. J Agr Sci 124: 173–194.

55. Burnett FJ, Hughes G (2004) The development of a risk assessment method to identify wheat crops at risk from eyespot. HGCA Project Report: 87 pp.

56. Agrios GN (2005) Plant pathology. Fifth edition. Elsevier Academic Press. San Diego, USA. 948 pp.

57. Datnoff LE, Elmer WH, Huber DM (2007) Mineral nutrition and plant disease. American Phytopathological Society Press. St Paul, Minessota, USA. 278 pp.

58. Loyce C, Meynard JM, Bouchard C, Rolland B, Lonnet P, et al. (2008) Interaction between cultivar and crop management effects on winter wheat diseases, lodging, and yield. Crop Prot 27: 1131–1142.

59. Zhang XY, Loyce C, Meynard JM, Savary S (2006) Characterization of multiple disease systems and cultivar susceptibilities for the analysis of yield losses in winter wheat. Crop Prot 25: 1013–1023.

60. Parnell S, Gilligan CA, Lucas JA, Bock CH, van den Bosch F (2008) Changes in fungicide sensitivity and relative species abundance in Oculimacula yallundae and O. acuformis populations (eyespot disease of cereals) in Western Europe. Plant Pathol 57: 509–517.

61. Colbach N, Lucas P, Cavelier N (1994) Influence of crop succession on foot and root diseases of wheat. Agronomie 14: 525–540.

62. Colbach N, Huet P (1995) Modelling the frequency and severity of root and foot diseases in winter wheat monocultures. Eur J Agron 4: 217–227.

63. Dore T, Clermont-Dauphin C, Crozat Y, David C, Jeuffroy MH, et al. (2008) Methodological progress in on-farm regional agronomic diagnosis. A review. Agron Sustain Dev 28: 151–161.

64. Cavelier A, Cavelier N, Colas AY, Montfort F, Lucas P (1998) ITITECH: a survey to improve the evaluation of relationships between cultural practices and cereal disease incidence. Brighton Crop Protection Conference: Pests & Diseases. Volume 3: Proceedings of an International Conference, Brighton, UK, 16–19 November. 1023–1028.

65. Wallach D, Makowski D, Jones J (2006). Working Dynamic Crop Models: Evaluation, Analysis, Parameterization and Application. Elsevier. 447p.

66. Nash JE, Sutcliffe JV (1970) River flow forecasting through conceptual models part I — A discussion of principles. J Hydrol 10: 282–290.

67. Sadok W, Angevin F, Bergez JE, Bockstaller C, Colomb B, et al. (2009) MASC, a qualitative multi-attribute decision model for ex ante assessment of the sustainability of cropping systems. Agron Sustain Dev 29: 447–461.

68. Pelzer E, Fortino G, Bockstaller C, Angevin F, Lamine C, et al. (2012) Assessing innovative cropping systems with DEXiPM, a qualitative multi-criteria assessment tool derived from DEXi. Ecol Indic 18: 171–182.

69. Bohanec M, Messéan A, Scatasta S, Angevin F, Griffiths B, et al. (2008) A qualitative multi-attribute model for economic and ecological assessment of genetically modified crops. Ecol Model 215: 247–261.

70. Vilichmeller V (1992) Mixed cropping of cereals to suppress plant-diseases and omit pesticide applications. Biol Agric Hortic 8: 299–308.

71. Mundt CC, Brophy LS, Schmitt MS (1995) Choosing crop cultivars and cultivar mixtures under low versus high disease pressure - a case-study with wheat. Crop Prot 14: 509–515.

72. Saur L, Mille B (1997) Disease progress of Pseudocercosporella herpotrichoides in mixed stands of winter wheat cultivars. Agronomie 17: 113–118.

73. Colbach N (1995) Modélisation de l'influence des systèmes de culture sur les maladies du pied et des racines du blé tendre d'hiver. Doctorat de l'INA P-G, Paris. 258 p.

10

Bumblebee Venom Serine Protease Increases Fungal Insecticidal Virulence by Inducing Insect Melanization

Jae Su Kim[1], Jae Young Choi[2], Joo Hyun Lee[2], Jong Bin Park[2], Zhenli Fu[2], Qin Liu[2], Xueying Tao[2], Byung Rae Jin[3], Margaret Skinner[4], Bruce L. Parker[4], Yeon Ho Je[2,5]*

1 Department of Agricultural Biology, College of Agricultural & Life Sciences, Chonbuk National University, Jeonju, Korea, 2 Department of Agricultural Biotechnology, College of Agriculture & Life Sciences, Seoul National University, Seoul, Korea, 3 Department of Applied Biology, College of Natural Resources and Life Science, Dong-A University, Busan, Korea, 4 Entomology Research Laboratory, University of Vermont, Burlington, Vermont, United States of America, 5 Research Institute for Agriculture and Life Sciences, Seoul National University, Seoul, Korea

Abstract

Insect-killing (entomopathogenic) fungi have high potential for controlling agriculturally harmful pests. However, their pathogenicity is slow, and this is one reason for their poor acceptance as a fungal insecticide. The expression of bumblebee, *Bombus ignitus,* venom serine protease (VSP) by *Beauveria bassiana* (ERL1170) induced melanization of yellow spotted longicorn beetles (*Psacothea hilaris*) as an over-reactive immune response, and caused substantially earlier mortality in beet armyworm (*Spodopetra exigua*) larvae when compared to the wild type. No fungal outgrowth or sporulation was observed on the melanized insects, thus suggesting a self-restriction of the dispersal of the genetically modified fungus in the environment. The research is the first use of a multi-functional bumblebee VSP to significantly increase the speed of fungal pathogenicity, while minimizing the dispersal of the fungal transformant in the environment.

Editor: Alfredo Herrera-Estrella, Cinvestav, Mexico

Funding: This work was supported by the grant from the Next-Generation BioGreen 21 Program (number PJ008198), Rural Development Administration, Republic of Korea. Joo Hyun Lee, Jong Bin Park, and Zhenli Fu were supported by 2nd stage of the Brain Korea 21 project. The funders had no role in study design, data collection and analysis, decision to publish, or preparation of the manuscript.

Competing Interests: The authors have declared that no competing interests exist.

* E-mail: btrus@snu.ac.kr

Introduction

Insect killing (entomopathogenic) fungi have high potential in controlling agriculturally harmful pests [1]. Some products have been industrialized as follows: *Beauveria bassiana* (e.g., BotaniGard® (BioWorks), Mycotrol® (Koppert), and Boverin® (Biodron)), *Beauveria brongniartii* (Betel® (Natural Plant Protection)), *Lecanicillium longisporum* (Vertalec® (Koopert)), *Metarhizium acridum* (Green Muscle® (CABI Bioscience)), *Metarhizium flavoviride* (Biogreen® (Becker Underwood)), and *Isaria fumosorosea* (PreFeRal® (Biobest) and Priority® (T. Stanes & Company)) [2]. The main active components of these commercial products are conidia (asexual spores) with high variability in virulence and slow pathogenesis [3], thus having difficulties in the expansion of fungal insecticide market [4].

So far, some efforts have been given to the expression of pathogenesis-related genes, such as *B. bassiana* chitinase gene [5] and *Bacillus thuringiensis* vegetative insecticidal protein (VIP) gene [6] in *B. bassiana* and insect-specific scorpion neurotoxin (AaIT) gene [7] in *M. anisopliae* to increase fungal virulence. These proteins were previously reported expressed in baculovirus expression vector system (BEVS) with the assessment of insecticidal potentials. Additionally, insect cuticle-degrading fungal own Pr1 protease and fusion protein of Pr1 and chitinase gene were over-expressed in *M. anisopliae* and *B. bassiana*, respectively [8,9]. But much more virulent entomopathogenic fungi need to be developed for efficacious pest management.

Melanization was studied as a rapid insect response to challenges of its immune system and as a novel strategy to accelerate host mortality. When arthropods encounter an immune challenge, they initiate a serine protease cascade that, in turn, leads to the activation of prophenoloxidase (proPO)-activating factors (PPAFs) [10]. These factors are activated by cleavage between clip of PPAFs and serine protease domains. Once activated, PPAFs catalyze the conversion of proPO to phenoloxidase (PO). This causes the conversion of phenols to diphenol, quinine and, finally, melanin.

Recently we found that bumblebee (*Bombus ignitus*) venom serine protease (Bi-VSP) has an arthropod PPAF function and fibrinolytic activity [11]. In some insects, Bi-VSP triggers the phenoloxidase (PO) cascade by the activation of PPAF. Injection of purified Bi-VSP induces a lethal melanization response in target insects by modulating the innate immune response. In mammals, Bi-VSP acts similarly to snake venom serine protease [12], which exhibits mammalian fibrinolytic activity. The fibrinolytic activity of Bi-VSP, possibly inhibiting blood coagulation, can facilitate the spread of toxic components throughout the bloodstream. Blood coagulation disorders are a global and frequently lethal medical disease. When clots are not dissolved, they accumulate in blood vessels and cause thrombosis leading to myocardial infraction and other cardiovascular diseases [13].

It is hypothesized that the fungi-based expression of Bi-VSP induces fast melanization of whole insect bodies, and the transformants possibly have much higher insecticidal potency than the previous achievements. In this work, we integrated multi-functional Bi-VSP to the insect killing fungus, *B. bassiana* ERL1170 by restriction enzyme-mediated integration method, which was confirmed by RT-PCR and western blotting, followed by a fibrinolysis assay. For the extracellular secretion of Bi-VSP protein, the active domain of the *vsp* gene was tailed to the signal sequence of *B. bassiana* chitinase. A selected transformant was injected to yellow spotted longicorn beetle larvae to confirm insect melanization and sprayed on beet armyworms to examine mortality. Our work is the first fungus-based expression of Bi-VSP, which was not available in an insect cell-mediated BEVS.

Results

Integration of *vsp* Gene into a Transformation Vector

For integration of the *vsp* gene into *B. bassiana* ERL1170 and the extracellular secretion of VSP protein, the active domain of the *vsp* gene was tailed with *B. bassiana* signal (*Bbs*) sequence for chitinase (**Figure 1a**) and inserted into a fungal transformation vector, yielding the binary plasmid pAB-Bbs-VSP (9.9 kb) (**Figure 1b**). A shuttle vector, pBluscript II KS(+)-egfp cassette (expression fragment) was used to insert the *Bbs-vsp* PCR products to the fungal transformation vector, pABeG provided by Dr. Feng Ming-Guang in Zhezhang University (**Figure S1**). The reason of the use of shuttle vector, pBluscript II KS(+) was that pABeG had the same promoters and terminators for *bar* and *egfp* genes, thus having difficulties in cloning. The pABeG has phosphinothricin (PPT) resistant *bar* and *egfp* genes, and each gene is expressed under the control of the same *gpdA* promoter in the same transcriptional direction. The binary plasmid, linearized by cutting with *Hind*III, was transformed into ERL1170 by the restriction enzyme-mediated integration based on blastospores [15].

Expression of Bi-VSP in *B. bassiana*

Transformation of the competent blastospores of wild type *B. bassiana* with the *Bbs-vsp*-vectoring binary plasmid produced 422 colonies on plates of Czapek's solution agar medium containing 600 µg ml^{-1} PPT. After three rounds of subculturing on PPT-free fourth-strength Sabouraud dextrose agar (SDA/4) plates, one of the putative transformants, BbsVSP-#181 was selected. BbsVSP-#181 grew similarly to the wild type on SDA/4 plates (**Figure 2a**). White mycelial growth and in 7 days a similar number of conidial production (Wt: $1.9 \times 10^8 \pm 3.1 \times 10^7$ and BbsVSP-#181:$1.8 \times 10^8 \pm 2.4 \times 10^7$ conidia per cm^2) was observed. From an economic standpoint, additional efforts are not necessary to increase BbsVSP-#181 conidial production to the level of the wild type. The transformant was consistently found to express the *Bbs-vsp* gene as determined by RT-PCR analysis (**Figure 2b**). In the Western blot, ~27 kDa (744 bp) VSP was detected in the supernatant concentrates by its polyclonal antibody, but not from the supernatant of the wild-type strain (**Figure 2c**). Transcription of *Bbs-vsp* gene and translation followed by extracellular secretion of VSP was confirmed. Secreted VSP is possibly ready to induce PPAFs in insects after the hyphal penetration, resulting in faster pathogenesis.

To determine the biological activity of the transformant (BbsVSP-#181), a fibrin plate assay and a bioassay against yellow spotted longicorn beetle larvae were conducted. In fibrin plate assay in a 60-mm dish, supernatant concentrate (25-fold) of BbsVSP-#181 strain made a clear area (degradation of fibrin) on the fibrin plate but no corresponding clearance with the wild type (**Figure 2d**). BbsVSP-#181 strain had supernatant dosage-dependent fibrinogen degradation activity when fibrinogen clots were submerged to the supernatant (**Figure S2**). Given the

previous report describing fibrinolytic activity of VSP, secreted VSP from the BbsVSP-#181 strain proved to maintain its own biological functions.

When fungal conidia (40 µl of 1×10^7 conidia ml^{-1}) were injected to second instars of yellow spotted longicorn beetle larvae, BbsVSP-#181-injected larvae completely turned black in 4 days, compared to the light pink color of the wild type-injected larvae in which fungal growth proceeded slowly (**Figure 2e**). In the BbsVSP-#181 treatment, small dark brown spots were observed 2 days post-injection, followed by complete insect melanization without fungal outgrowth in 7 days, but the wild type-injected larvae turned light pink as the mycosis developed (without forming dark spots on the host cuticle) until the hosts were covered by a sporulating mass of mycelium (**Figure S3**). Consequently, production of VSP and its use as an activator for PPAFs were confirmed.

Virulence of BbsVSP-#181 against Beet Armyworms

A transformant, BbsVSP-#181 conidial suspension (1×10^7 conidia ml^{-1}) was sprayed on 2nd instar of beet armyworm (*Spodoptera exigua*) larvae under laboratory conditions to assess its pest control activity. The transformant had significantly faster virulence than the wild type in controlling beet armyworms (**Figure 3a**). To achieve 50% mortality against beet armyworms at 1×10^7 conidia ml^{-1} dose, BbsVSP-#181 required 4.2 (± 0.8) days, compared to more than 7 days for the wild type (estimated by 9.3 (± 2.5) days) (**Table S1**). Thus, it takes 2.2-fold shorter time for controlling beet armyworms. BbsVSP-#181-treated beet armyworms turned black in 4 days and no further development was observed (**Figure 3b**). The melanization in beet armyworms began during the early stages of fungal pathogenesis, and soon after fungal germination and hyphal penetration. However, beet armyworms in the wild type treatment developed to 4th instar larvae with mycosis in 7 days. The wild type fungus had more time to achieve mechanical penetration and enzymatic degradation for complete mortality. Non-treated control beet armyworms developed to 5th instar in 10 days. Secondly, 7 days after the spray treatments (1×10^5, 1×10^6, 1×10^7, and 1×10^8 conidia ml^{-1}), LC$_{50}$ (lethal concentration causing 50% mortality) of BbsVSP-#181 was 3.6 (± 0.8)$\times 10^5$ conidia ml^{-1}, which was significantly lower than that of wild type (41.3 (± 17.1)$\times 10^5$ conidia ml^{-1}).

Discussion

The fungus-based expression of bumblebee serine protease significantly increased the virulence of wide type, which may be compared to the expression of scorpion neurotoxin (AaIT) [7] and the over-expression of fungal own Pr1 protease [8,9] as mentioned above. In the BbsVSP-#181 treatment, it took 2.2-fold shorter time for controlling beet armyworms. In the expression of neurotoxin, AaIT59 transformant required 4.5 days (wild type: 6.3 days) to achieve 50% mortality against Tobacco cutworms and 6.1 days (wild type: 9.9 days) against yellow fever mosquitoes. Approximately it took 1.4 to 1.6-fold shorter time for controlling the cutworms and mosquitoes. Similarly, in the over-expression of Pr1 protease, transformants required 93–96 h (wild type: 128 h) against gypsy moths and 98–121 h (wild type: 131 h) against green peach aphids. Approximately it took 1.1 to 1.4-fold shorter time for controlling the moths and aphids.

VSP-integrated insect-killing fungi have some advantages in pest management. They control agriculturally harmful insects in a short time compared to the wild type. Activation of melanization cascade is very sensitive to initiators and proceeds very quickly [17]. Thus, low levels of hyphal penetration may be enough to

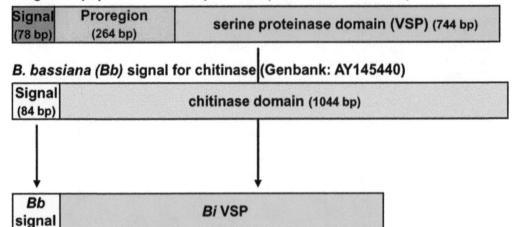

a

B. ignitus (Bi) venom serine protease (Genbank: FJ159443)

| Signal (78 bp) | Proregion (264 bp) | serine proteinase domain (VSP) (744 bp) |

B. bassiana (Bb) signal for chitinase (Genbank: AY145440)

| Signal (84 bp) | chitinase domain (1044 bp) |

| Bb signal | Bi VSP |

```
   1  ATGGCTCCTT TTCTTCAAAC CAGCCTCGCGCTCCTTCCAT TGTTGGCTTC CACCATGGTC  ┐
  61  AGCGCCTCGCCCCTTGGCGCCGCGAGTGGTCGGTGGTAAGC CAGCTGTACT TGGTGCTTGG  } Bb signal (84 bp)
 121  CCATGGATTG CTGCATTAGG TTTTCGTTAT CCCCGAAACCCAGCTCTTGA ACCACTATGG  ┐
 181  AAGTGCGGAG GTTCCCTGAT ATCGTCTAGG CATGTTTTAA CTGCAGCACA TTGTGCAGAA  │
 241  ATCAATGAAT TGTACGTGGT TCGTATCGGT GACTTAAATC TAGTACGAAA TGACGACGGA  │
 301  GCGCATCCTG TTCAAATAGA AATCGAATCT AAAATAATAC ATCCTGATTA TATTTCCGGA  │
 361  GTAACCAAAC ATGATATCGC CATTCTTAAA TTGGTGGAGG AGGTGCCATT TTCGGAGTAC  │
 421  GTATATCCCA TTTGTCTTCC CGTAGAGGAT AACCTTCGAA ATAACAATTT CGAGCGCTAT  } Bi VSP (744 bp)
 481  TACCCCTTCG TTGCTGGATG GGGATCACTA GCACATCATG GACCAGGTAG TGACGATTTA  │
 541  ATGGAAGTAC AAGTGCCAGT GATTAGCAAC ACCGAATGCA AGAACTCTTA TGCCAGATTT  │
 601  GCTGCTGCAC ATGTTACCGA TACTGTATTA TGCGCCGGATACACTCAAGG CGGAAAGGAT  │
 661  GCTTGTCAAG GTGACAGCGG AGGACCACTG ATGCTACCAA AGAAATTCAC CTTCTATCAA  │
 721  ATAGGTGTTG TGTCTTATGG TCATAAGTGC GCCGCAGCTGGATATCCCGG CGTTTACACT  │
 781  AGGGTCACGT CGTACCTCGACGACTTTATT CTCCCAGCGATGCAATAAGG ATCCCG       ┘
```

b

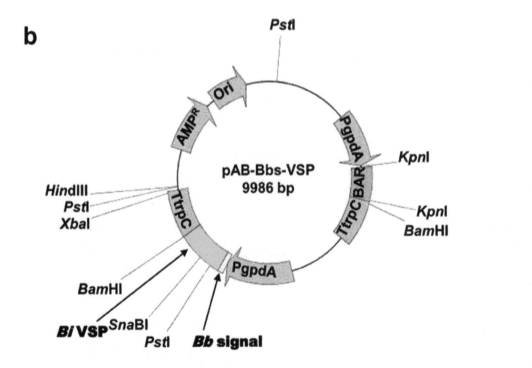

Figure 1. Tailing of *B. ignitus* (*Bi*) VSP fragment with *B. bassiana* (*Bb*) chitinase signal for extracellular secretion and integration of the fusion form into a fungal transformation vector. (a) A schematic diagram of the fusion of *Bb* chitinase signal and *Bi* VSP fragments. The 5′-end of *Bi* serine protease domain was tailed by 3′-end of *Bb* signal by four-round serial PCR. The fusion form of PCR product was confirmed by sequencing. *Bb* signal, shadowed; and *Bi* VSP, not-shadowed. (b) A map of fungal transformation vector, pAB-BbsVSP including the fusion form of *Bb* signal and *Bi* VSP fragment. The plasmid retains the *BAR* and *AMP* resistance genes of the parent plasmid pABeG (*BAR* is a selectable marker providing resistance to glufosinate).

induce the melanization cascade when VSP-integrated fungus is applied. From an economic standpoint, VSP-integrated fungi do not need to spend great deal of energy for hyphal penetration into the haemocoel. However, in research on the expression of chitinase, vegetative insecticidal protein and insect-specific scorpion toxin (expressed in haemocoel), hyphal growth and penetration should be fully accomplished for the expression of integrated genes. Fungal penetration and expression of VSP is a strategically well combined tactic to achieve fast control with high biological performance. Particularly, VSP-integrated fungi can be more useful in controlling pests with a short-term life cycle.

Another merit of VSP-integrated insect-killing fungi is the self-restriction of further reproduction and dispersal in the environments by the genetically modified fungus. Because VSP-mediated insect melanization quickly kills target insects and the fungal pathogen that introduced this toxin, no further development by the fungus should be possible. The rapid killing of its host insects and the death of treated fungus explains why no mycoses were found in the VSP-integrated fungus treatments. Melanization

improves the effectiveness of other immune responses that promote arthropod resistance to microbial infection [18] and suppresses the infection of parasitoids [19]. Dispersal of genetically modified fungi can be naturally inhibited in the environment so that the registration process may be minimized although fundamental safety tests are required. But, genetically modified fungi, even those whose reproduction or dispersal is self-limited, may or may not be easy or even possible to register in many countries merely because they are GMOs, regardless of their beneficial properties. It may take more times in the industrialization of this VSP-integrated insect-killing fungus than expectation. However, in other cases (expression of other functional insecticidal proteins such as chitinase, vegetative insecticidal protein, and insect-specific scorpion toxin), dispersal of genetically modified fungi may be usual events, so it should be carefully controlled in the environment.

VSP-integrated insect-killing fungi inherit any fungal own host spectrum, by which VSP can be expressed in potentially many insects. Among microbial pest control agents, *Bacillus thuringiensis*

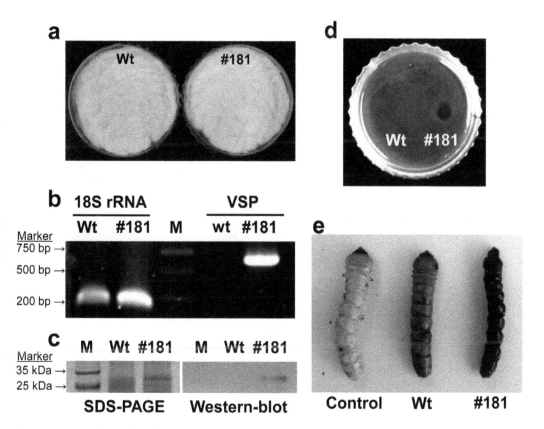

Figure 2. Expression of bumble bee venom serine protease (VSP) in an insect-killing fungus, *B. bassiana* ERL1170 (transformant: BbsVSP-#181). (a) Solid culture of wild type (Wt) and BbsVSP-#181 on fourth-strength Sabouraud dextrose agar (SDA/4) for 7 days. (b) RT-PCR analysis of VSP in BbsVSP-#181 (#181). M, Marker. (c) Western blot analysis of liquid-cultured BbsVSP-#181 supernatant using an antiserum to bumble bee VSP. (d) Fibrinolytic activity of BbsVSP-#181 supernatant. Serine proteases are known to have fibrinolytic activity. (e) Melanization activity of BbsVSP-#181 spores (conidia) against yellow spotted longicorn beetles 4 days post injection. Beetles were injected with conidia at 40 µl $(1 \times 10^7$ conidia ml$^{-1})$ per larva. Phosphate buffered saline (PBS) solution was used as a base for all the treatments.

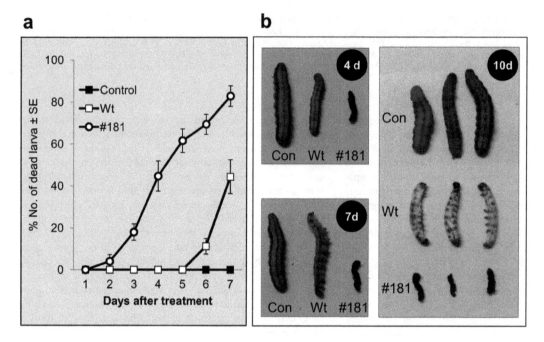

Figure 3. Insecticidal activity of wild type (Wt) and transformant BbsVSP-#181 (#181) against beet armyworm larvae in laboratory conditions. (a) Percentage (%) of dead beet armyworm larvae after the spray of Wt and BbsVSP-#181 spores at 1×10^7 conidia ml^{-1} ($N = 27$). Siloxane solution (0.03%) as a surfactant was used as a base for all the treatments. (b) Symptoms of beet armyworms in 4, 7 and 10 days after the treatment. BbsVSP-#181-treated beet armyworms turned black in 4 days and no stage development was observed. But beet armyworms in the wild type treatment developed to fourth instars with mycosis in 7 days and completely mycotized in 10 days.

(Bt) and baculoviruses can be alternatively considered, but host spectrums are mainly limited to lepidopteran pests (moths) and BEVS is not available in VSP expression as described above. Insect-killing fungi, particularly *B. bassiana* virulent to many agricultural pests such as moths, aphids, mites, stink bugs, whiteflies, thrips and soil-dwelling beetles.

Mass production of entomopathogenic fungi has been effectively developed and is cost-effective [20]. Thanks to the development of industrialization technology, VSP-integrated fungal spores (conidia) or VSP proteins can be easily mass-produced and harvested. A great deal of effort has been given to increase the shelf life of fungal spores during distribution and after application. Some studies are necessary to investigate whether any directed genetic modification of a particular fungal strain does or does not affect such critically important properties of the fungus.

In conclusion, for the first time bumble bee venom serine protease (Bi-VSP) has been successfully expressed in the insect-killing fungus, *B. bassiana* ERL1170 and has caused the melanization and rapid death of yellow spotted longicorn beetle larvae and beet armyworm larvae as well as supernatant-mediated mammalian fibrinolysis. This research highlights the expression of multi-functional Bi-VSP (not available in BEVS) in a fungal platform that is especially relevant for agricultural (fungal application) and pharmacological (purified proteases) fields with much stronger biological activities. These results could significantly increase the economic value of entomopathogenic fungi for at least some specific application.

Materials and Methods

Microbial Strains

The wild type strain *B. bassiana* ERL1170 (ARSEF2060 in USDA-ARS in Itheca) [14] was provided by Entomology Research Laboratory, University of Vermont, USA, and main-

tained on fourth-strength Sabouraud dextrose agar (SDA/4) in darkness at 25°C for colony growth. *Escherichia coli* TOP10 (Invitrogen, Carlsbad, CA), used for DNA manipulation, were cultured in Luria-Bertani (LB) medium containing 50 g ml^{-1} ampicillin [15].

Vector Construction

A fungal transformation vector, pABeG expressing *egfp* gene (provided by Dr. Feng Ming-Guang in Zhezhang University, China) was used as a plasmid backbone and its availability in *B. bassiana* was confirmed. The pABeG has phosphinothricin (PPT) resistant *bar* and *egfp* genes, and each gene is expressed under the control of *gpdA* promoter in the same transcriptional direction. Active domain of *vsp* gene was tailed with *B. bassiana* signal (*Bbs*) sequence and inserted into a fungal transformation vector, yielding the binary plasmid pAB-Bbs-VSP (9.9 kb). The full-length, 744 bp active domain *vsp* gene (GenBank FJ159443) was amplified by PCR of pGEM-Bi-VSP donated by Dr. Byung Rae Jin in Dong-A University, S. Korea. For extracellular secretion [5], the 5'-end of *Bi* serine protease active domain was tailed by 3'-end of 84 bp *Bb* signal fragment for chitinase (GenBank AY145440) by four-round serial PCR (**Table S2**), finally flanked with *Bam*HI at 3'-end (*Bbs-vsp*).

A PCR product of *Bbs-vsp* was integrated into the fungal transformation vector, pABeG containing *egfp* expression cassette by exchanging *egfp* gene with *Bbs-vsp* gene with the help of the shuttle vector, pBluscript II KS(+). The 3,668 bp *egfp* expression cassette including *gpdA* promoter (P*gpdA*) and *trpC* terminator (T*trpC*) was cut from pABeG using *Bgl*II and *Hind*III and integrated into pBluscript II KS(+), which was previously cut using *Bam*HI (compatible end to *Bgl*II) and *Hind*III. The ligated plasmid was designated as pBluscript II KS(+)-egfp cassette. To integrate *Bbs-vsp* into the position of *egfp* in pBluscript II KS(+)-egfp cassette, the *Bbs-vsp* PCR product was cut using *Bam*HI and pBluscript II KS(+)-egfp cassette was cut using *Nco*I/blunted and *Bam*HI to remove *egfp* region. The insert and the vector was ligated and designated as pBluscript II KS(+)-Bbs-vsp cassette. Lastly, to integrate the *Bbs-vsp* cassette from pBluscript II KS(+)-Bbs-vsp cassette into the fungal transformation vector, pABeG, pBluscript II KS(+)-Bbs-vsp cassette was cut using *Spe*I and *Hind*III and pABeG was cut using *Xba*I (compatible end to *Spe*I) and *Hind*III, finally yielding the binary plasmid pAB-Bbs-VSP (9.9 kb) (**Figure S1**).

Fungal Transformation

The binary plasmid, linearized by cutting with *Hind*III, was transformed into *B. bassiana* ERL1170 by the restriction enzyme-mediated integration based on blastospores [16]. Transformants were grown on Czapek's solution agar containing 600 µg ml^{-1} PPT. Putative transformants were sub-cultured three times on PPT-free SDA/4 at 25°C. Genomic DNAs were extracted from 5-day old fungal mycelial mass by the quick fungal genomic DNA extraction method [21] and the presence of *bar* and *Bbs-vsp* was examined by PCR with primers Bar-F and Bar-R (5′-AGTCGACCGTGTACGTCTCC-3′ and 5′GAAGTC-CAGCTGCCAGAAAC-3′) and primers Bbs-vsp-F and Bbs-vsp-R (5′- ATGGCTCCTTTTCTTCA-3′ and 5′-TCCGCTGTCACCTTGAC-3′).

Verification of Expression

Transcription of *Bbs-vsp* in the transformants was examined by the extraction of RNAs from 5-day old fungal mycelial mass, produced in SDA/4 in darkness at 25°C, using TRIZOL (Invitrogen) method and reverse transcription PCR (RT-PCR) with the primers Bbs-vsp-F and Bbs-vsp-R. For western blotting, transformants and wild type were cultured in fourth strength Sabouraud dextrose broth (SDB/4) at 25°C and 150 rpm of shaking for 5 days. Cultured broth was filtered using 3M filter papers and syringe filters (0.25 µm) and concentrated by ultrafiltration using Amicon tubes (Millipores, MA, USA). The concentrates were subjected to 12% SDS-polyacrylamide gel electrophoresis (PAGE) and electrophoretically transferred to a polyvinylidene difluoride (PVDF) membrane. A polyclonal antibody against VSP expressed in Sf-9 insect cells through baculovirus expression vector system (BEVS), provided by Dr. Byung Rae Jin, was derived from mouse and used to detect the expression of VSP secreted from mycelia of transformants. The PVDF membrane was incubated with a 1,000-fold dilution of the polyclonal antibody and a 2,000-fold dilution of goat anti-rabbit IgG horseradish peroxidase (HRP) as the second antibody. Visualization was performed using the luminol reagent SC-2048 (Santa Cruz Biotechnology Co.).

Fibrinolytic Activity Assay

The original fibrin plate assay [22] was slightly modified for use here in measuring the fibrinolytic activity of the supernatants. An aliquot of 5 ml fibrinogen (Sigma-Aldrich, from human plasma) (0.25%) solution in PBS (phosphate buffered saline, pH 7.4) was mixed with 10 units of thrombin (Sigma-Aldrich, from human plasma) (1 unit/50 µl) in a 60-mm Petri dish and incubated at 37°C for 15 min to speed up the clotting. A supernatant concentrate (10 µl) was dropped onto the fibrin plate. The plates were then incubated at 37°C for 2 h and visually inspected for liquefaction.

Secondly, to investigate the degradation of fibrin in supernatant solution, an aliquot of 100 µl fibrinogen (0.25%, in PBS) solution was mixed with 10 units of thrombin (1 unit/50 µl) in an 1.5-ml Eppendorf tube and incubated at 37°C for 15 min for clotting. Supernatant was loaded at 200, 400, and 800 µl tube^{-1} and the tubes were incubated at 37°C for 3 hr. After the incubation, solutions in the tubes were completely removed using a pipette and the amount of remaining fibrin was observed.

Bioassay

Yellow spotted longicorn beetle (*Psacothea hilaris*) for fungal injection and beet armyworm (*Spodoptera exigua*) for fungal spray were supplied by the Department of Agricultural Biology, National Academy of Agricultural Science, Republic of Korea. They were reared as previously described [23,24] and subjected to bioassays [25]. To produce test fungal spores (conidia), a transformant and wild type fungi were inoculated on SDA/4 at 100 µl (1×10^7 conidia ml^{-1}) per 60-mm diam Petri-dish and incubated in darkness at 25°C for 10 days. As inocula for injection, conidial suspensions, where hyphae were removed, were adjusted to 1×10^7 conidia ml^{-1} using PBS for injection and 0.03% (v/v) siloxane solution (Silwet L-77) as a wetting agent. PBS and siloxane solution served as controls.

For injection, second instars of yellow longicorn beetle larvae were placed at 4°C for 20 min. Second instars of yellow longicorn beetle larvae were injected with 40 µl of hypha-free conidial suspensions filtered using cheese clothes, where a sterile needle was promptly pierced under the epidermis of soft membranous cuticle between the sixth and seventh abdominal segments. Injected larvae were placed in 60-mm diam. Petri dishes that contained artificial diets (mulberry leaf and branch powder 100 g, carrageenan 5 g, distilled water 300 ml) ($1 \times 1 \times 0.5$ cm^3 piece per dish). The dishes were covered with lids and held in an incubator at 25±1°C and 16:8 (L/D). Petri dishes were not stacked to keep from excess moisture from forming inside of the dishes. Symptom of melanization and mycosis was observed daily for 10 days.

In spray test, a group of 10 larvae was placed in a 60-mm Petri dish (3 dishes/treatment), and all dishes were covered with lids and held at 4°C for 20 min to reduce mobility. Fungal suspensions were sprayed at 10 ml per dish using a microsprayer, and dishes were covered with lids and sealed with Parafilm. They were held in an incubator at 25±1°C and 16:8 (L/D). Petri dishes were not stacked to keep from excess moisture from forming inside of the dishes. Mortality was assessed by counting the number of live and dead larvae per dish daily for 7 days. This entire bioassay was repeated twice using different batches of conidial suspensions on different days. Secondly, to determine lethal concentration causing 50% mortality (LC$_{50}$), conidial suspensions were adjusted to 1×10^5, 1×10^6, 1×10^7, and 1×10^8 conidia ml^{-1} using 0.03% (v/v) siloxane solution (Silwet L-77) and subjected to the same spray test as described above. Data on the percentage of live larvae was analyzed by a general linear model, followed by Tukey's honestly significant difference, and median survival time and lethal concentration were determined with probit analysis using a SPSS ver. 17.0 (SPSS Inc., 2009) at the 0.05 (α) level.

Supporting Information

Figure S1 Flow chart of pAB-BbsVSP construction. (a) Construction of pBluscript II KS(+)-egfp cassette. The 3.7 kb *egfp* expression cassette was cut from pABeG and inserted to pBluscript II KS(+). **(b)** Construction of pBluscript II KS(+)-Bbs-vsp cassette. The *Bbs-vsp* PCR product was inserted to the position of *egfp* in pBluscript II KS(+)-egfp cassette. **(c)** Construction of the binary plasmid pAB-Bbs-VSP. The *Bbs-vsp* expression cassette from pBluscript II KS(+)-Bbs-vsp cassette was inserted to the position of *egfp* expression cassette in pABeG.

Figure S2 Degradation of fibrin in the wild type (Wt) and the BbsVSP-#181 transformant (#181) supernatant solutions 3 h of post-incubation at 37°C. Supernatant was loaded at 200, 400, and 800 μl tube^{-1}, where 100 μl fibrinogen (0.25%, in PBS) solution was clotted by 10 units of thrombin (1 unit/50 μl). Treated supernatant solution was completely removed and the amount of remaining fibrin was observed.

Figure S3 Yellow spotted longicorn beetles injected with wild type (Wt) and BbsVSP-#181 transformant (#181) conidia at 40 μl (1×10^7 conidia ml^{-1}) per larva 2, 4 and 7 days after injection. Phosphate buffered saline (PBS) solution was used as a base for all the treatments. In the BbsVSP-#181 treatment, small dark brown spots (arrows) were observed 2 days post-injection, followed by complete insect melanization without fungal outgrowth in 7 days, but the wild type-injected larvae turned pink as mycosis without dark spots and finally covered with fungal mycelial mass.

Table S1 Comparison of virulence between wild type and BbsVSP-#181 against beet armyworm larvae in laboratory conditions.

Table S2 Primers used for four-round serial PCR to tale *Bi* serine protease domain (vsp) with *B. bassiana* signal (Bbs) fragment for chitinase, finally flanked with *Bam*HI at 3′-end.

Acknowledgments

We are grateful to Richard A. Humber (USDA-ARSEF, USA) and Stefan T. Jaronski (USDA-ARS NPARL, USA) for their critical readings and comments on this manuscript before submission, and to Ming-Guang Feng in Zhezhang University for providing the fungal transformation vector, pABeG.

Author Contributions

Conceived and designed the experiments: JSK JYC BRJ YHJ. Performed the experiments: JSK JYC JHL JBP ZF QL XT. Analyzed the data: JSK JYC MS BLP YHJ. Contributed reagents/materials/analysis tools: JHL JBP ZF QL XT. Wrote the paper: JSK JYC MS BLP YHJ.

References

1. Hajek AE (1997) Ecology of terrestrial fungal entomopathogens. Adv Microb Ecol 15: 193–249.
2. Copping LG (2004) Manual of Biocontrol Agents, 3rd ed. Hampshire: BCPC Publications.
3. Inglis GD, Johnson DL, Goettel MS (1997) Effects of temperature and sunlight on mycosis of *Beauveria bassiana* (Hyphomycetes: Sympodulosporae) of grasshoppers under field conditions. Environ Entomol 26: 400–409.
4. Yatin BT, Venkataraman NS, Parija TK, Panneerselvam D, Govindanayagi P, et al. (2006) The New Biopesticide Market. Denver: Business Communications Research.
5. Fang W, Leng B, Xiao Y, Jin K, Ma J, et al. (2005) Cloning of *Beauveria bassiana* chitinase gene Bbchit1 and its application to improve fungal strain virulence. Appl Environ Microbiol 71: 363–370.
6. Qin Y, Ying SH, Chen Y, Shen ZC, Feng MG (2010) Integration of insecticidal protein Vip3Aa1 into *Beauveria bassiana* enhances fungal virulence to *Spodoptera litura* larvae by cuticle and *Per Os* infection. Appl Environ Microbiol 76: 4611–4618.
7. Wang C, St. Leger RJ (2007) A scorpion neurotoxin increases the potency of a fungal insecticide. Nature Biotechnol 25: 1455–1456.
8. St. Leger RJ, Joshi L, Bidochka MJ, Roberts DW (1996) Construction of an improved mycoinsecticide overexpressing a toxic protease. Proc Natl Acad Sci USA 93: 6349–6354.
9. Fang W, Feng J, Fan Y, Zhang Y, Bidochka MJ, et al. (2009) Expressing a fusion protein with protease and chitinase activities increases the virulence of the insect pathogen *Beauveria bassiana*. J Invertebr Pathol 102: 155–159.
10. Cerenius L, Söderhäll K (2004) The prophenoloxidase-activating system in invertebrates. Immunol Rev 198: 116–126.
11. Choo YM, Lee KS, Yoon HJ, Kim BY, Sohn MR, et al. (2010) Dual function of a bee venom serine protease: prophenoloxidase-activating factor in arthropods and fibrin(ogen)olytic enzyme in mammals. PLoS ONE 5: e10393.
12. Swenson S, Markland FS (2005) Snake venom fibrin(ogen)olytic enzymes. Toxicon 45: 1021–1039.

13. Toombs CF (2001) New directions in thrombolytic therapy. Curr Opin Pharmacol 1: 164–168.
14. Kim JS, Parker BL, Skinner M (2010) Effects of culture media on hydrophobicity and thermotolerance of *Bb* and *Ma* conidia, with description of a novel surfactant based hydrophobicity assay. J Invertebr Pathol 105: 322–328.
15. Bloom M, Freyer G, Micklos D (1996) Laboratory DNA Science. California: Benjamin Cummings.
16. Ying SH, Feng MG (2006) Novel blastospore-based transformation system for integration of phosphinothricin resistance and green fluorescence protein genes into *Beauveria bassiana*. Appl Microbiol Biotechnol 72: 206–210.
17. Jiang H, Kanost MR (2000) The clip-domain family of serine proteinases in arthropods. Insect Biochem Mol Biol 30: 95–105.
18. Tang H, Kambris Z, Lemaitre B, Hashimoto C (2006) Two proteases defining a melanization cascade in the immune system of drosophila. J Biol Chem 281: 28097–28104.
19. Vass E, Nappi AJ (2000) Developmental and immunological aspects of Drosophila-parasitoid relationships. J Parasitol 86: 1259–1270.
20. Kim JS, Kassa A, Skinner M, Hata T, Parker BL (2011) Production of thermotolerant entomopathogenic fungal conidia on millet grains. J Indus Microbiol Biotechnol 38: 697–704.
21. Chi MH, Park SY, Lee YH (2009) A quick and safe method for fungal DNA extraction. Plant Pathol. J. 25: 108–111.
22. Astrup T, Mullertz S (1952) The fibrin plate method for the estimation of fibrinolytic activity. Arch Biochem Biophys 40: 346–351.
23. Goh HG, Lee SG, Lee BP, Choi KM, Kim JH (1990) Simple mass-rearing of beet armyworm, *Spodoptera exigua* (Hübner) (Lepidoptera: Noctuidae), on an artificial diet. Korean J Appl Entomol 29: 180–183.
24. Scrivener AM, Watanabe H, Noda H (1997) Diet and carbohydrate digestion in the yellow-spotted longicorn beetle *Psacothea hilaris*. J Insect Physiol 43: 1039–1052.
25. Butt TM, Goettel MS (2000) Bioassays of Entomopathogenic Microbes and Nematodes. Wallingford: CABI Publishing. 141–196.

11

An Epidemiological Framework for Modelling Fungicide Dynamics and Control

Matthew D. Castle*, Christopher A. Gilligan

Department of Plant Sciences, University of Cambridge, Cambridge, United Kingdom

Abstract

Defining appropriate policies for controlling the spread of fungal disease in agricultural landscapes requires appropriate theoretical models. Most existing models for the fungicidal control of plant diseases do not explicitly include the dynamics of the fungicide itself, nor do they consider the impact of infection occurring during the host growth phase. We introduce a modelling framework for fungicide application that allows us to consider how "explicit" modelling of fungicide dynamics affects the invasion and persistence of plant pathogens. Specifically, we show that "explicit" models exhibit bistability zones for values of the basic reproductive number (R_0) less than one within which the invasion and persistence threshold depends on the initial infection levels. This is in contrast to classical models where invasion and persistence thresholds are solely dependent on R_0. In addition if initial infection occurs during the growth phase then an additional "invasion zone" can exist for even smaller values of R_0. Within this region the system will experience an epidemic that is not able to persist. We further show that ideal fungicides with high levels of effectiveness, low rates of application and low rates of decay lead to the existence of these bistability zones. The results are robust to the inclusion of demographic stochasticity.

Editor: Simon Gubbins, Institute for Animal Health, United Kingdom

Funding: MDC acknowledges the support of a Biotechnology and Biological Sciences Research Council (BBSRC) PhD studentship. CAG acknowledges the support of a BBSRC Professorial Fellowship. The funders had no role in study design, data collection and analysis, decision to publish, or preparation of the manuscript.

Competing Interests: The authors have declared that no competing interests exist.

* E-mail: mdc31@cam.ac.uk

Introduction

Fungicide use is an essential aspect of disease management in modern agriculture [1]. There are, however, increasing restrictions upon the use of fungicides (and other forms of chemical control) [2,3] and the trend is for reduced usage, to avoid undue release to the environment or to minimise the risk of fungicide resistance. Reduction in the availability and use of fungicides imposes greater demands upon efficient use of the limited resources available. One way to approach this is by the use of mathematical models to investigate the effects of a given fungicide on the host crop and the pathogen to aid the design of appropriate disease management strategies [4–6]. The benefits of modelling diseases are well acknowledged: it allows theoretical optimal control strategies to be developed that minimise economic costs and maximise crop returns [7–10], which can then be tested experimentally [11]. Furthermore the continued widespread use of fungicides is threatened by the emergence of resistant pathogen strains, often as a direct consequence of the application strategy itself [1,2,12] and in order to implement effective resistance management strategies it is often necessary to model those strategies first [9,13–16].

Most models that include fungicides often account for fungicide dynamics by simply modifying the parameters of the underlying epidemiological models (i.e. reducing the infection rates and/or increasing the host recovery rates) [9,14,17]. Much work has been done to investigate the consequences of using different underlying models for both the host population and pathogen dynamics (see [6] for a review) but relatively little work has been carried out to investigate the dynamics of the fungicides themselves and how the timing of initial infection relative to host population growth affects invasion and persistence in a chemically controlled system. Generally, previous work has implicitly made one or more of the following assumptions; that there is complete coverage by the fungicide for either the entire host population or a fixed subset of the host population [10,14,16,18,19]; that a generic, multi-purpose, fungicide has been applied to the hosts (i.e. they have not readily separated out the effects of different fungicide types such as protectants, curatives or eradicants) [9,20,21]; or that the fungicides are permanent (i.e. that the chemicals do not decay or that their effects do not change over time) [18].

Here we propose a parsimonious model framework that is designed to integrate epidemiological and fungicide dynamics. Specifically we consider purely protectant fungicides, and we distinguish the effects of the rate of application, the decay in activity and the partial effectiveness of the fungicide on the epidemic dynamics. We first use the framework to construct a deterministic compartmental model that incorporates host growth and explicitly allows for the dynamic application of a purely protectant fungicide through the use of multiple susceptible host classes. We compare this model with a conventional model that implicitly incorporates the protectant fungicide dynamics through modification of the underlying infection rates for a single susceptible host class. For brevity, we refer to these two models as the explicit and implicit models respectively. We then extend the explicit model to include demographic stochasticity.

Given that the application of fungicides is usually designed to prevent invasion of a pathogen or to eliminate the pathogen, we use the two models to ask the following questions:

- does the explicit inclusion of fungicide dynamics affect the criteria for invasion and persistence?
- does the timing of initial infection affect the criteria for invasion and persistence in a chemically controlled system?
- how are the differences manifested?
- does failure to account for fungicide dynamics lead to erroneous predictions of control effectiveness in epidemiological models?
- are there critical regions of parameter space concerned with coverage, decay and effectiveness of fungicides for which these differences are most exaggerated?
- are the differences maintained when allowance is made for stochastic variability in the transmission of infection?

Methods

Deterministic Model Derivation

Consider a pathogen spreading through a population of hosts such as a single field of a crop, in which the epidemiological host unit is an amount of plant tissue i.e. a leaf, stem, or root [4] and there is a simple density dependent growth of the host up to a carrying capacity. In the absence of any form of chemical control the population is divided into two classes, susceptible ($\hat{S}$) and infected ($\hat{I}$). We assume that infected host tissue consumes resources but does not contribute to new host growth [22], that infection is adequately described as a mass action process (this assumption holds well for pathogens within areas smaller than their average dispersal distance [23]) and that hosts cannot be re-infected after they cease to be infectious, and are removed. We investigate the effects of applying a purely protectant fungicide to the system. Here we assume that a protectant fungicide affects susceptible hosts, reducing their capacity to become infected.

We first consider a conventional model that only includes implicit fungicide dynamics, we assume that all susceptible hosts are protected. All factors affecting the effectiveness of the protectant are subsumed into a single parameter (ρ) which represents the average decrease in the infection rate for the host population. The dynamics of this model are represented in Figure 1 and by the equations:

$$\frac{d\hat{S}}{d\hat{t}} = \hat{\lambda}\hat{S}\left(1 - \frac{\hat{S}+\hat{I}}{\hat{\kappa}}\right) - \frac{\rho\hat{\beta}\hat{S}\hat{I}}{\hat{\kappa}}, \qquad (1)$$

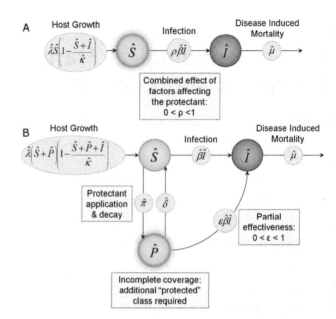

Figure 1. Model transtion diagrams. (A) The conventional model (the 'implicit' model) incorporates fungicide dynamics implicitly and therefore assumes uniform, permanent coverage by the protectant. (B) The model with explicit fungicide dynamics (the 'explicit' Model) allows for partial coverage and decay in activity of the protectant as well as different infection rates depending on whether the hosts have a protectant fungicide applied or not. The variables and parameters are explained in Table 2.

$$\frac{d\hat{I}}{d\hat{t}} = \frac{\rho\hat{\beta}\hat{S}\hat{I}}{\hat{\kappa}} - \hat{\mu}\hat{I}. \qquad (2)$$

Here, density dependent host growth is governed by the rate $\hat{\lambda}$ and a carrying capacity $\hat{\kappa}$. $\hat{\beta}$ represents the infection rate in the absence of chemical control. The parameter ρ ($0 \leq \rho \leq 1$) is a measure of fungicide effectiveness. It reduces the infection rate and represents the combined effects of the partial coverage, decay in activity and incomplete effectiveness of the protectant fungicide (possible expressions for ρ that reflect different combinations of these effects are summarised in Table 1). Infected hosts are removed at a per capita rate ($\hat{\mu}$).

We choose an expression for ρ that takes account of the uninfected population being an aggregation of completely unprotected and partially protected hosts. Given a protectant application rate ($\hat{\pi}$) and a protectant decay rate ($\hat{\delta}$) the long term

Table 1. Expressions for ρ. Potential expressions for ρ, the reduction in infection rate in the implicit model, and their interpretations in terms of assumptions made about the behaviour of the protectant.

Functional Form For ρ	Interpretation
ε	1. Complete coverage of hosts. 2. No decay of protectant. 3. Protectant is partially effective
$1-\pi$	1. Fixed proportion of hosts covered 2. No decay of protectant 3. Protectant is completely effective
$\left(\dfrac{\delta+\varepsilon\pi}{\delta+\pi}\right)$	1. Fixed equilibrium proportion of hosts covered 2. Protectant is allowed to decay 3. Protectant is partially effective

proportions for unprotected and protected hosts are $\frac{\hat{\delta}}{\hat{\delta}+\hat{\pi}}$ and $\frac{\hat{\pi}}{\hat{\delta}+\hat{\pi}}$ respectively. In the absence of explicit treatment of fungicide dynamics, we assume these proportions are maintained throughout the infection process, and that the protected hosts are infected at a reduced per capita rate $(\varepsilon\hat{\beta}, 0 \leq \varepsilon \leq 1)$, where ε is a measure of the effectiveness the protectant fungicide $(\varepsilon = 0$ and $\varepsilon = 1$ correspond to completely effective and ineffective protectant fungicides respectively). The total rate of infection is therefore

$$\begin{pmatrix} \text{Infection rate of} \\ \text{unprotected proportion} \end{pmatrix} + \begin{pmatrix} \text{Infection rate of} \\ \text{protected proportion} \end{pmatrix}$$

$$= \hat{\beta}\left(\frac{\hat{\delta}}{\hat{\delta}+\hat{\pi}}\hat{S}\right)\frac{\hat{I}}{\kappa} + \varepsilon\hat{\beta}\left(\frac{\hat{\pi}}{\hat{\delta}+\hat{\pi}}\hat{S}\right)\frac{\hat{I}}{\kappa}$$

$$= \left(\frac{\hat{\delta}+\varepsilon\hat{\pi}}{\hat{\delta}+\hat{\pi}}\right)\frac{\hat{\beta}\hat{S}\hat{I}}{\kappa}.$$

As such we take $\rho = \frac{\hat{\delta}+\varepsilon\hat{\pi}}{\hat{\delta}+\hat{\pi}}$, which yields our conventional, 'implicit' model:

$$\frac{\mathrm{d}\hat{S}}{\mathrm{d}\hat{t}} = \hat{\lambda}\hat{S}\left(1 - \frac{\hat{S}+\hat{I}}{\hat{\kappa}}\right) - \left(\frac{\hat{\delta}+\varepsilon\hat{\pi}}{\hat{\delta}+\hat{\pi}}\right)\frac{\hat{\beta}\hat{S}\hat{I}}{\hat{\kappa}}, \qquad (3)$$

$$\frac{\mathrm{d}\hat{I}}{\mathrm{d}\hat{t}} = \left(\frac{\hat{\delta}+\varepsilon\hat{\pi}}{\hat{\delta}+\hat{\pi}}\right)\frac{\hat{\beta}\hat{S}\hat{I}}{\hat{\kappa}} - \hat{\mu}\hat{I}. \qquad (4)$$

In order to incorporate fungicide dynamics explicitly we propose an alternative model with an additional protected host class $(\hat{P})$. Here, the per capita protectant application rate $(\hat{\pi})$ and the per capita protectant decay rate $(\hat{\delta})$ allow individuals to move between the susceptible $(\hat{S})$ and protected $(\hat{P})$ states. The protected state experiences reduced infection rates $(\varepsilon\hat{\beta})$ whilst the unprotected susceptible state experiences normal infection rates $(\hat{\beta})$. New host units are assumed to be unprotected which is consistent with a non-systemic protectant fungicide. The dynamics of this model are represented in Figure 1 and by the equations:

$$\frac{\mathrm{d}\hat{S}}{\mathrm{d}\hat{t}} = \hat{\lambda}(\hat{S}+\hat{P})\left(1 - \frac{\hat{S}+\hat{P}+\hat{I}}{\hat{\kappa}}\right) - \hat{\pi}\hat{S} + \hat{\delta}\hat{P} - \frac{\hat{\beta}\hat{S}\hat{I}}{\hat{\kappa}}, \qquad (5)$$

$$\frac{\mathrm{d}\hat{P}}{\mathrm{d}\hat{t}} = \hat{\pi}\hat{S} - \hat{\delta}\hat{P} - \frac{\varepsilon\hat{\beta}\hat{P}\hat{I}}{\hat{\kappa}}, \qquad (6)$$

$$\frac{\mathrm{d}\hat{I}}{\mathrm{d}\hat{t}} = \frac{\hat{\beta}\hat{I}(\hat{S}+\varepsilon\hat{P})}{\hat{\kappa}} - \hat{\mu}\hat{I}. \qquad (7)$$

Non-dimensionalisation. We introduce the dimensionless variables

$$S = \frac{\hat{S}}{\hat{\kappa}}, \quad P = \frac{\hat{P}}{\hat{\kappa}}, \quad I = \frac{\hat{I}}{\hat{\kappa}}, \quad t = \hat{\beta}\hat{t},$$

and parameters

$$\lambda = \frac{\hat{\lambda}}{\hat{\beta}}, \quad \pi = \frac{\hat{\pi}}{\hat{\beta}}, \quad \delta = \frac{\hat{\delta}}{\hat{\beta}}, \quad \mu = \frac{\hat{\mu}}{\hat{\beta}}.$$

The conventional model equations (1) and (2) with implicit fungicide dynamics become

$$\frac{\mathrm{d}S}{\mathrm{d}t} = \lambda S(1-(S+I)) - \left(\frac{\delta+\varepsilon\pi}{\delta+\pi}\right)SI, \qquad (8)$$

Implicit Model

$$\frac{\mathrm{d}I}{\mathrm{d}t} = \left(\frac{\delta+\varepsilon\pi}{\delta+\pi}\right)SI - \mu I, \qquad (9)$$

and the alternative model equations (5), (6) and (7) with explicit fungicide dynamics become

$$\frac{\mathrm{d}S}{\mathrm{d}t} = \lambda(S+P)(1-(S+P+I)) - \pi S + \delta P - SI, \qquad (10)$$

$$\frac{\mathrm{d}P}{\mathrm{d}t} = \pi S - \delta P - \varepsilon PI, \quad \text{Explicit Model} \qquad (11)$$

$$\frac{\mathrm{d}I}{\mathrm{d}t} = I(S+\varepsilon P) - \mu I. \qquad (12)$$

Henceforth we refer to the conventional model as the implicit model and the alternative model as the explicit model. The parameters, their definitions, and typical values used for the numerical simulations are summarised in Table 2. The default values used hereafter are chosen primarily to highlight the characteristic properties of the model system, however the parameter ratios are consistent with those used in disease management practices for both barley and wheat crops [24–27].

Stochastic Model

In order to demonstrate that existence of the bistability zone is robust to demographic stochasticity and that it is not just a property of the deterministic nature of the explicit model, a continuous-time Markov process version of the explicit model is constructed.

Again $\hat{S}$, $\hat{I}$ and $\hat{P}$ represent the actual numbers of hosts in each of the susceptible, infected and protected classes. Given a carrying capacity, $\hat{\kappa}$, the infinitesimal transition probabilities that describe the Markov process are given in Table 3. For computational efficiency a continuous-time Gillespie algorithm [28] is used to generate the sequence of transition event times for each simulation and thus obtain the trajectories for each class.

Table 2. Parameter Summary. Summary of dimensionless state variables, initial conditions, equilibria values and system parameters for both the implicit and explicit models.

Variable	Definition	Description	Default Value
S	$\hat{S}\hat{\kappa}^{-1}$	Density of susceptible hosts	-
P	$\hat{P}\hat{\kappa}^{-1}$	Density of protected hosts	-
I	$\hat{I}\hat{\kappa}^{-1}$	Density of infected hosts	-
t	$\hat{\beta}\hat{t}$	Time	-
$S_{\infty:0}$	$\hat{S}_{\infty:0}\hat{\kappa}^{-1}$	Disease free equilibrium density of susceptible hosts	0.0099
$P_{\infty:0}$	$\hat{P}_{\infty:0}\hat{\kappa}^{-1}$	Disease free equilibrium density of protected hosts	0.9900
S_∞	$\hat{S}_\infty\hat{\kappa}^{-1}$	Endemic equilibrium density of susceptible hosts	varies
P_∞	$\hat{P}_\infty\hat{\kappa}^{-1}$	Endemic equilibrium density of protected hosts	varies
I_∞	$\hat{I}_\infty\hat{\kappa}^{-1}$	Endemic Equilibrium density of infected hosts	varies
S_0	$\hat{S}_0\hat{\kappa}^{-1}$	Initial density of susceptible hosts at infection	varies
P_0	$\hat{P}_0\hat{\kappa}^{-1}$	Initial density of protected hosts at infection	varies
I_0	$\hat{I}_0\hat{\kappa}^{-1}$	Initial density of infected hosts	varies
λ	$\hat{\lambda}\hat{\beta}^{-1}$	Growth rate of hosts	0.5
μ	$\hat{\mu}\hat{\beta}^{-1}$	Removal rate	varies
π	$\hat{\pi}\hat{\beta}^{-1}$	Protectant application rate	0.1
δ	$\hat{\delta}\hat{\beta}^{-1}$	Protectant decay rate	0.001
ε	-	Protectant effectiveness	0.1
ρ	$\dfrac{\hat{\delta}+\varepsilon\hat{\pi}}{\hat{\delta}+\hat{\pi}}$	Reduction in infection rate (implicit model)	0.109
$\hat{\kappa}$	-	Disease free carrying capacity	2000
R_0	$\dfrac{\delta+\varepsilon\pi}{\mu(\delta+\pi)}$	Basic reproductive number	varies

Results

Equilibrium Analyses

Both the implicit model and the explicit model have the same basic reproductive number for the pathogen:

$$R_0 = \frac{1}{\mu}\left(\frac{\delta+\varepsilon\pi}{\delta+\pi}\right), \qquad (13)$$

with the disease free equilibria $(S_{\infty:0},I_{\infty:0})$ and $(S_{\infty:0},P_{\infty:0},I_{\infty:0})$ for the models given by

$$(S_{\infty:0},I_{\infty:0})=(1,0) \quad \text{Implicit Model} \qquad (14)$$

$$(S_{\infty:0},P_{\infty:0},I_{\infty:0})=\left(\frac{\delta}{\delta+\pi},\frac{\pi}{\delta+\pi},0\right) \quad \text{Explicit Model} \qquad (15)$$

It is trivial to show that both disease free equilibria are stable for $R_0<1$ and unstable for $R_0>1$. The implicit model only admits a single endemic equilibrium (S_∞,I_∞), given by:

$$(S_\infty,I_\infty)=\left(\frac{1}{R_0},\left(\frac{\lambda}{\lambda+\mu R_0}\right)\left(1-\frac{1}{R_0}\right)\right), \qquad (16)$$

whereas the explicit model admits multiple endemic equilibria,

given by:

$$(S_\infty,P_\infty,I_\infty)=\left(\mu\left(\frac{\delta+\varepsilon I_\infty}{\delta+\varepsilon\pi+\varepsilon I_\infty}\right),\mu\left(\frac{\pi}{\delta+\varepsilon\pi+\varepsilon I_\infty}\right),I_\infty\right), \qquad (17)$$

where I_∞ is any root to the cubic equation:

$$
\begin{aligned}
&\varepsilon^2(\lambda+1)I^3 + \\
&\varepsilon((\delta+\varepsilon\pi)(\lambda+2)+\lambda(\delta+\pi)+\varepsilon\lambda(\mu-1))I^2 + \\
&((\delta+\varepsilon\pi)^2+\lambda(\delta+\varepsilon\pi)(\pi+\delta-\varepsilon)-\varepsilon\lambda(\delta+\pi)(2\mu-1))I - \\
&\lambda(\pi+\delta)((\delta+\varepsilon\pi)-\mu(\delta+\pi))=0,
\end{aligned} \qquad (18)
$$

that satisfies $I_\infty\in[0,1]$.

It can be shown that the endemic equilibrium for the implicit model is both biologically meaningful and stable only for $R_0>1$. This model gives rise to a classical epidemiological bifurcation diagram (a graph of I_∞ as a function of R_0 showing a transcritical bifurcation at $R_0=1$) given by Figure 2 and we see a single invasion threshold at $R_0=1$ as expected.

Allowance for fungicide dynamics in the explicit model yields zero, one, or two biologically realistic endemic equilibria depending on the values of the model parameters ε, π, δ and λ. In the region where $R_0>1$ there is only ever one single, stable, endemic equilibrium. However, in the region where $R_0<1$ the other parameters affect the number of possible equilibria. Let

Table 3. Transition Probabilities. Infinitesimal transition probabilities for the stochastic Markov process version of the explicit model.

Transition Event	Infinitesimal Transition Probability
Net Birth	$\lambda\left(\hat{S}+\hat{P}\right)\left(1-\dfrac{\left(\hat{S}+\hat{P}+\hat{I}\right)}{\hat{\kappa}}\right)\Delta t$
Susceptible Infection	$\dfrac{\hat{S}\hat{I}}{\hat{\kappa}}\Delta t$
Protected Infection	$\dfrac{\varepsilon\hat{P}\hat{I}}{\hat{\kappa}}\Delta t$
Protectant Application	$\pi\hat{S}\Delta t$
Protectant Decay	$\delta\hat{P}\Delta t$
Removal	$\mu\hat{I}\Delta t$
No Event Happens	$1-\left[\lambda\left(\hat{S}+\hat{P}\right)\left(1-\dfrac{\left(\hat{S}+\hat{P}+\hat{I}\right)}{\hat{\kappa}}\right)+\dfrac{\left(\hat{S}+\varepsilon\hat{P}\right)\hat{I}}{\hat{\kappa}}+\pi\hat{S}+\delta\hat{P}+\mu\hat{I}\right]\Delta t$

$$\gamma=(\lambda+1)\delta^2+(\lambda(1+\varepsilon)+2\varepsilon)\pi\delta+\varepsilon(\lambda+\varepsilon)\pi^2-(1-\varepsilon)\varepsilon\lambda\pi. \quad (19)$$

If $\gamma>0$ then there is no biologically meaningful (stable or unstable) endemic equilibrium for $R_0<1$. This again gives rise to a classical epidemiological bifurcation diagram. If, however, $\gamma<0$, then a second threshold R_c exists and it is possible for the system to have two endemic equilibria for R_0 between R_c and 1 (Figure 2). The lower branch of the endemic equilibrium curve is always unstable and the upper branch stable [29]. It follows that failure to account explicitly for fungicide dynamics can lead to an erroneous understanding of the nature of the system. In particular, explicitly accounting for fungicide dynamics means that for values of R_0 less than 1, it is still possible for infection to persist within the system.

The threshold $\gamma=0$ can be interpreted as a surface in the fungicide parameter ($\pi\delta\varepsilon$) space with λ fixed (see Figure 3). Here we note that if the fungicide parameters lie in the region below this surface then the model exhibits bistability. The surface encloses the region of fungicide parameter space near to the origin, i.e. low

values of the scaled parameters (π, δ and ε). It is a key result to note that these values correspond to highly effective fungicides (low ε), with long lifespans (low δ) that are applied infrequently (low π) and so it follows that a move towards using fungicides with these properties can have extremely undesirable consequences in terms of effective control.

Invasion and Persistence Criteria

We now consider how invasion and persistence criteria depend on the choice of model and on the host growth state of the system at the time of initial infection. In particular we determine how these criteria depend on both the basic reproductive ratio R_0 and the initial levels of infection (I_0). Invasion is related to the immediate behaviour of the system, and a pathogen is considered to have invaded if there is an immediate increase in the initial

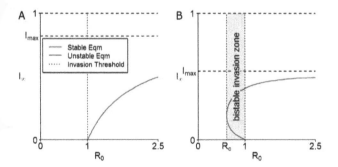

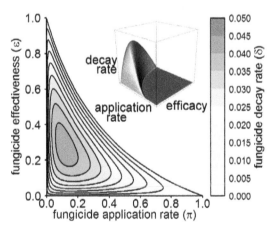

Figure 2. Model bifurcation diagrams. Bifurcation diagrams for the implicit and explicit models showing the effect of explicitly including fungicide dynamics on the endemic equilibria of the system. (A) The implicit model only admits bifurcation diagrams with a single invasion threshold at $R_0=1$. (B) The explicit model can additionally admit a bistable invasion zone. Within this zone invasion depends upon the initial level of infection (I_0). The default parameters used for these plots are given in Table 2.

Figure 3. Critical fungicide properties required for bifurcation. A contour plot for critical fungicide properties (efficacy, application rate and decay rate) that defines the region where there is a bistable invasion zone. For values of π, δ and ε that lie within this region, the system exhibits a backwards bifurcation and a pathogen is able to invade and persist for values of $R_0<1$. For parameter values outside this region the system exhibits a traditional bifurcation diagram and a pathogen is only able to invade for $R_0>1$. Inset is a 3D perspective plot of the same surface. For these plots the growth rate of the host, $\lambda=0.5$.

levels of infection. Persistence is related to the long term behaviour of the system, and a pathogen is considered to persist if the infection levels reach stable endemic equilibrium values.

Disease Free Host Growth. In the absence of infection the implicit and explicit models become

$$\frac{dS}{dt} = \lambda S(1-S). \quad \text{Implicit Model} \quad (20)$$

$$\frac{dS}{dt} = \lambda(S+P)(1-(S+P)) - \pi S + \delta P,$$

$$\text{Explicit Model} \quad (21)$$

$$\frac{dP}{dt} = \pi S - \delta P.$$

Both models assume the same logistic growth of uninfected hosts up to a disease free equilibrium carrying capacity. However for the explicit model the relative levels of susceptible and protected hosts will vary during this growth phase (see Figure 4). Analytic solutions exist to both of these sets of equations.

Infection Choice. For each value of I_0 we must choose which uninfected hosts become infected. For the implicit model there is a single uninfected class and so the initial infected individuals must come from this class. If, prior to infection, the system has S_i susceptible individuals, where $0 < S_i < 1$ then immediately post infection the system must be in the state:

$$(S_0, I_0) = (S_i - I_0, I_0) \quad 0 \le I_0 \le S_i \quad (22)$$

For the explicit model, with two uninfected classes, a choice is required. We assume that the proportion of I_0 that comes from each uninfected class is proportional to their relative susceptibilities (i.e. more initial infections come from the susceptible class than the protected class) and their relative densities. i.e. If, prior to infection, the system is in state $(S_i, P_i, 0)$ and immediately post

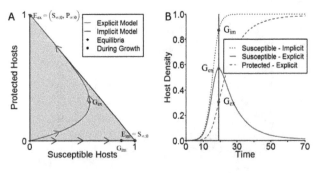

Figure 4. Disease-free growth dynamics. (A) A state space plot of the relative densities of susceptible (S) and protected (P) hosts during the disease free growth period for the explicit and implicit models. The points E_{ex} and E_{im} correspond to the disease-free equilibrium (DFEQ) densities for the explicit and implicit models respectively, and the points G_{ex} and G_{im} correspond to non-equilibrium densities (i.e. densities during the growth stage) of the disease-free states of the explicit and implicit models. (B) The equivalent plot of the densities of susceptible and protected hosts against time for the explicit model along with the equivalent growth curve of susceptible host density for the implicit model. The parameter values used here are given in Table 2.

infection the system is in state (S_0, P_0, I_0) then

$$\text{proportion of } I_0 \text{ from susceptible class} = S_i - S_0 \propto 1 \times S_i$$

$$\text{proportion of } I_0 \text{ from protected class} = P_i - P_0 \propto \varepsilon \times P_i$$

Consequently, for an initial level of infection, I_0, the initial post infection state (S_0, P_0, I_0) is given by:

$$S_0 = \begin{cases} S_i\left(1 - \dfrac{I_0}{S_i + \varepsilon P_i}\right) & 0 \le I_0 \le S_i + \varepsilon P_i \\ 0 & S_i + \varepsilon P_i < I_0 < S_i + P_i \end{cases} \quad (23)$$

$$P_0 = \begin{cases} P_i\left(1 - \dfrac{\varepsilon I_0}{S_i + \varepsilon P_i}\right) & 0 \le I_0 \le S_i + \varepsilon P_i \\ S_i + P_i - I_0 & S_i + \varepsilon P_i < I_0 < S_i + P_i \end{cases} \quad (24)$$

This gives us a method for moving from a pre-infection system state $(S_i, 0)$ or $(S_i, P_i, 0)$ to an initial post-infection state (S_0, I_0) or (S_0, P_0, I_0) for the implicit and explicit models respectively.

Invasion and Persistence Thresholds. For a pathogen to invade we require that the level of infection increases immediately from the post-infection system state. For the implicit model this requires

$$\left.\frac{dI}{dt}\right|_{I_0} > 0 \quad \Rightarrow \quad I_0(\rho S_0 - \mu) > 0. \quad (25)$$

We use the relationship between pre-infection levels and post-infection levels to create a threshold in $R_0 - I_0$ parameter space for each pre-infection susceptible host level S_i. The threshold is given by:

$$I_0 = S_i - \frac{1}{R_0} \quad (26)$$

This reduces to the classic invasion threshold if S_i is taken to be the disease free equilibrium level $S_{\infty:0} = 1$. However for the explicit model we require:

$$\left.\frac{dI}{dt}\right|_{I_0} > 0 \quad \Rightarrow \quad I_0((S_0 + \varepsilon P_0) - \mu) > 0. \quad (27)$$

This gives the following threshold in $(R_0 - I_0)$ parameter space for each pre-infection host state (S_i, P_i):

$$I_0 = \begin{cases} \dfrac{(S_i + \varepsilon P_i)^2}{(S_i + \varepsilon^2 P_i)}\left(1 - \left(\dfrac{S_{\infty:0} + \varepsilon P_{\infty:0}}{S_i + \varepsilon P_i}\right)\dfrac{1}{R_0}\right) & 0 \le I_0 \le S_i + \varepsilon P_i \\ (S_i + \varepsilon P_i)\left(1 - \left(\dfrac{S_{\infty:0} + \varepsilon P_{\infty:0}}{S_i + \varepsilon P_i}\right)\dfrac{1}{R_0}\right) & S_i + \varepsilon P_i < I_0 < S_i + P_i \end{cases} \quad (28)$$

In the absence of analytical tractability persistence thresholds for the two models were determined numerically for a range of pre-infection states. The parameter values used for the numerical calculation are summarised in Table 2.

Using the results from the growth curves in section 0 and Figure 4 we illustrate how both the invasion and persistence thresholds depend upon the inclusion of fungicide dynamics for two distinct pre-infection states: the disease free equilibrium (points E_{ex} and E_{im} in Figure 4) and a pre-infection state during where the uninfected hosts are still growing (points G_{ex} and G_{im} in Figure 4). The thresholds obtained are shown in Figure 5. We conclude that the explicit inclusion of fungicide dynamics leads to a zone of bistability for values of R_0 less than 1, and that this zone is not simply an artefact of the equilibrium analysis but is robust to the system's transient dynamics. Furthermore we show that for pre-infection states where the host population is still growing the zone of bistability is extended and lower initial levels of infected hosts (I_0) are able to invade and persist within the same range of R_0 values. In addition an extra invasion zone now exists for a range of values of R_0 below 1. Within this invasion zone, a pathogen is able to invade but not persist (see Figure 5).

Deterministic Realisations. Invasion trajectories for both models are shown in Figure 6. For initial conditions at point A ($R_0 = 0.4$, $I_0 = 0.06$), a value of R_0 is chosen so that $R_i < R_0 < R_c$ for the explicit model lies within the invasion zone for a pre-infection state with growing hosts. By considering the trajectories that result from this point it can clearly be seen that for the implicit model the choice of pre-infection state affects the pathogen's ability to invade. For initial conditions at points B ($R_0 = 0.8$, $I_0 = 0.06$) and C ($R_0 = 0.8$, $I_0 = 0.05$), a value of R_0 is chosen so that $R_c < R_0 < 1$ for the explicit model parameters. By considering the trajectories that result from these two points it can clearly be seen that for the implicit model the choice of initial infection level does not affect the ability of the pathogen to persist, whereas for the explicit model the initial infection level does affect the ability of the pathogen to persist. For initial conditions at point D ($R_0 = 1.8$, $I_0 = 0.05$) we can see that both models predict invasion for $R_0 > 1$,

but that they do not agree on the final endemic level of infection, with the explicit model predicting higher levels.

Stochastic Realisations. Using a nominal carrying capacity of 2000 ($\hat{\kappa} = 2000$), 5000 simulations are performed for every initial condition in the $I_0 - R_0$ parameter space ($I_0 \in [0, 0.01, \ldots, 1]$, $R_0 \in [0, 0.05, \ldots, 2.50]$, starting from a disease-free equilibrium level of uninfected hosts only) and the proportion of simulations that result in pathogen persistence is recorded and shown in Figure 7. The bistability zone still exists when allowance is made for demographic stochasticity. Figure 8 shows probability density plots obtained from the 5000 simulations for the initial conditions corresponding to points B ($I_0 = 0.06$, $R_0 = 0.8$), C ($I_0 = 0.05$, $R_0 = 0.8$) and D ($I_0 = 0.05$, $R_0 = 1.8$). It can be seen that for a stochastic framework, starting at initial conditions near to the deterministic threshold, individual trajectories may tend towards either equilibrium (endemic or disease free) regardless of whether the initial condition is above or below the deterministic persistence threshold and so we see distinctly bimodal probability density functions as a result. This is due to demographic stochastic effects allowing individual trajectories to enter different basins of attraction and so tend towards either equilibrium. Crucially, this means that by creating a stochastic version of the explicit model we have shown that not only is the bistability zone still present but the range of initial infection levels that can lead to persistence is extended. The robustness of the effects was tested for a range of population sizes ($\hat{\kappa} \in [200, \ldots, 20000]$) and shown to be sustained.

Discussion

In this paper we develop a simple model to investigate the effects of fungicide dynamics on pathogen invasion and persistence. Specifically we incorporate protectant fungicide dynamics into a conventional model in a way that separates out the effects of partial coverage, incomplete effectiveness and decay in activity of a protectant fungicide to create an alternative model that takes

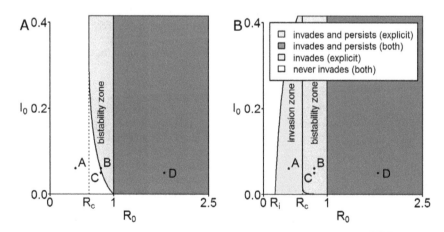

Figure 5. Invasion and persistence plots. Here we show the effects of R_0 and initial level of infection at invasion (I_0) on invasion and persistence of the pathogen. (A) Pathogen invades system at disease free equilibrium (points E_{ex} and E_{im} in Figure 4): for given R_0 and I_0 values the plot shows the long-term behaviour of both models in terms of the ability of a pathogen to invade and persist. For the implicit model R_0 completely characterises the long-term behaviour of the system (with $R_0 = 1$ being the threshold). For the explicit model, the ability of a pathogen to invade and persist is also determined by initial level of infection, I_0, for a range of R_0 values ($R_c < R_0 < 1$). (B) Pathogen invades system during growth phase (points G_{ex} and G_{im} in Figure 4): For the implicit model R_0 still completely characterises the long-term behaviour of the system. For the explicit model the bistability zone is extended and lower initial levels of infected hosts (I_0) are able to invade and persist within the same range of R_0 values ($R_c < R_0 < 1$). In addition, an extra invasion zone now exists for a range of values of R_0 ($R_i < R_0 < 1$) where here $R_i < R_c$. Within this zone the pathogen is able to invade but is not able to persist. Initial conditions for selected numerical simulations are shown. A:($R_0 = 0.4$, $I_0 = 0.06$) is within the invasion zone for the explicit model starting during the growth phase. B:($R_0 = 0.8$, $I_0 = 0.06$) is just above the invasion and persistence threshold for the explicit model staring from the disease free equilibrium. C:($R_0 = 0.8$, $I_0 = 0.05$) is just below the same threshold. D:($R_0 = 1.8$, $I_0 = 0.05$) is significantly above the invasion and persistence thresholds for both models. The default parameter values are given in Table 2.

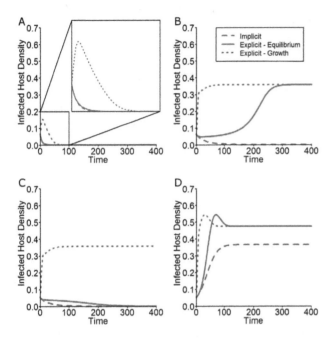

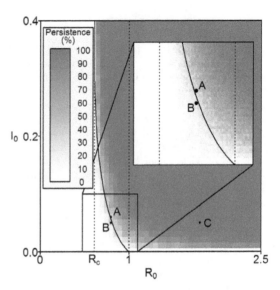

Figure 7. Persistence plot for the stochastic version of the explicit model. For given R_0 and I_0 values, the plot shows the proportion of stochastic simulations that resulted in pathogen invasion and persistence (500 simulations per point). It can be seen that persistence is still determined by I_0 for a range of R_0 values ($R_c < R_0 < 1$) and furthermore that simulations with initial conditions below the deterministic threshold are still able to persist (see inset). Initial conditions for selected numerical simulations are shown. A:($R_0 = 0.8$, $I_0 = 0.06$) is just above the deterministic invasion and persistence threshold for the explicit model. B:($R_0 = 0.8$, $I_0 = 0.05$) is just below the same threshold. C:($R_0 = 1.8$, $I_0 = 0.05$) is significantly above the invasion and persistence thresholds for both models. The default parameter values are given in Table 2.

Figure 6. Deterministic realisations for the implicit and explicit models. These plots correspond to the initial conditions in the I_0-R_0 parameter space shown in Figure 5. (A) For a trajectory starting at point A ($R_0 = 0.4$, $I_0 = 0.06$) both models predict the long term extinction of the pathogen but the explicit model predicts invasion of the pathogen for a pre-infection state given by point G_{ex} in Figure 4. (B) For a trajectory starting at point B ($R_0 = 0.8$, $I_0 = 0.06$) the implicit model predicts extinction, whereas the explicit model predicts persistence of the pathogen for both pre-infection states. (C) Both models predict the extinction of the pathogen for a system at disease-free equilibrium but the explicit model predicts the long-term persistence of the pathogen for a pre-infection state given by point G_{ex} in Figure 4. (D) Both models predict the persistence of the pathogen sarting from point D ($R_0 = 1.8$, $I_0 = 0.05$), regardless of the pre-infection state. The default parameter values are given in Table 2.

explicit account of the fungicide dynamics. Previous work has utilised simple models of fungicide application to investigate a number of issues e.g. fungicide resistance [9,13,19,30] and optimal control strategies [8–10], but the models used have not incorporated explicit fungicide dynamics (as the convention is to take implicit account of fungicide dynamics by mapping the effect onto the transmission rate).

The allowance for explicit fungicide dynamics markedly changes the inferences on both the invasion and persistence of the pathogen. Both frameworks appear superficially to have the same epidemiological properties and thresholds (R_0 is identical for both) but the explicit model exhibits a bistability zone for values of R_0 less than 1, a range that the implicit model considers completely safe from invasion. Previous work on human and animal vaccination models [29,31], re-infection models [32] and models of sexually transmitted diseases (with high and low transmissibility groups) [33] have exhibited similar bistability properties, but this is the first time that this property has been demonstrated for chemical control in agricultural systems. In addition previous epidemiological research has only considered the existence of a bistability zone for pathogens invading a system already at equilibrium. The extension of this result to systems where a pathogen invades a growing host state is a novel result with some striking repercussions. Not only does the inclusion of fungicide dynamics affect the criteria for invasion and persistence,

(increasing the risk of a pathogen successfully invading and persisting) when compared with models that only implicitly include fungicide dynamics, but if the initial infection occurs at a time before the host population has reached its equilibrium state then this risk is exacerbated. Now it is even more likely that a pathogen will be able to invade and persist, and also there is an opportunity for a pathogen to simply invade in the short term, causing an early epidemic before eventually dying out.

The existence of both a bistability zone and an invasion zone below the traditionally accepted invasion threshold has several important implications both in the use of models to guide disease control, and in the selection of fungicide traits to promote effective disease control. Firstly, for a non-negligible, initial level of infection, a pathogen may be able to invade a system with a fungicidal control regime that an implicit model would predict to be adequate. For example, protectant application rates, fungicide efficacies and fungicide activity decay rates may be chosen to reduce the R_0 value of the system below 1 yet unknowingly still remain above the critical thresholds $R_0 = R_c$, or $R_0 = R_i$. The system lies either in the bistability zone and is at risk from invasion by a high enough level of initial infection or it lies in the invasion zone and is at risk from an invasion during the host growth phase. The second implication is that for values of R_0 just above 1, the two models predict very different endemic levels of infection; for the implicit model, having an R_0 value just above 1 would always result in an invading pathogen achieving a negligible endemic equilibrium (see Figure 2) whereas for the explicit model an equivalent R_0 value will always lead to a much higher endemic equilibrium (see Figure 2). Consequently it would be possible for a system to experience sudden large invasions from small initial infections as a result of a small change in value of R_0 (from just

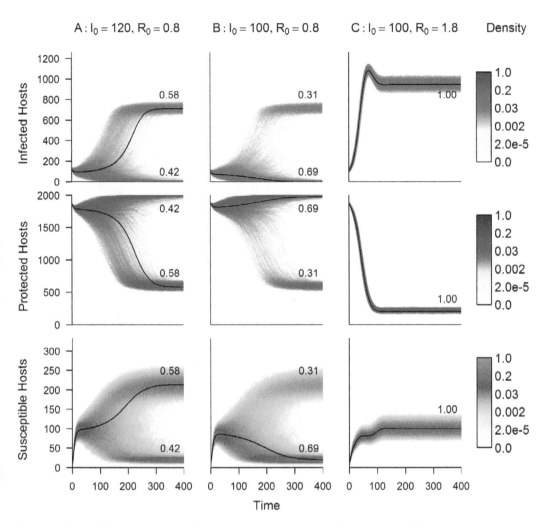

Figure 8. Probability density plots from the stochastic model. Here we show 5000 stochastic realisations of the explicit model that correspond to the initial condition points in the I_0-R_0 parameter space shown in Figure 7. Here the host carrying capacity, $\hat{\kappa}$ is 2000. The left hand figures show the probability density plots for all classes with initial conditions given by Point A ($\hat{I}_0 = 120$, $R_0 = 0.8$) i.e. just above the deterministic invasion threshold, the middle column shows the probability density plots for Point B ($\hat{I}_0 = 100$, $R_0 = 0.8$), just below the invasion threshold, and the right hand column shows the probability density plots for Point C ($\hat{I}_0 = 100$, $R_0 = 1.8$), which is significantly above the invasion threshold. The deterministic trajectory for the corresponding class and initial condition is overlaid on each plot. The numbers within each plot indicate the proportion of realisations that reach those final distinct host levels. Note that the density scale is non-linear. The default parameter values are given in Table 2.

below 1 to just above 1). Finally, in order to control an existing outbreak, it is necessary to reduce R_0 not just to below 1 but to below the critical threshold value R_c. The predicted effort required to eliminate an established pathogen from a system therefore depends on the choice of model. A model without explicit fungicide dynamics will underestimate the effort required and consequently allow the pathogen to persist.

Of particular concern for disease management strategies is the drive in the agrochemical industry to produce longer lasting, more effective fungicides, which necessitate lower application rates. These aspirational fungicide properties are exactly those that correspond to the parameter region for the explicit model in which bistability zones occur (Figure 3).

A continuous-time, Markov process version of the explicit model is constructed and used to demonstrate that the bistability result is not just a property of the deterministic models. The simulations show that the probability distributions for the stochastic model exhibit bimodal behaviour in the region of bistability, with one mode at the endemic equilibrium and one at the disease free equilibrium. It is clear that the effects of fungicide dynamics on invasion and persistence criteria are still maintained when allowance is made for stochastic variability in the various transmission processes.

Taking all of this into account it is reasonable to conclude that failure to account for fungicide dynamics naturally leads to erroneous predictions of control effectiveness.

Author Contributions

Conceived and designed the experiments: MDC. Performed the experiments: MDC. Analyzed the data: MDC. Wrote the paper: MDC CAG.

References

1. De Waard MA, Georgopoulos SG, Hollomon DW, Ishii H, Leroux P, et al. (1993) Chemical controls of plant diseases: problems and prospects. Annual Review of Phytopathology 31: 403–421.

2. Russell PE (2005) A century of fungicide evolution. Journal of Agricultural Science 143: 11–25.

3. Zilberman D, Schmitz A, Casterline G, Lichtenberg E, Siebert JB (1991) The economics of pesticide use and regulation. Science 253: 518–522.

4. Gilligan CA (2002) An epidemiological framework for disease management. Advances in Botanical Research 38: 1–64.

5. Gilligan CA, van den Bosch F (2008) Epidemiological models for invasion and persistence of pathogens. Annual Review of Phytopathology 46: 385–418.

6. van den Bosch F, Gilligan CA (2008) Models of fungicide resistance dynamics. Annual review of phytopathology 46: 123–147.

7. Dybiec B, Kleczkowski A, Gilligan CA (2004) Controlling disease spread on networks with incomplete knowledge. Phys Rev E 70: 66145.

8. Dybiec B, Kleczkowski A, Gilligan CA (2005) Optimising control of disease spread on networks. Acta Physica Polonica B 36: 1509–1526.

9. Hall RJ, Gubbins S, Gilligan CA (2007) Evaluating the performance of chemical control in the presence of resistant pathogens. Bulletin of Mathematical Biology 69: 525–537.

10. Forster GA, Gilligan CA (2007) Optimizing the control of disease infestations at the landscape scale. Proceedings of the National Academy of Sciences, USA 104: 4984–4989.

11. Reinink K (1986) Experimental verification and development of EPIPRE, a supervised disease and pest management system for wheat. European Journal of Plant Pathology 92: 3–14.

12. Brent KJ, Hollomon DW (2007) Fungicide Resistance: The assessment of Risk. Technical report, FRAC.

13. Birch CPD, Shaw MW (1997) When can reduced doses and pesticide mixtures delay the build-up of pesticide resistance? A mathematical model. Journal of Applied Ecology 34: 1032–1042.

14. Gubbins S, Gilligan CA (1999) Invasion thresholds for fungicide resistance: deterministic and stochastic analyses. Proceedings of the Royal Society of London Series B-Biological Sciences 266: 2539–2549.

15. Holt J, Chancellor TCB (1999) Modelling the spatio-temporal deployment of resistant varieties to reduce the incidence of rice tungro disease in a dynamic cropping system. Plant Pathology 48: 453–461.

16. Hall RJ, Gubbins S, Gilligan CA (2004) Invasion of drug and pesticide resistance is determined by a trade-off between treatment efficacy and relative fitness. Bulletin of Mathematical Biology 66: 825–840.

17. Chin KM (1987) A simple model of selection for fungicide resistance in plant pathogen populations. Phytopathology 77: 666–669.

18. Parnell S, Gilligan CA, van den Bosch F (2005) Small-scale fungicide spray heterogeneity and the coexistence of resistant and sensitive pathogen trains. Phytopathology 95: 632–639.

19. Parnell S, van den Bosch F, Gilligan CA (2006) Large-scale fungicide spray heterogeneity and the regional spread of resistant pathogen strains. Phytopathology 96: 549–555.

20. Levy Y, Levi R, Cohen Y (1983) Buildup of a pathogen subpopulation resistant to a systemic fungicide under various control strategies: a exible simulation model. The American Phytopathological Society 73: 1475–1480.

21. Shaw MW (1989) Independent action of fungicides and its consequences for strategies to retard the evolution of fungicide resistance. Crop Protection 8: 405–411.

22. Cunniffe NJ, Gilligan CA (2010) Invasion, persistence and control in epidemic models for plant pathogens: the effect of host demography. Journal of the Royal Society Interface 7: 439–451.

23. Gilligan CA, Truscott JE, Stacey AJ (2007) Impact of scale on the effectiveness of disease control strategies for epidemics with cryptic infection in a dynamical landscape: an example for a crop disease. Journal of the Royal Society Interface 4: 925–934.

24. Spink J, Blake J, Bingham I, Hoad S, Foulkes J, et al. (2006) The barley growth guide. Technical report, HGCA.

25. Sylvester-Bradley R, Berry P, Blake J, Kindred D, Spink J, et al. (2008) The wheat growth guide. Technical report, HGCA.

26. Blake J, Paveley N, Fitt BDL, Oxley S, Bingham I, et al. (2012) Barley disease management guide 2012. Technical report, HGCA.

27. Paveley N, Blake J, Gladders P, Cockerell V (2012) Wheat disease management guide 2012. Technical report, HGCA.

28. Gillespie DT (1977) Exact Stochastic Simulation of Coupled Chemical Reactions. The Journal of Physical Chemistry 81: 2340–2361.

29. Kribs-Zaleta CM, Velasco-Hernandez JX (2000) A simple vaccination model with multiple endemic states. Mathematical Biosciences 164: 183–201.

30. Milgroom M (1990) A stochastic Model for the intial occurence and development of fungicide resistance in plant pathogen populations. Phytopathology 80: 410–416.

31. Brauer F (2004) Backward bifurcations in simple vaccination models. Journal of Mathematical Analysis and Applications 298: 418–431.

32. Safan M, Heesterbeek H, Dietz K (2006) The minimum effort required to eradicate infections in models with backward bifurcation. Journal of mathematical biology 53: 703–718.

33. Hadeler KP, Castillo-Chavez C (1995) A core group model for disease transmission. Mathematical Biosciences 128: 41–55.

12

The Emergence of Resistance to Fungicides

Peter H. F. Hobbelen[1], Neil D. Paveley[2], Frank van den Bosch[1]*

1 Rothamsted Research, Harpenden, Hertfordshire, United Kingdom, **2** ADAS UK Ltd, High Mowthorpe, Duggleby, Malton, North Yorkshire, United Kingdom

Abstract

Many studies exist about the selection phase of fungicide resistance evolution, where a resistant strain is present in a pathogen population and is differentially selected for by the application of fungicides. The emergence phase of the evolution of fungicide resistance - where the resistant strain is not present in the population and has to arise through mutation and subsequently invade the population - has not been studied to date. Here, we derive a model which describes the emergence of resistance in pathogen populations of crops. There are several important examples where a single mutation, affecting binding of a fungicide with the target protein, shifts the sensitivity phenotype of the resistant strain to such an extent that it cannot be controlled effectively ('qualitative' or 'single-step' resistance). The model was parameterized for this scenario for *Mycosphaerella graminicola* on winter wheat and used to evaluate the effect of fungicide dose rate on the time to emergence of resistance for a range of mutation probabilities, fitness costs of resistance and sensitivity levels of the resistant strain. We also evaluated the usefulness of mixing two fungicides of differing modes of action for delaying the emergence of resistance. The results suggest that it is unlikely that a resistant strain will already have emerged when a fungicide with a new mode of action is introduced. Hence, 'anti-emergence' strategies should be identified and implemented. For all simulated scenarios, the median emergence time of a resistant strain was affected little by changing the dose rate applied, within the range of doses typically used on commercial crops. Mixing a single-site acting fungicide with a multi-site acting fungicide delayed the emergence of resistance to the single-site component. Combining the findings with previous work on the selection phase will enable us to develop more efficient anti-resistance strategies.

Editor: Joy Sturtevant, Louisiana State University, United States of America

Funding: Rothamsted Research receives support from the Biological and Biotechnological Research Council of the United Kingdom. Part of this research was funded by the Department for Environment, Food and Rural Affairs of the United Kingdom (project number PS2712). The funders had no role in study design, data collection and analysis, decision to publish, or preparation of the manuscript.

Competing Interests: The authors have declared that no competing interests exist.

* E-mail: frank.vandenbosch@rothamsted.ac.uk

Introduction

The evolution of fungicide resistance can be divided into an emergence phase and a selection phase [1,2,3]. In the emergence phase, the resistant strain has to arise through mutation and subsequently invade the pathogen population. In this phase, the number of fungicide resistant lesions is very small and the resistant strain may become extinct due to stochastic variation, in spite of fungicide applications providing the resistant strain with a higher fitness than the sensitive strain. The length of the emergence phase (emergence time) can be defined as the time from the introduction of a new fungicide mode of action until the resistant strain succeeds in building up a large enough sub-population so that it is unlikely to die out due to chance. The evolution of resistance then enters the selection phase in which the application of fungicides increases the frequency of the resistant strain in the pathogen population [1,3].

Fungicide resistance management strategies aim to delay the evolution and spread of resistance in a sensitive pathogen population, while ensuring effective disease control. Due to the differences in the dynamics of the resistant strain between the emergence phase and the selection phase, the usefulness of resistance management strategies may also differ between these two phases. For example, in the selection phase, the frequency of resistance in the pathogen population will generally increase faster for higher dose rates of the fungicide [3]. However, in the emergence phase, there are two opposing effects of dose on resistance evolution: A high dose rate of a fungicide (close to, or at the label recommended dose) may delay the emergence of resistance by reducing the size of the sensitive pathogen population and therefore the number of resistant mutants produced per unit time. However, the smaller pathogen population will reduce the competition between the sensitive and the resistant strain for healthy host tissue to infect and may therefore increase the probability that the resistant mutant invades the pathogen population (Fig. 1). We therefore hypothesize that the choice of dose rate of a fungicide in the emergence phase may change the emergence time in a number of different ways (Fig. 2). If the emergence time is most sensitive to changes in the number of mutations produced per time unit, the emergence time will increase with increasing dose rate of the fungicide. However, if the emergence time is most sensitive to changes in the strength of competition for healthy leaf area, the emergence time will decrease with increasing dose rate of the fungicide.

There is a range of experimental studies on the development of resistance in response to the dose rates of a fungicide and the mixing or alternation of fungicides [4,5,6,7,8,9,10,11,12]. However, in many of these studies resistant strains were either introduced [4,6,13] or were already present at a significant frequency at the start of experiments [7,9,14]. As even a frequency of 1% represents a large population of resistant lesions, these studies describe the selection phase in the evolution of fungicide

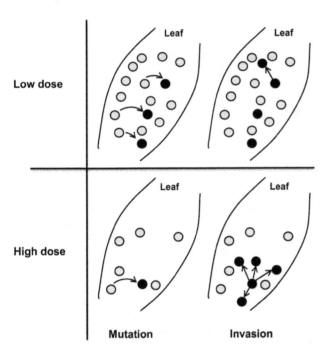

Figure 1. Fungicide dose and resistance emergence. The effect of the dose rate of a fungicide on the rate at which resistant lesions (black circles) arise through mutation and subsequently invade a sensitive pathogen population (grey circles). Curved arrows in the left subfigures represent mutation events and straight arrows in the right subfigures represent the colonization of new leaf area by the resistant lesions that arose through mutation in the left subfigures. This figure was adapted from figure 8 in [3].

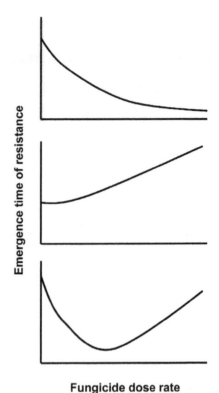

Figure 2. The shape of dose-emergence time curves. Possible ways in which the dose rate of a high-risk fungicide may affect the emergence time of resistance in a sensitive pathogen population. This figure was adapted from figure 9 in [3].

resistance. The effect of fungicide treatment strategies on emergence time can therefore not be determined from the experimental literature.

There are some papers in the biomathematical literature studying the emergence of, what is called, escape mutants [15,16,17]. The models and methods developed give insight into the life-cycle parameters that are of key relevance for the emergence of new pathogen strains. It is however not possible to use these results to study the emergence of fungicide resistance due to the absence of seasonality in host density, and the fact the selection pressure (the fungicide) is not constant through time. Previously we have shown that in the selection phase this periodicity of the host density and the time dependence of the selection pressure are key to understand both the qualitative and quantitative relation between selection for fungicide resistance and fungicide application regimes [1,18,19,20].

To our knowledge, no models have been published that account for the time dependence of key processes as well as the stochastic nature of resistant mutants arising and reproducing to invade a sensitive pathogen population in the emergence phase of the evolution of fungicide resistance. This also holds for models of insecticide and herbicide resistance. The aim of this study was therefore to develop a model for the emergence phase in the evolution of fungicide resistance, which describes the effect of fungicides on mutation and invasion. The model was derived from a successfully tested fungicide resistance model [20] describing the section phase and then parameterized for *Mycosphaerella graminicola* on winter wheat (*Triticum aestivum*).

To show how this model could be used to evaluate resistance management strategies, we determined the effect of the dose rate of a high resistance risk fungicide on the emergence time of resistance in a population of *M. graminicola* on winter wheat for different mutation probabilities, fitness costs of resistance and sensitivity levels of the resistant strain. We also evaluated the usefulness of mixing a high-risk fungicide with a low-risk fungicide for delaying the emergence of resistance. For the analyses in this paper we define a high-risk fungicide as a fungicide prone to substantial efficacy reduction due to a single mutation in the pathogen strain, such that selection for the resistant strain will eventually result in ineffective disease control by the high-risk fungicide used alone. We define a low-risk fungicide as a fungicide for which resistance does not evolve in the pathogen population in the time frame under consideration, but with efficacy that is too low to provide sufficient disease control on its own. These high and low-risk fungicides might typically represent single-site and multi-site acting substances.

Materials and Methods

Type of resistance described by the model

We developed a model to describe the emergence of resistance to high-risk fungicides which are prone to substantial efficacy reduction due to a single mutation in the pathogen strain. We assume that this single mutation decreases the sensitivity of the resistant pathogen strain to such an extent that the high-risk fungicide loses its ability to provide sufficient disease control of a pathogen population dominated by the resistant strain. The average difference in sensitivity between the sensitive and resistant pathogen populations is then much larger than the difference in sensitivity within these two populations. In that case, it is reasonable to represent the pathogen population as consisting of one sensitive and one resistant strain. This type of qualitative resistance development has occurred in response to, for example, methyl benzimidazole carbamate (MBC) [21] and quinone outside inhibitor (QoI) fungicides [22,23].

An overview of the model structure

The model describes a resistant strain arising by mutation and reproduction in a sensitive population of *M. graminicola* on winter wheat in response to solo use of a high-risk fungicide or mixtures of a high-risk and a low-risk fungicide. The main part of the model, which describes the emergence of resistance within growing seasons, consists of a deterministic and a stochastic sub-model (Fig. 3). The deterministic sub-model describes the dynamics of the crop canopy, the sensitive pathogen strain and the variation in the fungicide concentrations in time. This part of the model was derived from a successfully tested fungicide resistance model for the selection phase of fungicide resistance development [20]. It was not necessary to use a stochastic sub-model to describe the seasonal dynamics of the sensitive strain, because the sensitive stain is always present in high enough densities to prevent extinction due to random processes and is well represented by the mean of the process. Using a deterministic sub-model had the advantage of a much shorter simulation times. A stochastic sub-model was used to describe the dynamics of the resistant strain, because the population of resistant lesions is very small and random processes may lead to the extinction of this strain. We assume that the frequency of the resistant strain in the pathogen population is too small to affect the dynamics of the sensitive strain through competition for healthy leaf area.

The structure of the deterministic sub-model

See Figure 3 for a graphical presentation of the model structure.

The dynamics of the crop canopy. The model predicts the seasonal dynamics of the canopy in order to account for the availability of healthy leaf area on the growth of the pathogen population. The canopy consists of the combined areas of leaves 1–3 (counting down from the flag leaf, which is designated leaf 1), because this leaf area intercepts the sprayed fungicides. Hereafter, we refer to leaves 1–3 as the "upper leaves" and refer to leaves further down the stem as "lower leaves". We use the term "density" to refer to leaf area per area of ground. The density of the total leaf area (A) is the sum of the densities of healthy, infected and dead leaf area and increases according to the monomolecular equation [24]:

$$\frac{dA}{dt} = k(A_{max} - A) \tag{1}$$

In the absence of disease, the seasonal dynamics of the healthy leaf area (H) consist of a growth phase, followed by a plateau and subsequently a senescence phase. The growth phase ends when the flag leaf has completely emerged (GS 39 on Zadoks' scale). Senescence starts at anthesis (GS 61) and is complete at the end of grain filling (GS 87). The density of healthy leaf area in the absence of disease is described by the equation

$$\frac{dH}{dt} = k(A_{max} - A) - \sigma(t)H \tag{2}$$

where $\sigma(t)$ represents the senescence rate. The senescence rate increases exponentially from approximately 0 ($<10^{-7}$) at GS 61 to a maximum value of 0.105 at GS 87 according to the function:

$$\sigma(t) = \begin{cases} 0, & t < t_{GS61} \\ 0.005\left(\dfrac{t - t_{GS61}}{t_{GS87} - t_{GS61}}\right) + 0.1e^{-0.02(t_{GS87} - t)}, & t \geq t_{GS61} \end{cases} \tag{3}$$

This reduces the healthy leaf area at GS 87 to $<1\%$ of the maximum leaf area, which approximates complete senescence.

The sensitive pathogen population. The lifecycle of the sensitive pathogen strain is divided into a latent stage (L_s) with length $1/\delta_s$ and an infectious stage (I_s) with length $1/\mu$. During the latent stage, the pathogen grows within the intercellular space in leaf tissues and senescence decreases the density of latent leaf area. At the start of the infectious stage, the pathogen kills the host cells and starts spore production. The rate at which infectious leaf area generates latent leaf area is determined by the product of i) the spore production rate per unit of infectious leaf area (ϕ), ii) the probability of a spore landing on the upper leaves of the canopy, iii) the probability of landing on healthy leaf area, given that a spore lands on the upper leaves (H/A), iv) the infection efficiency, and v) the area occupied by a lesion, which develops after the successful infection of healthy leaf area by one spore (ψ). Points ii and iv are incorporated in compound parameter ε_s. Hereafter, we refer to this parameter as the infection efficiency.

At the beginning of a growing season, the canopy becomes infected by spores from infectious lesions on lower leaves. The density of leaf area occupied by the infectious stage of the sensitive strain () decreases according to the function

$$\frac{dF_s}{dt} = -\lambda F_s \tag{4}$$

with parameter λ representing the loss rate of infectious leaf area on lower leaves. To constrain complexity, the model is not spatially explicit, hence spores produced by sensitive lesions have the same probability of landing on upper leaves (included in compound parameter) whether they are produced on lower or upper leaves. This leads to the following equations to describe the dynamics of the sensitive pathogen population:

$$\frac{dH}{dt} = \kappa(A_{max} - A) - \phi\varepsilon_s\psi\left(\frac{H}{A}\right)(F_s + I_s) - \sigma(t)H \tag{5}$$

$$\frac{dL_s}{dt} = \phi\varepsilon_s\psi\left(\frac{H}{A}\right)(F_s + I_s) - \delta_s L_s - \sigma(t)L_s \tag{6}$$

$$\frac{dI_s}{dt} = \delta_s L_s - \mu I_s \tag{7}$$

The impact of fungicides on the sensitive pathogen strain. Both the low-risk and high-risk fungicides were represented as having protectant activity, which reduced the infection efficiency of the sensitive strain (ε_s). The high-risk fungicide was also represented as having eradicant activity. Eradicant activity was defined here as the ability of the fungicide to slow fungal growth during the latent period (fungistatic activity), rather than converting latent leaf area back into healthy tissue. Thus, symptom expression was delayed or prevented during the life of the crop canopy. This representation fits well with the observation that fungicides with 'eradicant' activity can only provide effective

Deterministic submodel   Stochastic submodel

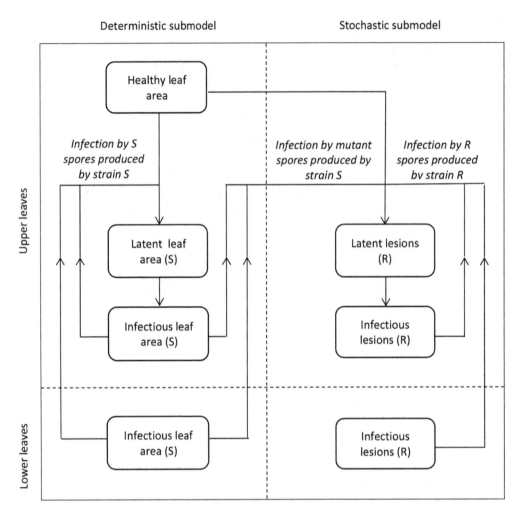

Figure 3. The structure of the simulation model. The model describes the emergence of a resistant pathogen strain (R) in a sensitive population (S) of *M. graminicola* on winter wheat in response to applications of a high-risk fungicide.

control of visible symptoms if treatments are applied prior to half way through the latent period. We therefore assumed that the eradicant activity of the high-risk fungicide increased the length of the latent stage of the sensitive strain $(1/\delta_s)$. The infection efficiency depends on the concentrations of the low-risk (C_A) and high-risk fungicides (C_B) according to the function:

$$\varepsilon_s = \varepsilon\left(1 - \alpha_A\left(1 - e^{-\beta_A C_A}\right)\right)\left(1 - \alpha_{B,s}\left(1 - e^{-\beta_B C_B}\right)\right) \quad (8)$$

The length of the latent stage depends on the concentration of the high-risk fungicide according to the function:

$$\frac{1}{\delta_s} = \frac{1}{\delta}\left(1 - \alpha_{B,s}\left(1 - e^{-\beta_B C_B}\right)\right) \quad (9)$$

In these equations, parameters ε and $1/\delta$ represent the infection efficiency and the length of the latent stage of the sensitive pathogen strain in the absence of fungicides, respectively. Parameter α_A represents the maximum possible reduction of the infection efficiency by the low-risk fungicide and parameter β_A determines the curvature of the dose-response curve. Parameter

$\alpha_{B,s}$ represents the maximum possible reduction of the life-cycle parameters of the sensitive strain by the high-risk fungicide and parameter β_B determines the curvature of the dose-response curve. The concentrations of the low-risk (C_A) and high-risk fungicides (C_B) decay in time according to the functions:

$$\frac{dC_A}{dt} = -v_A C_A \quad (10)$$

$$\frac{dC_B}{dt} = -v_B C_B \quad (11)$$

with v_A and v_B representing the decay rates of the low-risk and high-risk fungicides, respectively.

This leads to an exponential decay of the activity of the fungicides in time, with the loss of activity highest just after application of the fungicides. To explore the effect of the type of function used to describe the decay of the activity of fungicides on the emergence time, we also determined emergence times using a gamma distribution for the rate of fungicide decay in time. (Text S1). This change did not affect the qualitative conclusions about the effect of the dose rate of a high-risk fungicide on the emergence of resistance or the usefulness of mixtures of a low-risk and a high-risk fungicide for delaying the emergence of resistance to the high-risk fungicide (Text S1).

Fungicide dose response curves

Figure 4 shows the model predictions for the loss of healthy area duration, an indicator of yield [25], due to an average epidemic *M. graminicola* on winter wheat in the United Kingdom as a function of the dose rate in case of solo use of the low-risk fungicide and solo use of the high-risk fungicide for the scenario assuming exponential decay of fungicides.

The structure of the stochastic sub-model

We used a modified version of Gillespie's stochastic simulation algorithm [26] to simulate the dynamics of the number of lesions of the resistant strain within a growing season. To use this algorithm, firstly, all possible events in the model which change the number of resistant lesions are labelled. Secondly, equations must be derived for the average rates at which these events occur at a given time t within a growing season. If N represents the total number of events, we labelled each event and corresponding event rate as E_i and R_i, respectively, with subscript $i \in [1, N]$. The modified version of Gillespie's stochastic simulation algorithm calculates the size of the next time step (t_{step}) as the maximum step that can be taken while satisfying the following condition for the change in event rates (R_i) during a time step:

$$|R_i(t + t_{step}) - R_i(t)| \leq \eta |R_i(t)| \qquad (12)$$

This condition limits the absolute change in any event rate R_i during a time step to a certain fraction η of the value of this event

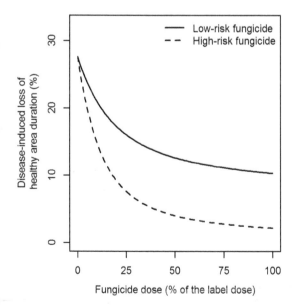

Figure 4. Fungicide dose response curves. The effect of the dose rate of the low-risk and high-risk fungicide (solo use) on the disease-induced loss of healthy area duration for a sensitive population of *M. graminicola* on winter wheat. Healthy area duration was calculated as the area under the green leaf area curve from anthesis to the end of the growing season and is an indicator of the yield loss of winter wheat [25]. We assumed an average epidemic of *M. graminicola* for the United Kingdom in the absence of fungicide applications. Fungicides were applied twice during a growing season (see text) at a constant dose rate.

rate at time t in order to ensure the accuracy of the stochastic simulation algorithm [26]. Parameter η is conventionally set to 0.03 [26] and we used this value in our simulations. The number of times that an event E_i occurs during a time step is subsequently determined by drawing from a Poisson distribution with mean $R_i(t) t_{step}$. Below, we describe the events which change the number of lesions of the resistant strain and derive equations for the average rates of these events.

Events and event rates. The stochastic sub-model describes the variation in the number of lesions of the resistant pathogen strain in time. The life-cycle of the resistant strain is divided into a latent (L_r) stage with length ($1/\delta_r$) and an infectious stage (I_r) with length ($1/\mu$). Latent lesions may die as a result of senescence. New latent lesions form as a result of infection by mutant spores from the sensitive strain and by spores from infectious lesions of the resistant strain on upper (I_r) and lower leaves (F_r). This leads to a total of six events, which can change the number of lesions of the resistant strain on upper and lower leaves.

The first event (E_1) is the successful infection of healthy leaf area by a mutant spore produced by the sensitive strain. Mutant spores can be produced by the sensitive pathogen population on upper and lower leaves. The number of new latent lesions generated per time unit is determined by the product of i) the leaf area occupied by infectious lesions of the sensitive strain, ii) the spore production rate per unit of infectious leaf area (ϕ), iii) the probability of a spore having the resistant genotype (θ) iv) the probability of a spore landing on the upper leaves of the canopy, v) the probability of landing on healthy leaf area, given that a spore lands on the upper leaves (H/A), and vi) the infection efficiency of a resistant spore. The amount of infectious leaf area is determined by the product of the density of infectious leaf area ($F_s + I_s$) and the size of the wheat growing area (o). Points iv and vi are incorporated in compound parameter ε_r. We assumed that spores produced by resistant lesions have the same probability of landing on upper leaves (included in compound parameter) whether they are produced on lower or upper leaves. The average number of new latent lesions generated per time unit by infectious leaf area on both upper and lower leaves is then:

$$R_1 = o\phi\theta\varepsilon_r \left(\frac{H}{A}\right)(F_s + I_s) \qquad (13)$$

If this event occurs, the number of latent lesions of the resistant strain increases by one.

The second event (E_2) is the successful infection of healthy leaf area by a spore from an infectious lesion of the resistant strain. The average number of latent lesions generated per time unit by infectious lesions of the resistant strain on both upper and lower leaves is determined by the product of i) the total number of infectious lesions of the resistant strain ($F_r + I_r$), ii) the spore production rate per infectious lesion ($\phi\psi$), iii) the probability of a spore landing on the upper leaves of the canopy, iv) the probability of landing on healthy leaf area, given that a spore lands on the upper leaves (H/A) and v) the infection efficiency of a resistant spore. Points iii and v are incorporated in compound parameter ε_r. The average number of new latent lesions generated per time unit by infectious lesions of the resistant strain is then:

$$R_2 = \phi\psi\varepsilon_r \left(\frac{H}{A}\right)(F_r + I_r) \qquad (14)$$

If this event occurs, the number of latent lesions of the resistant strain increases by one.

The third event (E_3) is the senescence of a latent lesion (L_r). The average number of latent lesions which die per time unit as a result of senescence of green leaf area is

$$R_3 = \sigma(t)L_r \tag{15}$$

with senescence rate $\sigma(t)$. If this event occurs, the number of latent lesions of the resistant strain decreases by one.

The fourth event (E_4) is the transition from a latent (L_r) to an infectious lesion (I_r). The average number of latent lesions which becomes infectious per unit of time is

$$R_4 = \delta_r L_r \tag{16}$$

with development rate δ_r equal to the inverse of the length of the latent period. If this event occurs, the number of latent lesions of the resistant strain decreases by one and the number of infectious lesions of the resistant strain on upper leaves increases by one.

The fifth event (E_5) is the death of an infectious lesion on lower leaves (F_r). The average number of infectious lesions on lower leaves dying per unit of time as a result of reaching the end of the infectious period is

$$R_5 = \mu F_r \tag{17}$$

with mortality rate μ equal to the inverse of the length of the infectious period.

If this event occurs, the number of infectious lesions of the resistant strain on lower leaves decreases by one.

The sixth and last possible event (E_6) is the death of an infectious lesion on upper leaves (I_r). The average number of infectious lesions on upper leaves dying per time unit as a result of reaching the end of the infectious period is

$$R_6 = \mu I_r \tag{18}$$

with mortality rate μ equal to the inverse of the length of the infectious period. If this event occurs, the number of infectious lesions of the resistant strain on upper leaves decreases by one.

Fitness costs of resistance and the impact of fungicides on the resistant strain. In the stochastic sub-model, which describes the dynamics of the resistant strain, fitness costs of resistance to the high-risk fungicide were assumed to reduce the infection efficiency of the resistant strain by a fraction ω. The protectant low-risk fungicide was assumed to reduce the infection efficiency of the resistant strain (ε_r). When resistance to the high-risk fungicide was represented as being partial (incomplete), the protectant activity of the high-risk fungicide reduced the infection efficiency and the eradicant activity of the high-risk fungicide increased the length of the latent stage of the resistant strain, $1/\delta_r$. The infection efficiency of the resistant strain depends on the fitness costs of resistance and the concentrations of the low-risk (C_A) and high-risk fungicides (C_B) according to the function:

$$\varepsilon_r = \varepsilon(1-\omega)\left(1-\alpha_A\left(1-e^{-\beta_A C_A}\right)\right)\left(1-\alpha_{B,r}\left(1-e^{-\beta_B C_B}\right)\right) \tag{19}$$

The length of the latent stage of the resistant strain depends on the concentration of the high-risk fungicide according to the function:

$$\frac{1}{\delta_r} = \frac{1}{\delta}\left(1-\alpha_{B,r}\left(1-e^{-\beta_B C_B}\right)\right) \tag{20}$$

In these equations, parameter $\alpha_{B,r}$ represents the maximum possible reduction of the life-cycle parameters of the partially resistant strain by the high-risk fungicide. In case of complete resistance, $\alpha_{B,r}=0$.

The number of infectious lesions of the resistant strain at the start of a new growing season. The epidemic on the upper leaves is initiated by spores from infectious lesions on lower leaves. Which of these infectious lesions are resistant to the fungicide is determined by drawing from a binomial distribution $B(n,p)$ with mean np and variance $np(1-p)$. Parameter n of the binomial distribution is the total number of infectious lesions on lower leaves at the start of a growing season:

$$n = F_0\left(\frac{o}{\psi}\right) \tag{21}$$

Parameter p of the binomial distribution is the probability of an infectious lesion on lower leaves being resistant to the fungicide, at the start of a growing season. This probability was set to the fraction of infectious lesions at the end of the previous growing season, which was resistant to the high-risk fungicide:

$$p = \frac{I_r^-}{I_s^-(o/\psi)+I_r^-} \tag{22}$$

In the emergence phase the frequency of resistance (p) is very low and $np(1-p)\approx np$. As a result the mean and variance of the binomial distribution are approximately the same and amount to:

$$np = F_0(o/\psi)\left(\frac{I_r^-}{I_s^-(o/\psi)+I_r^-}\right) = F_0\left(\frac{I_r^-}{I_s^-+\frac{I_r^-}{(o/\psi)}}\right) \approx F_0\left(\frac{I_r^-}{I_s^-}\right) \tag{23}$$

This simplification can be made, because the number of resistant lesions at the end of the previous growing season (I_r^-) is much smaller than the ratio of the total wheat growing area and the area occupied by a single lesion, o/ψ. It follows that the size of the wheat growing area (o) has a negligible effect on the number of infectious lesions of the resistant strain at the start of the next growing season, because it does not occur in the approximation on the right side of the equation.

Parameter values

The definitions, values and dimensions of the model parameters are given in Table 1. The values of all parameters were the same as we have used before [18,19,20], except for the spore production rate per infectious leaf area (ϕ), the area occupied by one *M. graminicola* lesion (ψ), the infection efficiency (ε), (previously multiplied to give the transmission rate, but for the stochastic sub-model included as separate parameters), the size of the wheat growing area (o) and the probability that a spore produced by the sensitive strain carries a resistant mutation (θ). The spore production rate per infectious leaf area was calculated by dividing the total number of spores produced during the infectious period per infectious leaf area [27] by the length of the infectious period in degree- days [18]. The area occupied by one *M. graminicola* lesion (ψ) was taken from [28]. The product of the infection efficiency (ε), the spore production rate per infectious leaf area (ϕ) and the area occupied by one *M. graminicola* lesion (ψ) has the same value as the transmission rate parameter (ρ) in [18]. The infection efficiency (ε) was therefore calculated as $\varepsilon = \rho/(\phi\psi)$. We set the size of the wheat growing area to 350,000 km^2, which reflects the size of the winter wheat growing area in Europe during the years 2000–2008 (Eurostat, the statistical office of the European Union). We subsequently chose the value of parameter θ such that the median emergence time at half dose rates of the fungicide amounted to ten years for the default scenario (see below). This resulted in mutation probability θ amounting to $1.13 \cdot 10^{-16}$. We used ten years, because this is an average time for the period from the introduction of a high-risk fungicide on the market until the first detection of resistant strains in European crops [29]. It should be noted that the emergence of resistance is influenced by the product of the size of the parameters o and θ (Eq. 13). There are many combinations of the values of parameters o and θ which result in the same product. To avoid confusion, it should be noted that the mutation probability θ is not the same as the mutation rate. In this paper, we define the mutation rate as the number of mutant spores with a resistant genotype produced per time by the sensitive pathogen population.

Emergence criterion

We define the emergence time as the number of growing seasons from the introduction of a new fungicide mode of action on the market until the resistant subpopulation has reached a size that makes extinction due to stochastic processes unlikely. We used a fixed emergence threshold of 30 resistant lesions at the start of a growing season to determine whether the size of the resistant subpopulation was large enough to have emerged. At or above this emergence threshold, the probability of the resistant strain becoming extinct during a period of 100 years in the absence of new mutations varied between 0–3%, depending on the simulated scenario (see below). To assess the accuracy of this emergence threshold we defined an alternative emergence threshold and compared results. For the alternative, we calculated emergence thresholds specific for each scenario as the lowest possible number of resistant lesions at the start of a growing season for which the probability of the resistant strain becoming extinct during a period of 100 years in the absence of new mutations was < 5%. Emergence times were very similar for both types of emergence thresholds (Text S1). The emergence times for all simulations in the main text are determined using a fixed threshold of 30 resistant lesions at the start of a growing season. This corresponds to a frequency of resistance in the total pathogen population at the start of the growing season of $2.4 \cdot 10^{-11}$%.

Simulations

Having parameterized the model, we first simulated the dynamics of the resistant strain during a 1000-year period before the introduction of the fungicide to study the possibility that a resistant strain might already be present in the pathogen population at numbers above the emergence threshold, prior to the introduction of the fungicide on the market. We subsequently determined the effect of the dose rate of the high-risk fungicide on the emergence time of the resistant strain, for different values of the mutation probability, fitness costs of resistance and the sensitivity of the resistant strain to the fungicide. In all scenarios, the high-risk fungicide was applied twice during each growing season which corresponds to a standard UK treatment pro-

gramme [30]. The first spray each season was applied at the full emergence of leaf 3 (approximately GS 32) and the second spray was applied at complete emergence of leaf 1 (GS 39), counting down from the flag leaf. For each scenario, we varied the dose rate of the high-risk fungicide from 10% to 100% of the label recommended dose in steps of 10% and performed 5000 simulations per dose rate. The median emergence time was stable for this number of repetitions (Text S1).

We first determined the effect of the dose rate of the high-risk fungicide on the emergence time for the default scenario. In the default scenario, i) fitness costs were assumed to reduce the infection efficiency of the resistant strain by 10%, ii) resistance to the high-risk fungicide was assumed to be complete, and iii) the mutation probability was chosen such that the median emergence time at half dose rates was 10 years. To determine the effect of variations in the mutation probability on the emergence time, we performed simulations with a mutation probability amounting to 0.1, 0.2, 5 and 10 times the mutation probability in the default scenario. To determine the effect of variation in the fitness costs of resistance on the emergence time, we performed simulations with the reduction of the infection efficiency increasing from 2.5 to 15% in steps of 2.5%, while keeping the values of other parameters the same as the default scenario. Finally, to determine the effect of variation in the sensitivity of the resistant strain to the high-risk fungicide, we performed simulations with the maximum reduction of life-cycle parameters by the high-risk fungicide increasing from 5 to 20% in steps of 5% ($\alpha_{B,r}$ varying from 0.05 to 0.2 in steps of 0.05, respectively), while keeping the values of other parameters the same as the default scenario.

We also determined the effect of mixing a low-risk with a high-risk fungicide on the emergence time of resistance for the default scenario only. Both fungicides were applied twice during each growing season, as described above. We varied the dose rates of both the low-risk and high-risk fungicides in the mixture from 10% to 100% of the label recommended dose in steps of 10%. We determined the emergence time for all possible combinations of these dose rates and performed 5000 simulations per combination of dose rates.

Results

The dynamics of the resistant strain in the absence of fungicides.

In the absence of fungicide, the resistant strain arises in the pathogen population through mutation, but fitness costs of resistance prevent it from building up a large number through drift. The temporal dynamics of the resistant strain in the absence of fungicides is characterized by the alternation of short periods during which the resistant strain is present, with periods during which the resistant strain is absent (Fig. 5). Increasing the mutation probability increases the percentage of time that the resistant strain is present in the pathogen population, while increasing the fitness costs of resistance decreases this percentage. The probability of the resistant strain having already emerged when a new fungicide mode of action is introduced equals the probability that the number of resistant lesions at the start of a growing season exceeds the emergence threshold (30 resistant lesions) in the absence of fungicides. This probability amounted to $\leq$ 0.16% for all simulated values of the mutation probability and fitness costs of resistance.

Table 1. The definitions, values and dimensions of model parameters[a].

Parameters	Definition	Value	Dimension
Host			
γ	Growth rate of leaf area	$1.26 \cdot 10^{-2}$	t^{-1}[b]
A_{max}	Maximum density of leaf area	4.1	leaf area per area of ground
$\sigma(t)$	Senescence rate	Eq. 3	t^{-1}
	The size of the wheat growing area	$3.5 \cdot 10^5$	km^2
All pathogen strains			
λ	Loss rate of infectious leaf area/lesions on lower leaves[c]	$8.5 \cdot 10^{-3}$	t^{-1}
ψ	The area occupied by one lesion	$0.3 \cdot 10^{-10}$	km^2
ϕ	Spore production rate per unit of infectious leaf area	$7.3 \cdot 10^{12}$	$t^{-1} km^{-2}$
ε	Infection efficiency in the absence of fungicides and fitness costs of resistance[d]	$9.5 \cdot 10^{-5}$	_[e]
$1/\delta$	Length of the latent stage in the absence of fungicides	266	T
$1/\mu$	Length of the infectious stage	456	T
Sensitive pathogen strain			
F_0	Initial density of infectious lesions on lower leaves[c]	$1.09 \cdot 10^{-2}$	leaf area per area of ground
ε_s	Infection efficiency in the presence of fungicides	Equation 8	t^{-1}
$1/\delta_s$	Length of the latent stage in the presence of fungicides	Equation 9	T
θ	Mutation probability	Variable[f]	_[e]
Resistant pathogen strain			
ε_r	Infection efficiency in the presence of fungicides and/or fitness costs of resistance	Equation 18	t^{-1}
$1/\delta_r$	Length of the latent stage in the presence of fungicides	Equation 19	T
ω	The fraction by which the infection efficiency of the resistant strain is reduced due to fitness costs of resistance	Variable[f]	_[e]
Dose-response curve and decay rate parameters			
α_A	Maximum reduction of the infection efficiency of the sensitive and resistant strain by the low-risk fungicide	0.48	_[e]
$\alpha_{B,s}, \alpha_{B,r}$	Maximum reduction of the life-cycle parameters of the sensitive ($\alpha_{B,s}$) and resistant strain ($\alpha_{B,r}$) by the high-risk fungicide	1, variable[f]	_[e]
β_A, β_B	Curvature parameter of the dose-response curve for the low-risk (β_A) and high-risk fungicide (β_B)	9.9, 9.6	_[e]
ν_A, ν_B	Decay rate of the low-risk (ν_A) and high-risk fungicide (ν_B)	$6.9 \cdot 10^{-3}$, $1.1 \cdot 10^{-2}$	t^{-1}

[a]Parameter values were taken from [18], except for parameters , ψ, ϕ and θ. The estimation of the values of these parameters is described in the text.
[b]The character 't' represents degree-days.
[c]Lower leaves are leaves that emerged before leaf 3, when counting down from the flag leaf (flag leaf = 1).
[d]A compound parameter which combines the infection efficiency and the probability of a spore landing on the upper leaves of the canopy (see text).
[e]Dimensionless.
[f]See the text for the range of values of parameters θ, ω and $\alpha_{B,r}$ in the simulations.

The effect of the dose rate of the high-risk fungicide on the emergence time

We first determined the effect of dose rate of the high-risk fungicide on the emergence time for the default scenario, which assumes a 10% reduction of the transmission rate due to fitness costs of resistance, complete resistance to the fungicide and a mutation probability of $1.13 \cdot 10^{-16}$. For this scenario, the median emergence time initially decreased sharply with increasing fungicide dose, but was much less sensitive to changes in dose when fungicide dose rates increased above approximately 50% of the label recommended dose (Fig. 6A). The distribution of the emergence time for a given dose rate was skewed to the right (Fig. 6B) and the size of 95% confidence interval of the emergence time was large with upper boundaries at least 38 years higher than lower boundaries.

The sensitivity of the emergence time to changes in parameter values

We subsequently determined the sensitivity of the emergence time to changes in the mutation probability, the fitness costs of resistance and the sensitivity of the resistant strain to the high-risk fungicide. Multiplying the mutation probability with a factor 10 decreased the median emergence time by a factor 4.3–6.3 while dividing the mutation probability by a factor 10 increased the median emergence time by a factor 7.9–9.9, depending on the dose rate of the high-risk fungicide (Fig. 7A). Changing the size of the wheat growing area has a similar effect on the emergence time as changing the mutation probability, because the emergence time is influenced by the product of both parameters. Decreasing the fitness costs of resistance from 10 to 2.5% (reduction of the infection efficiency of the resistant strain) decreased the emergence

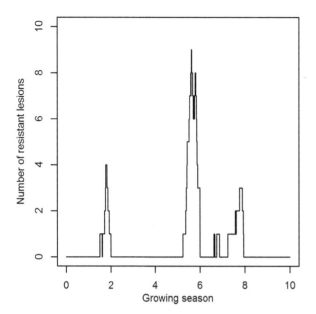

Figure 5. Temporal dynamics of the resistant sub-population. An example of the temporal dynamics of the number of resistant lesions in the absence of fungicide applications for the default scenario. In this scenario, the mutation probability amounts to $1.13 \cdot 10^{-16}$ and fitness costs of resistance reduce the infection efficiency by 10%.

time by a factor 1.3–1.9, while increasing fitness costs of resistance from 10 to 15% increased the median emergence time by a factor 1.1–2.5, depending on the dose rate of the high-risk fungicide (Fig. 7B). The effect of varying the sensitivity of the resistant strain to the high-risk fungicide on the emergence time was determined by increasing the maximum possible reduction of the life-cycle parameters of the resistant strain by the high-risk fungicide from 0 (complete resistance) to 20%. This increased the median emergence time by a factor 1.1–1.6, depending on the dose rate of the high-risk fungicide (Fig. 7C). The emergence time was most sensitive to changes in these parameters for low dose rates of the high-risk fungicide. The size of the 95% confidence interval of the emergence time increased or decreased when the median emergence time increased or decreased due to the variation in parameter values described above.

For all parameter settings described above, varying the dose rate of the high-risk fungicide had the same qualitative effect on the median and 95% confidence interval of the emergence time as for the default scenario.

The usefulness of mixing a low-risk with a high-risk fungicide to delay the emergence of resistance

Mixing a low-risk with a high-risk fungicide delayed the emergence of resistance to the high-risk fungicide in comparison to solo use of the high-risk fungicide at the same dose rate as in the mixture (Table 2). For a fixed dose rate of the high-risk fungicide, the delay in the emergence time initially increased with an increasing dose rate of the low-risk fungicide, but became much less sensitive when the dose rate of the low-risk fungicide increased above approximately 50% of the label recommended dose. For a fixed dose rate of the low-risk fungicide, the emergence time initially decreased sharply with an increasing dose rate of the high-risk fungicide, but was much less sensitive when dose rates of the high-risk fungicide increased above approximately 50% of the label recommended dose. The emergence time was delayed most

by mixing the lowest dose rate of the high-risk fungicide with a dose rate of the low-risk fungicide higher than approximately 50% of the label recommended dose.

Discussion

To our knowledge, this is the first time that a model structure is presented to describe the emergence of resistance to high-risk fungicides in a pathogen population. The model consists of a deterministic sub-model to describe the dynamics of the host and the sensitive pathogen population. The resistant strain occurs in very low densities during the emergence phase and stochastic processes determine when resistant mutants arise and whether they survive or not. A stochastic sub-model was therefore used to describe the dynamics of the resistant strain. Although the model structure is generic and could be applied to many foliar patho-systems on determinate crops, we have as an example parame-terized the model to describe the emergence of resistance in *M. graminicola* on winter wheat. For this specific system, we evaluated the effect of the dose rate of a high-risk fungicide on the emergence time of the resistant strain. We also determined the effect of mixing a high-risk fungicide with a low-risk fungicide on the emergence time of resistance to the high-risk fungicide. The model output suggests that the emergence time initially sharply decreases with increasing dose rate of the high-risk fungicide, but is virtually insensitive to changes within the range of higher dose rates typically needed for effective control of pathogens in commercial crops. This pattern was similar for a range of values for the mutation probability, fitness costs of resistance and sensitivities of the resistant strain to the high-risk fungicide. Mixing a high-risk fungicide with a low-risk fungicide delayed the emergence of resistance to the high-risk fungicide in comparison to solo use of the high-risk fungicide.

Explanation of the model output

The initial sharp decrease in emergence time with increasing dose rate of the high-risk fungicide shows that the emergence time is more sensitive to a reduction in the competition for healthy leaf between the resistant and sensitive strain than to a reduction in the number of mutants generated per time unit. The sensitivity of the emergence time to changes in the dose rate of the high-risk fungicide is virtually negligible at higher dose rates. This effect of dose on emergence time is caused by the high curvature of the dose-response curve resulting in the asymptote being reached at quite a low fungicide dose. As a result, the impact of the high-risk fungicide on the density of the sensitive strain will be similar for dosages above 0.4. The number of mutants generated per time unit and the availability of healthy leaf area which both depend on the size of the sensitive pathogen population, are therefore also approximately constant for dose rates larger than 0.4. This explains the approximately constant emergence time for these dose rates.

The delay in the emergence of resistance by mixing a high-risk fungicide with a low-risk fungicide occurs because the low-risk fungicide i) further decreases the size of the sensitive pathogen population and therefore the number of mutants generated per time unit, and ii) decreases the infection efficiency and therefore the survival probability of the resistant strain. The sensitivity of the emergence time to changes in the dose rate of the low-risk fungicide (for constant dose rates of the high-risk fungicide) is much less at higher dose rates, for similar reasons to those described for high-risk fungicides above.

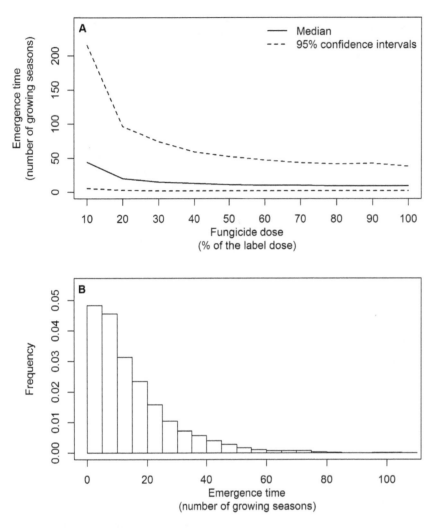

Figure 6. Emergence time and fungicide dose. The emergence time of a resistant strain in a sensitive population of *M. graminicola* on winter wheat in response to different dose rates of a high-risk fungicide for the default scenario (A). The bottom graph (B) shows the frequency distribution of the emergence time for a dose rate amounting to 50% of the label recommended dose. The shape of the distribution was the same for dose rates from 10 to 100% of the label recommended dose per spray. In the default scenario, the mutation probability was $1.13 \cdot 10^{-16}$, fitness costs of resistance reduced the infection efficiency by 10% and resistance to the high-risk fungicide was complete.

Comparison to the literature

Most experimental studies describe the development of fungicide resistance in response to different treatment strategies for the selection phase [4,5,6,7,8,9,10,12]. To our knowledge, there are four experimental studies which report the evolution of fungicide resistance in a sensitive laboratory population. In these studies fungicide resistance either did not evolve [31,32] or was already emerged (frequencies ≥1%) when detected [32,33,34]. The effect of fungicide treatment strategies on the time to the emergence of resistance can therefore not be determined from these studies.

There is modelling literature on the development of fungicide resistance in response to for example the dose rate, spray frequency and spray coverage of a fungicide [35,36,37,38,39,40] and in response to concurrent, sequential, alternating or mixture use of fungicides [18,19,35,40,41,42,43]. The models in virtually all of these studies were deterministic and are therefore unable to account for the stochastic nature of the dynamics of the resistant strain in the emergence phase. To our knowledge, there are two modelling studies which describe the dynamics of the resistant strain during a part of the emergence phase as well as during the

selection phase. In one of these studies [2], as is the case for our model, the dynamics of the resistant strain in the emergence phase were described using a stochastic model and the dynamics of the resistant strain during the selection phase were described using a deterministic model. However, contrary to our model, the emergence phase was assumed to last only until the first resistant mutant arose and the model did not therefore account for the possibility that a mutant may subsequently become extinct due to random processes. As a result, it was suggested that the length of the emergence phase increased with an increasing degree of control of the sensitive pathogen population by fungicides, which decreased the mutation rate. In our model, the mutation rate also decreases with increasing dose rates of fungicides (due to a smaller sensitive pathogen population), but the emergence time of resistance was predicted to decrease with increasing fungicide dose rates, due to the increased survival probability of resistant mutants at higher dose rates.

In the second modelling study [44], the resistant strain was introduced at the start of simulations and its dynamics in response to applications of a high-risk fungicide were described using a

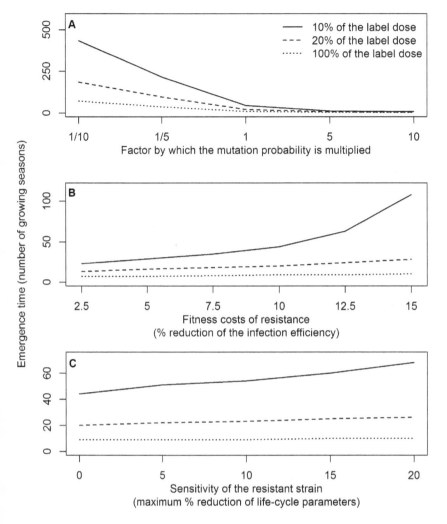

Figure 7. Emergence time and parameter values. The effect of the mutation probability (A), fitness costs of resistance (B) and the sensitivity of a resistant strain to a high-risk fungicide (C) on the emergence time of resistance in a sensitive population of *M. graminicola* on winter wheat. The emergence times are shown for dose rates of the high-risk fungicide amounting to 10, 20 and 100% of the label recommended dose per spray. The mutation probability is the probability that a spore produced by the sensitive pathogen population has a resistant genotype. By default, the mutation probability was $1.13 \cdot 10^{-16}$, fitness costs of resistance reduced the infection efficiency by 10% and resistance to the high-risk fungicide was complete.

stochastic model. No clear distinction was made between the emergence phase and the selection phase and the effect of the dose rate of a fungicide on the length of the emergence time was not determined. The output of the model suggested that the resistant strain can only invade a sensitive pathogen population when the fitness of the resistant strain is high enough compared to the fitness of the sensitive strain in the presence of fungicides. Our model results show that resistance always emerged in a population of *M. graminicola* on winter wheat for all simulated dose rates of the high-risk fungicide, and scenarios for fitness costs of resistance and partial resistance. However, our model results show that decreasing the fitness of the resistant strain relative to the sensitive strain, by increasing fitness costs of resistance or increasing the sensitivity of the resistant strain to the high-risk fungicide, increases the emergence time. Further decreasing the relative fitness of the resistant strain in our model simulations decreases the probability of emergence below 100% (results not shown).

Implications for resistance management

Hobbelen et al. [18] defined the effective life of a fungicide as the time from the start of a treatment until the loss of effective disease control. They subsequently used the term "effective life" to indicate the number of years that different treatment strategies were able to provide disease control in the selection phase only [18,19]. However, the time from the start of a treatment until the loss of effective disease control above includes both the emergence phase and the selection phase. Below, we use the term "effective life" to indicate the emergence time plus the number of years that a fungicide can provide effective disease control in the selection phase.

When resistance is detected in the field for the first time, it is likely that the resistant strain is already present at a frequency of one percent or more in certain areas. To reach that point, the pathogen population has been evolving by emergence and selection for many generations. Hence, much of the opportunity to slow evolution has already been lost if anti-resistance strategies are put in place in response to detection. Strategies need to be

Table 2. The effect of mixing a low-risk and a high-risk fungicide[a] on the number of growing seasons before resistance to the high-risk fungicide emerges[b,c] in a population of *M. graminicola* on winter wheat.

Dose rate of the low-risk fungicide[d]	Dose rate of the high-risk fungicide[d]									
	10	20	30	40	50	60	70	80	90	100
0	46[e]	21[e]	15[e]	13[e]	11	10	9	9	9	8
10	55[e]	23[e]	17[e]	14	13	11	11	10	10	9
20	65[e]	25[e]	18	14	13	12	11	11	10	9
30	73[e]	27[e]	19	15	13	12	11	11	10	10
40	78[e]	27[e]	19	16	14	13	12	11	11	10
50	78[e]	29[e]	20	16	14	13	12	11	11	10
60	82[e]	29	20	16	14	13	12	11	11	11
70	80[e]	28	20	16	14	13	12	12	11	11
80	81[e]	28	20	16	15	13	12	11	11	11
90	81[e]	28	20	16	14	13	12	12	11	11
100	81[e]	28	19	16	14	14	12	12	11	11

[a]The low-risk fungicide was assumed to be not at-risk of resistance development, but unable to provide sufficient disease control when used alone. The resistant strain was assumed to be completely insensitive to the high-risk fungicide.
[b]The resistant strain was considered to have emerged when the number of resistant lesions reaches or exceeds a threshold (see text).
[c]The emergence times in the table were calculated for the default scenario, which assumes that i) fitness costs of resistance reduce the infection efficiency of the resistant strain by 10%, ii) resistance to the high-risk fungicide is complete and iii) a mutation probability amounting to $1.13 \cdot 10^{-16}$.
[d]Fungicide doses are expressed as a fraction of the label recommended dose.
[e]Combinations of dose rates of the low-risk and high-risk fungicide that do not provide sufficient control of an average epidemic of *M. graminicola* on winter wheat. Effective disease control was defined as a disease-induced loss of healthy leaf area duration during the yield forming period equal to or below 5% [18].

implemented at introduction. This raises two new questions. Does the treatment strategy which is most effective at delaying resistance emergence differ from the treatment strategy which is most effective at slowing selection? If so, which of these two strategies should be used?

For solo use of a high resistance risk fungicide, the median emergence time of resistance was highest for the lowest dose rate of the high-risk fungicide that could provide effective control of an average epidemic of *M. graminicola* on winter wheat. Hobbelen et al. [18,19] determined the effect of the dose rate of a high-risk fungicide on the number of years that a high-risk fungicide can provide effective disease control in the selection phase for the same host-pathogen system. Their analyses showed that this number of years was constant or slightly decreased with increasing dose rate of the high-risk fungicide. This result was consistent for a range of fitness costs of resistance and for different degrees of partial resistance. For solo use of a high-risk fungicide, the model output thus suggests that both the part of the effective life spent in the emergence phase and the part of the effective life spent in the selection phase can be maximised, for a fixed number of fungicide applications per crop, by using the lowest dose which can provide effective disease control.

For mixtures of a high-risk and a low-risk fungicide, the median emergence time of resistance to the high-risk fungicide was highest when high dose rates of the low-risk fungicide were combined with the lowest possible dose rate of the high-risk fungicide necessary to provide sufficient disease control of an average epidemic of *M. graminicola* on winter wheat. Hobbelen et al. [18] determined the number of years that mixtures of a low-risk and a high-risk fungicide can provide sufficient disease control in the selection phase for the same host-pathogen system. Similar to the emergence phase, their analysis shows that this number of years is highest when high dose rates of the low-risk fungicide are combined with the lowest possible dose rate of the high-risk fungicide necessary to provide sufficient disease control. It can be

concluded that the dose and mixture treatment strategies which are most effective at delaying the evolution of fungicide resistance, do not differ between the emergence phase and the selection phase.

Generality of the model assumptions

The specific model in this paper describes the emergence of resistance to a high-risk fungicide in *M. graminicola* populations on winter wheat. However, the structure and assumptions underlying the model apply to many foliar fungal pathogens of cereal crops. For example, only parameter values would need to be changed to describe the development of the canopy of cereal crops other than winter wheat. Similarly, the division of the life cycle of fungal pathogens into latent and infectious stages is representative of all fungal pathogens.

The division of our model into deterministic and stochastic sub-models is to some extent artificial, because stochastic processes will not only influence the dynamics of the resistant strain, but also the dynamics of the host and the sensitive pathogen population. However, such a division is justified when the density of the host and the sensitive pathogen population are so high during most of the growing season that extinction due to stochastic processes is highly unlikely. The advantage of using a deterministic instead of a stochastic model to describe large populations is the much shorter simulation time.

There are also a number of limitations to the generality of the model. Firstly, the sensitivity of pathogen strains is assumed to be constant in time. As a result, the model cannot be used to describe a quantitative type of resistance development, characterised by a gradual decrease in sensitivity of the pathogen population due to the accumulation of mutations over time. The best strategy for delaying the emergence of strains with sharply decreased sensitivity due to a single mutation may not be the best strategy for delaying the emergence of strains in which the reduction in sensitivity due to each mutation is relatively small. A second limitation is that the model does not account for the spatial variation in the treatment programs for fungicides. In reality, spores which are resistant to the fungicide which is applied in one area may disperse to neighbouring areas treated with another fungicide. If the spore is sensitive to the fungicide applied in the neighbouring area, it is unlikely to survive. Accounting for spatial variation may therefore decrease the survival probability of resistant mutants and increase the emergence time of resistance. Thirdly, in the absence of peer-reviewed data, we have assumed that the mutation probability is not increased by the exposure to fungicides. Finally, when a low-risk and a high-risk fungicide are applied in a mixture, we have assumed that both fungicides act independently on the life-cycle parameters of the pathogen strain. The last two assumptions can however be changed by small adjustments to the model equations.

Priorities for future research

This initial analysis of fungicide resistance emergence opens several lines of future enquiry. Although several experiments have shown that environmental stress (caused for example by nutrient limitation, UV light, oxidative stress, antibiotic exposure or low pH) can increase the mutation rate in bacteria [45,46,47,48], a recent review [3] found no published studies on the effect of the dose rate of fungicides on the probability of mutations which decrease the sensitivity of pathogens to fungicides. Future work should test if there is a relationship between dose rate and the probability of such mutations, as this may change the current conclusion that mixing a low-risk with a high-risk fungicide increases the emergence time.

It would be useful to develop a stochastic model that describes both the emergence phase and the selection phase in the evolution of fungicide resistance. This would allow the calculation of a distribution for the time from the introduction of a fungicide on the market to the loss of effective disease control due to the evolution of resistance. In addition, a spatial version of such a model would provide insight into spatial differences in the evolution of resistance. At the end of the emergence phase the number of resistant lesions in the pathogen population is very small and large areas will still be occupied by a completely sensitive pathogen population. The time from the introduction of a fungicide on the market to the loss of effective control will therefore differ between wheat growing regions, depending on the rate of dispersal.

So far, we have used our model to analyse the effect of the dose rate of a high-risk fungicide on the emergence time of resistance and the usefulness of mixing a low-risk and a high-risk fungicide for delaying the emergence of resistance to the high-risk fungicide. The usefulness of other anti-resistance strategies remains to be evaluated. Finally, more research is needed to determine the effect of exposure to a mixture of fungicides on the life-cycle parameters of pathogen strains. In this model, we have assumed that fungicides act independently on life-cycle parameters. Deviations from this assumption will change the efficacy of fungicide mixtures and therefore the size of the sensitive pathogen population, which in turn influences the mutation rate and the ability of mutant spores to survive.

Model testing

In order to experimentally determine the emergence time of resistance, an emergence threshold must be defined above which the resistant strain is unlikely to become extinct if the fungicide treatment continues. The emergence threshold in our model was defined as the number of resistant lesions at the start of a growing season giving a specified low probability of extinction in the absence of new mutations. In experiments, it would not be possible to use this threshold as the generation of new mutations by the sensitive pathogen population cannot be stopped. It is therefore difficult to experimentally determine the length of the emergence phase in the evolution of resistance. However, it may be possible to design experiments that determine the time that it takes for a resistant strain to arise in a completely sensitive pathogen population and subsequently invade this population until it constitutes a specified very small threshold frequency in the pathogen population. It is important to note that the emergence threshold is a number of resistant lesions, which applies irrespective of the size of the sensitive pathogen population. For small pathogen populations, the emergence threshold corresponds to a higher frequency of the resistant strain in the pathogen population than for a large pathogen population and the required sample size to detect the resistant strain early may be less. However, for smaller populations, the time until a resistant mutant arises will be longer than for a large pathogen population.

Conclusion

In this study, we formulated a model structure to describe the emergence of resistance in a sensitive pathogen population. The resistance simulated was representative of observed cases where a mutation affecting the target protein results in a large shift in sensitivity. We subsequently showed how the model could be used to evaluate the usefulness of treatment strategies for delaying the emergence of such resistance. There are important conclusions from the model output which have implications for practical resistance management. In the absence of previous exposure to high-risk fungicides with the same mode of action, the model output suggests that resistance to high-risk fungicides is likely to emerge after their introduction on the market, making it important that anti-resistance strategies implemented at introduction are effective against both emergence and selection. Our analysis suggests that the dose and mixture treatment strategies which have been shown previously to reduce selection for resistance in the selection phase, may also be effective in prolonging the emergence phase in the evolution of resistance to fungicides.

Acknowledgments

Rothamsted Research receives support from the Biotechnology and Biological Sciences Research Council (BBSRC) of the United Kingdom.

Author Contributions

Conceived and designed the experiments: FB NP. Performed the experiments: FB PH. Analyzed the data: PH FB NP. Wrote the paper: PH FB NP.

References

1. van den Bosch F, Gilligan CA (2008) Models of fungicide resistance dynamics. Annu. Rev. Phytopathol. 46: 123–147.

2. Milgroom MG (1990) A stochastic model for the initial occurrence and development of fungicide resistance in plant pathogen populations. Phytopathology 80: 410–416.

3. van den Bosch F, Paveley N, Shaw M, Hobbelen P, Oliver R (2011) The dose rate debate: does the risk of fungicide resistance increase or decrease with dose? Plant Pathol. 60: 597–606.

4. Genet JL, Jaworska G, Deparis F (2006) Effect of dose rate and mixtures of fungicides on selection for QoI resistance in populations of Plasmopara viticola. Pest Manag. Sci. 62: 188–194.

5. Hunter T, Brent KJ, Carter GA, Hutchen JA (1987) Effects of fungicide spray regimes on incidence of dicarboxamide resistance in gray mold (Botrytis cinerea) on strawberry plants. Ann. Appl. Biol. 110: 515–525.

6. Mavroeidi VI, Shaw MW (2006) Effects of fungicide dose and mixtures on selection for triazole resistance in Mycosphaerella graminicola under field conditions. Plant Pathol. 55: 715–725.

7. McCartney C, Mercer PC, Cooke LR, Fraaije BA (2007) Effects of a strobilurin-based spray programme on disease control, green leaf area, yield and development of fungicide-resistance in Mycosphaerella graminicola in Northern Ireland. Crop Prot. 26: 1272–1280.

8. Sanders PL, Houser WJ, Parish PJ, Cole H (1985) Reduced-rate fungicide mixtures to delay fungicide resistance and to control selected turfgrass diseases. Plant Dis. 69: 939–943.

9. Thygesen K, Jorgensen LN, Jensen KS, Munk L (2009) Spatial and temporal impact of fungicide spray strategies on fungicide sensitivity of Mycosphaerella graminicola in winter wheat. Eur. J. Plant Pathol. 123: 435–447.

10. Turechek WW, Koller W (2004) Managing resistance of Venturia inaequalis to the strobilurin fungicides. Plant Health Prog Available: http://www.plantmanagementnetwork.org/pub/php/research/2004/strobilurin/. Accessed 11 December 2013.

11. Vali RJ, Moorman GW (1992) Influence of selected fungicide regimes on frequency of dicarboximide resistant and dicarboximide sensitive strains of Botrytis cinerea. Plant Dis. 76: 919–924.

12. Zhang CQ, Zhang Y, Zhu GN (2008) The mixture of kresoxim-methyl and boscalid, an excellent alternative controlling grey mould caused by Botrytis cinerea. Ann. Appl. Biol. 153: 205–213.

13. Metcalfe RJ, Shaw MW, Russell PE (2000) The effect of dose and mobility on the strength of selection for DMI fungicide resistance in inoculated field experiments. Plant Pathol. 49: 546–557.

14. LaMondia JA (2001) Management of euonymus anthracnose and fungicide resistance in Colletotrichum gloeosporioides by alternating or mixing fungicides. J. Environ. Hortic. 19: 51–55.

15. Serra MC (2006) On waiting time to escape. J. Appl. Probab. 43: 296–302.

16. Serra MC, Haccou P (2007) Dynamics of escape mutants. Theor. Popul. Biol. 72: 167–178.

17. Iwasa Y, Michor F, Nowak MA (2003) Evolutionary dynamics of invasion and escape. J. Theor. Biol. 226: 205–214.

18. Hobbelen PHF, Paveley ND, van den Bosch F (2011) Delaying selection for fungicide insensitivity by mixing fungicides at a low and high risk of resistance development: a modelling analysis. Phytopathology 101: 1224–1233.

19. Hobbelen PHF, Paveley ND, Oliver RP, van den Bosch F (2013) The usefulness of concurrent, alternating and mixture use of two high-risk fungicides for delaying the selection of resistance in populations of Mycosphaerella graminicola on winter wheat. Phytopathology 103: 690–707.

20. Hobbelen PHF, Paveley ND, Fraaije BA, Lucas JA, van den Bosch F (2011) Derivation and validation of a model to predict selection for fungicide resistance. Plant Pathol. 60: 304–313.

21. Sanoamuang N, Gaunt RE, Fautrier AG (1995) The segregation of resistance to carbendazim in sexual progeny of Monilinia fructicola. Mycol. Res. 99:677–680.

22. Fraaije BA, Cools HJ, Fountaine J, Lovell DJ, Motteram J, West JS, Lucas JA (2005) Role of ascospores in further spread of QoI-resistant cytochrome b alleles (G143A) in field populations of Mycosphaerella graminicola. Phytopathology 95:933–941.

23. Fraaije BA, Bayon C, Atkins S, Cools HJ, Lucas JA, Fraaije MW (2011) Risk assessment studies on succinate dehydrogenase inhibitors, the new weapons in the battle to control Septoria leaf blotch in wheat. Mol. Plant Pathol. 13:263–75.

24. Thornley JHM, Johnson IR (1990) Plant and Crop Modelling. Oxford: Clarendon Press.

25. Waggoner PE, Berger RD (1987) Defoliation, disease, and growth. Phytopathology 77:393–398.

26. Cao Y, Gillespie DT, Petzold LR (2006) Efficient step size selection for the tau-leaping simulation method. J. Chem. Phys. 124: 044109.

27. Eyal Z (1999) The septoria tritici and stagnospora nodorum blotch doseases of wheat. Eur. J. Plant Pathol. 105: 629–641.

28. Robert C, Fournier C, Andrieu B, Ney B (2008) Coupling a 3D virtual wheat (Triticum aestivum) plant model with a Septoria tritici epidemic model (Septo3D): a new approach to investigate plant-pathogen interactions linked to canopy architecture. Funct. Plant Biol. 35: 997–1013.

29. Grimmer M, van den Bosch F, Powers S, Paveley N (2011) Fungicide resistance risk assessment. In: Resistance 2011. Rothamsted Research. p 94.

30. Paveley ND, Blake J, Gladders P, Cockerell V (2011) The HGCA wheat disease management guide. United Kingdom: Home-Grown Cereals Authority.

31. Grabski C, Gisi U (1985) Mixtures of fungicides with synergistic interactions for protection against phenylamide resistance in phytophthora. In: Fungicides for crop protection. British Crop Protection Council. pp 315–318.

32. Horsten JAHM (1979) Acquired resistance to systemic fungicides of Septoria nodorum and Cercosporella herpotrichoides in cereals. Wageningen University, The NetherlandsPhD thesis

33. Carnegie SF, Cameron AM, Haddon P (2008) Effects of fungicide and rate of application on the development of isolates of Polyscytalum pustulans resistant to thiabendazole and on the control of skin spot. Potato Res. 51: 113–129.

34. Hoare FA, Hunter T, Jordan VWL (1986) Influence of spray programs on development of fungicide resistance in the eyespot pathogen of wheat, Pseudocercosporella herpotrichoides. Plant Pathol. 35: 506–511.

35. Doster MA, Milgroom MG, Fry WE (1990) Quantification of factors influencing potato late blight suppression and selection for metalaxyl resistance in Pytophthora infestans - a simulation approach. Phytopathology 80: 1190–1198.

36. Hall RJ, Gubbins S, Gilligan CA (2004) Invasion of drug and pesticide resistance is determined by a trade-off between treatment efficacy and relative fitness. Bull. Math. Biol. 66: 825–840.

37. Hall RJ, Gubbins S, Gilligan CA (2007) Evaluating the performance of chemical control in the presence of resistant pathogens. Bull. Math. Biol. 69: 525–537.

38. Parnell S, Gilligan CA, Van den Bosch F (2005) Small-Scale Fungicide Spray Heterogeneity and the Coexistence of Resistant and Sensitive Pathogen Strains. Phytopathology 95: 632–639.

39. Parnell S, van den Bosch F, Gilligan CA (2006) Large-Scale Fungicide Spray Heterogeneity and the Regional Spread of Resistant Pathogen Strains. Phytopathology 96: 549–555.

40. Shaw MW (2000) Models of the Effects of Dose Heterogeneity and Escape on Selection Pressure for Pesticide Resistance. Phytopathology 90: 333–339.

41. Birch CPD, Shaw MW (1997) When can reduced doses and pesticide mixtures delay the build-up of pesticide resistance? A mathematical model. J. Appl. Ecol. 34: 1032–1042.

42. Josepovits G (1989) A detailed model of the evolution of fungicide resistance for the optimization of spray applications. Crop Prot. 8: 106–113.

43. Shaw Mw (1993) Theoretical analysis of the effect of interacting activities on the rate of selection for combined resistance to fungicide mixtures. Crop Prot. 12: 120–126.

44. Gubbins S, Gilligan CA (1999) Invasion thresholds for fungicide resistance: deterministic and stochastic analyses. Proc. Biol. Sci 266: 2539–2549.

45. Bjedov I, Tenaillon O, Gerard B, et al. (2003) Stress-induced mutagenesis in bacteria. Science. 300: 1404–9.

46. Tenaillon O, Denamur E, Matic I (2004) Evolutionary significance of stress-induced mutagenesis in bacteria. Trends Microbiol. 12: 264–70.

47. Kang JM, Iovine NM, Blaser MJ (2006) A paradigm for direct stress-induced mutation in prokaryotes. FASEB J. 20: 2476–85.

48. Galhardo RS, Hastings PJ, Rosenberg SM (2007) Mutation as a stress response and the regulation of evolvability. Crit. Rev. Biochem. Mol. Biol. 42: 399–435.

13

NikA/TcsC Histidine Kinase Is Involved in Conidiation, Hyphal Morphology, and Responses to Osmotic Stress and Antifungal Chemicals in *Aspergillus fumigatus*

Daisuke Hagiwara[1]*, Azusa Takahashi-Nakaguchi[1], Takahito Toyotome[1], Akira Yoshimi[2], Keietsu Abe[2], Katsuhiko Kamei[1], Tohru Gonoi[1], Susumu Kawamoto[1]

1 Medical Mycology Research Center, Chiba University, Chiba, Japan, 2 New Industry Creation Hatchery Center, Tohoku University, Sendai, Japan

Abstract

The fungal high osmolarity glycerol (HOG) pathway is composed of a two-component system (TCS) and Hog1-type mitogen-activated protein kinase (MAPK) cascade. A group III (Nik1-type) histidine kinase plays a major role in the HOG pathway of several filamentous fungi. In this study, we characterized a group III histidine kinase, NikA/TcsC, in the life-threatening pathogenic fungus, *Aspergillus fumigatus*. A deletion mutant of *nikA* showed low conidia production, abnormal hyphae, marked sensitivity to high osmolarity stresses, and resistance to cell wall perturbing reagents such as congo red and calcofluor white, as well as to fungicides such as fludioxonil, iprodione, and pyrrolnitrin. None of these phenotypes were observed in mutants of the SskA response regulator and SakA MAPK, which were thought to be downstream components of NikA. In contrast, in response to fludioxonil treatment, NikA was implicated in the phosphorylation of SakA MAPK and the transcriptional upregulation of *catA*, *dprA*, and *dprB*, which are regulated under the control of SakA. We then tested the idea that not only NikA, but also the other 13 histidine kinases play certain roles in the regulation of the HOG pathway. Interestingly, the expression of *fos1*, *phkA*, *phkB*, *fhk5*, and *fhk6* increased by osmotic shock or fludioxonil treatment in a SakA-dependent manner. However, deletion mutants of the histidine kinases showed no significant defects in growth under the tested conditions. Collectively, although the signal transduction network related to NikA seems complicated, NikA plays a crucial role in several aspects of *A. fumigatus* physiology and, to a certain extent, modulates the HOG pathway.

Editor: Gustavo Henrique Goldman, Universidade de Sao Paulo, Brazil

Funding: This study was supported by a Grant-in-Aid for Scientific Research (to D.H.) from the Ministry of Education, Science, Sports and Culture, and partly by a Cooperative Research Grant of NEKKEN (2010–2012), a Cooperative Research Program of Medical Mycology Research Center, Chiba University (12-2), and MEXT Special Budget for Research Projects: The Project on Controlling Aspergillosis and the Related Emerging Mycoses. The funders had no role in study design, data collection and analysis, decision to publish, or preparation of the manuscript.

Competing Interests: The authors have declared that no competing interests exist.

* E-mail: dhagi@chiba-u.jp

Introduction

Aspergillus fumigatus is a major causative pathogen of invasive aspergillosis (IA) worldwide. This fungus infects immunocompromised patients, and IA is known for its relatively high mortality [1]. Despite recent progress in diagnostic and therapeutic modalities, IA is still one of the most life-threatening infectious diseases. One reason for this is that the antifungal options to combat this pathogen are very limited. Thus, the identification of molecular targets for new antifungal medications is an urgent issue.

The two-component system (TCS) is a signal transduction system that is conserved in a wide range of organisms from bacteria to higher plants, but not in mammals [2,3]. In general, it senses environmental stimuli by sensor histidine kinases (HKs), transmits signals, and leads to appropriate cellular responses with response regulators (RRs). The molecular basis of signaling is the His-to-Asp phosphorelay system, in which a phosphorus group directly and reversibly transfers between conserved histidine and asparagine residues in HK and RR domains, respectively. The fungal TCS has hybrid-histidine kinases (hHKs), which have both HK and RR domains, and a histidine-containing phosphotransfer protein (HPt) as an intermediate factor. As a consequence, the

fungal TCS functions as a multistep phosphorelay composed of hHKs, HPt, and RRs (His-Asp-His-Asp) [4].

Molecular analysis of the fungal TCS has been intensively performed in *Saccharomyces cerevisiae*, in which a single hHK, Sln1p, a single HPt, Ypd1p, and two response regulators, Ssk1p and Skn7p, have been shown to constitute TCS signaling. The TCS is directly linked to the stress-activated mitogen-activated protein kinase (MAPK; SAPK) cascade and regulates its activation in response to osmotic conditions in the extracellular environment, resulting in a high osmolarity glycerol (HOG) pathway [5,6]. To date, the link to the MAPK cascade has also been observed in several filamentous fungi including plant and human pathogens [7]. In model filamentous fungi such as *Neurospora crassa* and *Aspergillus nidulans*, the TCS plays a role in a variety of physiological cellular functions including conidiation, sexual development, oxidative stress response, osmotic adaptation, and sensitivity to fungicides [8–12]. In some plant pathogenic fungi, the TCS plays a crucial role in pathogenicity, and is supposedly the target of the phenylpyrrole and dicarboximide classes of fungicides that are widely used to protect crops in the agricultural industry [13–15]. The accumulation of these findings supports the idea that the TCS is a promising molecular target for new antifungal therapies against human pathogens, including *A. fumigatus* [16].

In this decade, genome data have become available for several fungi, which allowed us to quickly search for components of the TCS and examine the diversity and universality of the fungal TCS. Catlett et al. (2003) revealed that the number of hHKs in filamentous fungi is generally greater than 10, which presented a sharp contrast to the numbers observed in yeast such as *S. cerevisiae* (1 hHK), *Schizosaccharomyces pombe* (3 hHKs), and *Candida albicans* (3 hHKs) [17]. In *A. fumigatus*, 13 HKs have previously been identified, and three (Fos1, TcsB, and TcsC) of them have been investigated. A deletion mutant of the *fos1* gene showed a moderate resistance to fungicides and attenuated virulence [18,19]. The *tcsB* deletion mutant showed a slight sensitivity to sodium dodecyl sulfate (SDS) and growth inhibition under high temperature conditions [20,21]. TcsC, a group III histidine kinase, was recently characterized by McCormick et al. as described below [22]. Characterization of the other HKs would be helpful to improve the understanding of the TCS signaling circuitry in *A. fumigatus*.

Among the different types of HKs, a large amount of attention has been paid to group III (Nik1-type) HKs. This gene (*os-1/nikI*) was initially identified in *N. crassa* as an osmotic stress-sensitive mutant allele, and later it was identified as a dicarboximide-resistant mutant allele [12,23]. This HK possesses a characteristic motif in its N-terminal region, consisting of four to six repeats of the HAMP (histidine kinases, adenylyl cyclases, methyl-accepting chemotaxis proteins, and phosphatases) domain. Although the functions of the motif were obscure, a null mutation and deletion of the gene resulted in resistance to the dicarboximide and phenylpyrrole fungicides in all fungi that possess this type of HK in its genome [10,12,24–28]. Intriguingly, although *S. cerevisiae* has no Nik1-type HK, heterologous expression of Nik1-type HKs from other species made *S. cerevisiae* responsive to these fungicides [29–33]. These findings illustrated that Nik1-type HKs play a crucial role in the fungicide action and that the mode of action is convertible across some fungi. Furthermore, a recent striking finding for this type of HK is its involvement in dimorphic switching in dimorphic pathogens including *Penicillium marneffei*, *Histoplasma capsulatum*, and *Blastomyces dermatitidis* [28,34]. However, the detailed molecular mechanism has yet to be elucidated.

Recently, the characterization of TcsC, a Nik1-type HK of *A. fumigatus*, has been reported by McCormick et al. [22]. TcsC is required for hyper-osmotic stress adaptation and sensitivity to certain fungicides such as fludioxonil, as well as phosphorylation of the SakA MAPK in response to these stimuli. Despite these phenotypes *in vitro*, TcsC seemed to be dispensable for virulence. In this study, we characterized the Nik1-type HK of *A. fumigatus* (NikA/TcsC) with regard not only to its role in osmotic stress and fungicide responses, development, and morphology, but also with regard to its role in the signaling pathway associated with the SskA response regulator and SakA MAPK, which are downstream components of the HOG pathway. Furthermore, the involvement of the other HKs in the HOG pathway was investigated. Based on our findings, we discussed the molecular mechanism of the TCS circuitry and the stress response mechanism of the HOG pathway in *A. fumigatus*.

Results

NikA is Required for Normal Conidiation

To investigate the physiological role of the NikA/TcsC HK, we constructed a deletion mutant of the *nikA* gene. Afs35 (*akuA::loxP*) was used as a host strain (WT), and the *nikA* gene was replaced with the hygromycin-resistant marker [35]. Three independent transformants (Δ*nikA*) were obtained through a standard transformation procedure, and the deletion of the *nikA* gene was confirmed at both the genomic and expression levels (data not shown). First, colony growth and conidia production were examined in the *nikA* deletion mutant (named Δ*nikA* hereafter). On a 0.1% yeast extract-containing glucose minimal medium (YGMM) plate, the colony diameter of Δ*nikA* was 82.5±0.8% of the WT (Figure 1A, Table S1). Also, Δ*nikA* produced fewer conidia than WT after 48 h and 96 h of incubation (Figure 1B). The germination and viability rates of Δ*nikA* conidia were indistinguishable from those of WT conidia (data not shown). Scanning electron microscopy (SEM) observations showed that although the conidiophore structure was normal in Δ*nikA*, the number of conidiophores was reduced (Figure 1C). To confirm these defects, we constructed *nikA*-complemented strains (named Co-*nikA*). The fitness and number of conidia were restored in Co-*nikA* (Figure 1A to C), indicating that these phenotypes resulted from the absence of the *nikA* gene.

To confirm the observed differences in conidia production, we examined the expression level of the *nikA*, *brlA*, and *rodA* genes during asexual development. *brlA* encodes a master regulator of conidiation, whose expression is increased in early asexual development [36]. *rodA* encodes a hydrophobin, which is a major conidia coating protein in *A. fumigatus* [37]. We found that in the WT strain the expression level of the *nikA* gene was increased as asexual development proceeded (Figure 1D). Consistent with the reduction in conidia, the expression levels of *brlA* and *rodA* were lower in Δ*nikA* than in WT (Figure 1D). These results suggest that NikA plays an important role in asexual development and is required for a full level of conidia production.

NikA Affects the Resistance to Cell Wall–Perturbing Agents

Because the radial growth of Δ*nikA* was slightly disrupted, we compared the morphological trait of growing hyphae in Δ*nikA* and WT. Microscopy observations revealed that the hyphal shape of Δ*nikA* was obviously different from that of WT and Co-*nikA* under liquid culture conditions (Figure 2A). In fact, Δ*nikA* showed rough and aberrant hyphae. To gain more insight into the cell wall structure, we observed the hyphae by transmission electron microscopy (TEM). Hyphal diameter and cell wall thickness were determined from the images of hyphal cross-sections, which showed that the cell wall of Δ*nikA* was slightly thinner than that of WT and Co-*nikA* cells (Figure 2B). These abnormalities of Δ*nikA* hyphae led us to examine the sensitivity to cell wall-perturbing agents such as congo red (CR), calcofluor white (CFW), and micafungin (MCFG). Δ*nikA* showed significant resistance to CR and CFW compared with WT, whereas it only showed slight resistance to MCFG (Figure 3). Taken together these findings suggest that the cell wall structure of Δ*nikA* is different from that of WT.

We next examined the cell wall carbohydrate composition in WT and Δ*nikA* strains. The cell wall of both strains was fractionated into alkali-soluble (AS1 and AS2) and alkali-insoluble (AI) fractions. Alpha-1,3-glucan is largely fractionated in AS2, while beta-1,3-glucan and chitin are largely fractionated in AI. Galactomannan is fractionated in both AS2 and AI. The amounts of glucose or glucosamine in the AS2 and AI fractions were largely similar between Δ*nikA* and WT, indicating that the amounts of alpha-1,3-glucan, beta-1,3-glucan and chitin were indistinguishable (Table S2). Notably, there was a difference in the galactosamine contents between Δ*nikA* and WT. Polygalactosamine, alpha-1,3-glucan, and galactomannan are thought to compose an amorphous cement in cell wall structure [38]. Thus,

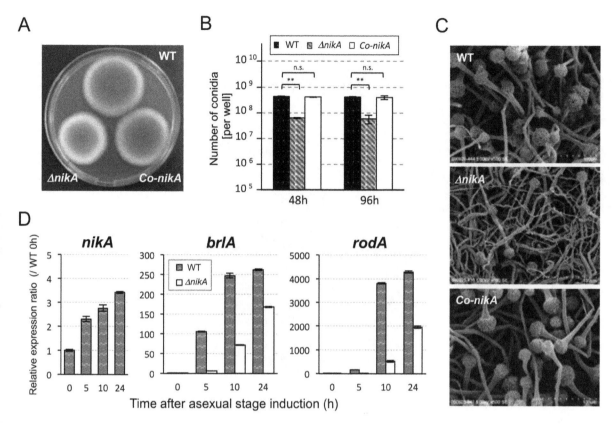

Figure 1. Deletion of the *nikA* gene results in a conidiation defect. (A) Conidial suspensions of the parental strain, Δ*nikA*, and the *nikA*-complemented strain (*Co-nikA*) were point-inoculated on 0.1% yeast extract containing glucose minimal medium (YGMM) agar plates and incubated for 68 h at 37°C. **(B)** The number of conidia in a 34 mm diameter well of a 6-well plate after 48 h and 96 h of incubation at 37°C was counted (See Materials and Methods). Error bars represent the standard deviations based on three independent replicates. **P<0.001 versus WT strain at any time point. *P*-values were calculated by the Student's *t*-test. (n.s.: not significant) **(C)** The colony surface was observed by scanning electron microscopy (SEM). The growth conditions were the same as those in (B): 48 h incubation at 37°C in 6-well plates. **(D)** The expression of *nikA*, *brlA*, and *rodA* during the asexual development stage was determined by real-time reverse transcriptase polymerase chain reaction (RT-PCR). To synchronize asexual development initiation, mycelia that were cultured in liquid YGMM for 18 h were harvested and transferred onto YGMM plates (the time point was set as 0 h of the asexual stage). Relative expression ratios were calculated relative to WT at 0 h. Error bars represent the standard deviations based on three independent replicates.

the reduction in galactosamine contents in Δ*nikA* may result in a modification of the cell wall structure that explains the resistances to cell wall-perturbing agents. However, the detailed function of galactosamine in the cell wall structure remains elusive. In any event, these results suggest that although the major cell wall carbohydrate composition of Δ*nikA* was comparable to that of WT, NikA affected hyphal morphology and resistance to cell wall stressors.

NikA is Required for Osmotic Stress Adaptation and Sensitivity to Fungicides

As reported in several studies, the Nik1-type HK is involved in fungicide action and osmotic stress adaptation in most fungi [24–31]. Thus, we sought to determine whether Δ*nikA* shows resistance to fungicides and sensitivity to osmotic stress. Prior to the evaluation of fungicide susceptibility of Δ*nikA*, the growth inhibitory concentration (IC) of the WT was determined. Fludioxonil, a phenylpyrrole fungicide, and iprodione, a dicarboximide fungicide, inhibited the radial growth on YGMM plates at concentrations above 0.01 μg/mL and 0.5 μg/mL, respectively, while pyrrolnitrin had a superior inhibiting activity (Figure S1). Based on these tests, the growth of Δ*nikA* was examined on plates containing appropriate concentrations of fungicides (roughly

estimated concentrations of IC50 and IC90 of the WT strain), revealing that Δ*nikA* had substantial resistance to the fungicides (Figure 4A). We next investigated the growth of Δ*nikA* on YGMM plates containing high concentrations of sorbitol, mannitol, NaCl, or KCl as high osmolarity stress conditions. Δ*nikA* showed impaired growth on the osmotic stress plates tested here (Figure 4B). Notably, resistance to fungicides and sensitivity to osmotic stresse of Δ*nikA* were also observed in liquid culture (data not shown), and these growth phenotypes were restored in *Co-nikA* (Figure S2). These results indicate that NikA plays an important role in the sensitivity to fungicides and osmotic stress adaptation in *A. fumigatus* as it does in other filamentous fungi.

SskA and SakA MAPK are not Required for Conidiation, Hyphal Morphology, and Sensitivity to the Fungicides

We next asked how NikA plays its role in the cellular functions and responses described above. Several lines of evidence have suggested that the NikA HK directly interacts with the HPt protein, YpdA, and modulates the SakA MAPK cascade via SskA RR. Thus, we decided to assess the involvement of the downstream components in conidia production, hyphal morphology, sensitivity to cell wall-perturbing reagents and fungicides, and osmotic stress resistance. We constructed mutant strains of the *sakA*

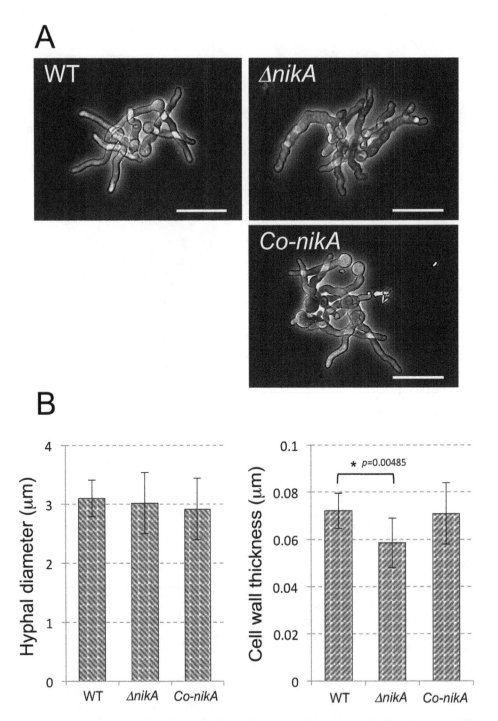

Figure 2. Defects of the cell wall structure in Δ*nikA*. (**A**) Conidia were incubated for 9 h in YGMM. The germlings were observed under optical microscopy. The bar represents 20 µm. (**B**) The cell wall thickness and the diameter of hyphae were examined from the hyphal cross-section. In each cross-section, the cell wall thickness was examined from at least 10 different positions. The largest and smallest measurements were eliminated, and the rest of values were averaged. The column indicates the mean value of the hyphal diameter (left) and cell wall thickness (right) from 10 different cross-sections (n = 10). Error bars represent the standard deviations. *P*-values versus WT were calculated by the Student's *t*-test.

and *sskA* genes (named Δ*sakA* and Δ*sskA*, respectively). Δ*sakA* and Δ*sskA* showed slightly faster radial growth (Δ*sakA*: 106.6±0.9% and Δ*sskA*: 104.0±1.2%) and comparable colony morphology and conidia production on YGMM with those of the WT (Figure 4B, Table S1, and data not shown). Light microscopy observations revealed that Δ*sakA* and Δ*sskA* displayed normal hyphae (data not shown). We also found that Δ*sakA* and Δ*sskA* were not as resistant to CR, CFW, and MCFG as Δ*nikA* was (Figure S3). With regard to fungicide resistance, Δ*sakA* and Δ*sskA* were as sensitive to fludioxonil and iprodione as the WT (Figure 4A and Table S1). Notably, Δ*sakA* and Δ*sskA* showed a slight resistance to pyrrolnitrin, suggesting that the SskA-SakA MAPK pathway is involved, to some extent, in the action of pyrrolnitrin (Figure 4A and Table S1). When grown on YGMM plates containing high osmotic

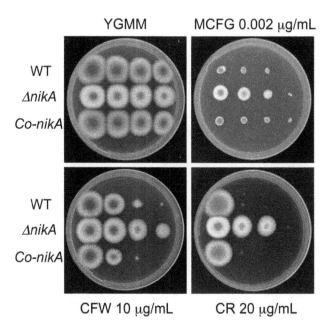

Figure 3. Growth of Δ*nikA* on plates containing cell wall-perturbing reagents. Series of diluted conidia suspensions (left to right: 10^4, 10^3, 10^2, and 10 conidia) of WT, Δ*nikA*, and Co-*nikA* were spotted onto YGMM plates with or without micafungin (MCFG), calcofluor white (CFW), or congo red (CR) at the indicated concentration. These plates were incubated at 37°C for 44 h and photographed.

stresses Δ*sakA* and Δ*sskA* showed significant, but less than that observed in Δ*nikA*, growth retardation (Figure 4B and Table S1). This result suggests that SskA and SakA play a partial role in the adaptation to high hyperosmotic environments, while NikA is crucial for it. Taken together, SskA and SakA were dispensable for normal conidiation, hyphal morphology, and resistance to cell wall-perturbing reagents and fungicides, suggesting that NikA plays a role in these cellular functions via a signaling pathway that is distinct from the SskA-SakA MAPK pathway.

SakA MAPK is Phosphorylated by Osmotic Shock and Fungicide Treatment via SskA

We next determined whether NikA is involved in the activation of the SakA MAPK cascade in response to environmental stimuli. It should be noted that besides osmotic stress and fungicides, Δ*nikA*, as well as Δ*sakA* and Δ*sskA*, displayed no significant growth differences when exposed to oxidative stress (hydrogen peroxide and menadione), antifungal chemicals (miconazole, fluconazole, and amphotericin B), and high temperature (42°C) compared with WT (data not shown). Thus, we focused on responses to osmotic stress and fungicides to define the signal transduction from the TCS to the SakA MAPK cascade in *A. fumigatus*. To monitor the SakA MAPK activation, we examined the phosphorylation level of SakA protein by immunoblot analysis with the commercial antibodies, anti-phospho-p38 MAPK (Thr180/Tyr182) and anti-Hog1 (y-215) (See Material and Methods). The phosphorylation of SakA was detected 10 min after the addition of sorbitol (1 M final concentration), and the levels decreased at 20 and 30 min (Figure 5A). When treated with fludioxonil or pyrrolnitrin, SakA was also phosphorylated (Figure 5A). These results suggest that SakA is rapidly activated after osmotic or fungicide treatments, and then the phosphorylated form gradually decreases.

Next, we investigated whether NikA and SskA are involved in the phosphorylation of the SakA MAPK. In Δ*sskA*, no phosphorylated SakA was detected with osmotic shock and fludioxonil treatment, while the total SakA protein was obviously present (Figure 5B and C). This indicates that SskA is indispensable for the phosphorylation of the SakA MAPK. In contrast, an increased level of SakA phosphorylation was detected in Δ*nikA* without stimulus, suggesting that the absence of NikA led to SakA phosphorylation (Figure 5D and E). Although the phosphorylation level was high in the absence of stimuli in Δ*nikA* cells, the addition of 1 M or 0.5 M sorbitol, or 10 μg/mL fludioxonil further increased the phosphorylation levels (Figure 5D and E). Interestingly, when treated with 1 or 0.1 μg/mL fludioxonil, there was no apparent increases in the phosphorylated form of SakA in Δ*nikA*, while phosphorylated SakA accumulated in WT. These results suggest that NikA plays a major role in SakA phosphorylation in response to fludioxonil treatment, but not in response to osmotic shock.

Osmotic Shock and Fungicide Induce Transcription of *catA*, *dprA*, and *dprB* through the HOG Pathway

To gain more insight into the role of NikA in mediating the SakA MAPK cascade, we examined the expression of genes that are possibly regulated under the control of the SakA MAPK. In our previous study on *A. nidulans*, *catA*, which encodes a conidia-specific catalase A, was regulated in a SakA/HogA MAPK cascade-dependent manner (unpublished data). Hence, we used the corresponding *A. fumigatus catA* to monitor the activation of the SakA MAPK cascade. In addition, the recently reported SakA-regulated genes *dprA* and *dprB*, both encoding a dehydrin-like protein, were also used for the expression analysis [39]. First, we investigated whether the expression of these genes was regulated in response to osmotic shock and fludioxonil treatment. After the addition of the stimuli, *catA*, *dprA*, and *dprB* were upregulated and the peak of expression was at 15–30 min (Figure S4).

Next, the upregulation of these genes (15 min after adding each stimulus) was investigated in Δ*nikA*, Δ*sakA*, and Δ*sskA*. In Δ*sakA* and Δ*sskA*, there was virtually no increase in the expression of any of the genes in response to osmotic shock (1 M sorbitol) or fludioxonil (10 μg/mL) treatment (Figure 6A and B). These results indicated that the upregulation of *catA*, *dprA*, and *dprB* was exclusively dependent on SskA and the SakA MAPK. In comparison, the expression of these genes significantly, but not completely, increased in response to osmotic shock and fludioxonil in Δ*nikA*, suggesting that the transcriptional responses of *catA*, *dprA*, and *dprB* are not fully dependent on NikA (Figure 6A and B). To further confirm this partial response, we examined the expression levels in WT and Δ*nikA* with various concentrations of sorbitol (0.25, 0.5, and 1 M) and fludioxonil (0.1, 1, and 10 μg/mL). When treated with 0.5 M sorbitol, the expression was significantly upregulated in WT as well as in Δ*nikA* (Figure 6C). In response to the serial concentrations of fludioxonil, Δ*nikA* was upregulated only with the 10 μg/mL treatment while upregulation was observed in WT even with the 0.1 μg/mL treatment. These results suggest that NikA is responsible for the transcriptional responses of *catA*, *dprA*, and *dprB* to the lower concentrations (0.1 and 1 μg/mL) of fludioxonil through the SskA-SakA pathway.

Expression of some Histidine Kinase Genes is Regulated through the SakA MAPK

The finding that the transcriptional responses to osmotic stress and fungicides were fully dependent on SskA and partly dependent on NikA led us to ask whether the other HKs play a compensative

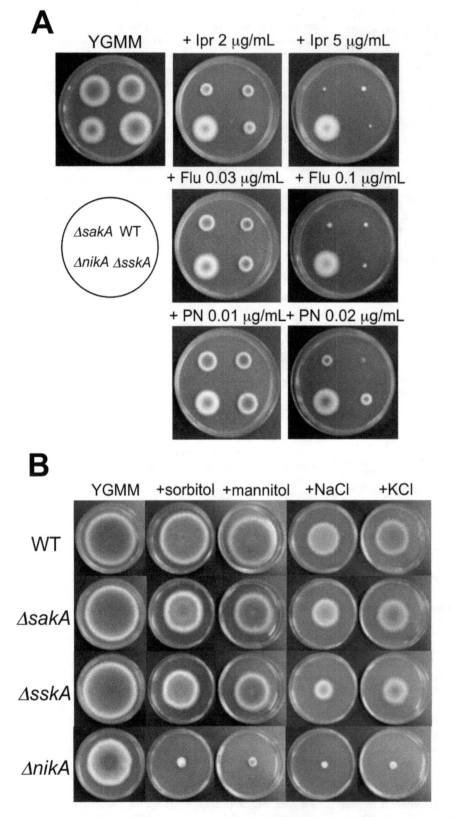

Figure 4. Growth of *sakA, sskA,* and *nikA* deletion mutants under high osmotic or fungicide stress conditions. (A) Growth on plates containing fungicides. Conidia of Δ*sakA*, Δ*sskA*, Δ*nikA*, and WT were inoculated onto YGMM containing iprodione (Ipr), fludioxonil (Flu), or pyrrolnitrin (PN) at the indicated concentrations and were incubated at 37°C for 48 h. **(B)** Growth on plates containing high osmotic stress. Conidia of Δ*sakA*, Δ*sskA*, Δ*nikA*, and WT were inoculated onto YGMM containing 1.2 M sorbitol, 1.2 M mannitol, 1 M NaCl, or 1 M KCl and were incubated at 37°C for 96 h.

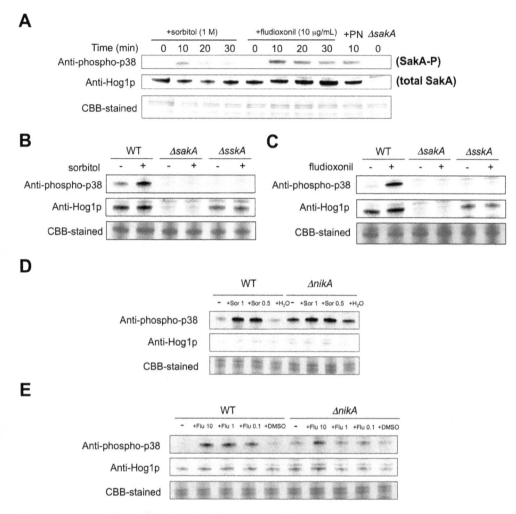

Figure 5. Immunoblot analysis for phosphorylation of the SakA MAPK in response to osmotic or fungicide stress. (A) The WT strain was grown for 18 h at 37°C. Then, sorbitol (1 M final concentration), fludioxonil (10 µg/mL final concentration), or pyrrolnitrin (1 µg/mL) was added. The mycelium was harvested at the indicated times, and total proteins were extracted. Anti-phospho-p38 was used to detect the phosphorylation of SakA, and anti-Hog1p was used to detect the total protein of SakA. A Coomassie Brilliant Blue (CBB)-stained gel is shown as a loading control. **(B and C)** WT, ΔsakA, and ΔsskA were grown with or without sorbitol (1 M final concentration) (B) or fludioxonil (10 µg/mL final concentration) (C). CBB-stained gels are shown as loading controls. **(D and E)** WT and ΔnikA were grown with or without sorbitol (0.5 and 1 M final concentrations) (D) or fludioxonil (0.1, 1, and 10 µg/mL final concentrations) (E). CBB-stained gels are shown as loading controls. The same volume of water (D) or DMSO (E) was added to the culture instead of sorbitol or fludioxonil as a control.

role for mediating the SakA MAPK cascade in ΔnikA cells. To examine the involvement of the other HKs in the HOG pathway, the expression level of all 13 HK genes in response to the osmotic stress and fludioxonil treatment was determined in WT and ΔsakA. fhk3 expression was not detected by our real-time reverse transcriptase-polymerase chain reaction (RT-PCR) system in several trials using different primer sets. In response to osmotic shock, the expression of fos1, phkA, phkB, and fhk6 was markedly upregulated (>4-fold), and this upregulation was dependent on the SakA MAPK (Figure 7A). Likewise, phkB and fhk5 were upregulated in a SakA MAPK-dependent manner by fludioxonil treatment (Figure 7B). These results suggest the possibility that Fos1, PhkA, PhkB, Fhk5, and Fhk6 play a certain role in the modulation of the SskA-SakA MAPK pathway in response to these stimuli.

To investigate whether Fos1, PhkA, PhkB, Fhk5, and Fhk6 are implicated in osmotic stress adaptation and fungicide action, we constructed the deletion mutants Δfos1, ΔphkA, ΔphkB, Δfhk5, and Δfhk6. Because ΔphkA could not be constructed in the Afs35

background, we created ΔphkA in the Af293 background and compared it to Af293. The colony growth of the mutants was investigated on a plate containing hyperosmotic stress or fungicides. Although ΔphkB colonies showed slight growth retardation on YGMM plates, these mutants showed a virtually wild-type growth rate on YGMM plates with or without hyperosmotic stress or fungicides (Figure S5). These results suggest that none of these HKs solely plays a role in the osmotic stress adaptation and fungicide action.

To reveal whether these HKs play a compensative role for NikA in modulating the SakA MAPK cascade, double deletion mutants should be constructed. Prior to this, we investigated the expression profiles of these HKs in ΔnikA cells that were treated with osmotic shock or fludioxonil. None of the HKs showed higher expression in ΔnikA than in WT (data not shown). This result led us to assume that none of these HKs were activated, at least at the expression level, in the cells lacking NikA. Therefore, we did not investigate the double deletion mutants in this study. Collectively, these results suggest that whereas NikA plays an important role in modulating

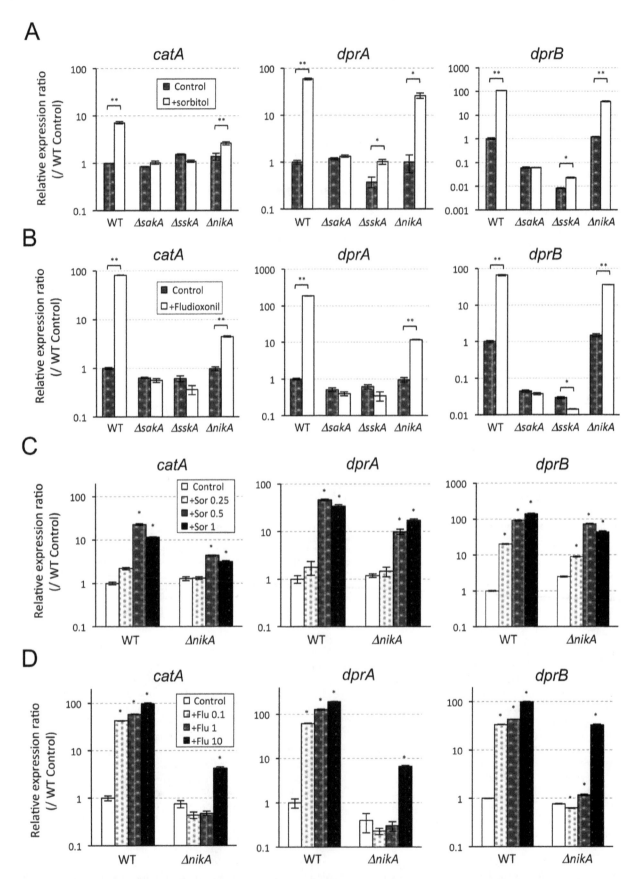

Figure 6. Transcriptional responses of *catA*, *dprA*, and *dprB* to osmotick shock and fungicide. (**A** and **B**) WT, Δ*sakA*, Δ*sskA*, and Δ*nikA* were grown in liquid YGMM for 18 h at 37°C. Then, sorbitol (A) or fludioxonil (B) was added (1 M or 10 μg/mL final concentrations, respectively). As a

control, an equivalent volume of water or DMSO was added to the cultures (indicated as Control). After 15 min, the mycelium was harvested, and the RNA was extracted. cDNAs was synthesized and used for real-time RT-PCR analysis. Relative expression ratios were calculated relative to the WT control. Error bars represent the standard deviations based on three replicates. P-values were calculated by the Student's t-test: *$P<0.005$; **$P<0.001$, compared to the Control in any strain. (**C** and **D**) WT and $\Delta nikA$ were grown in liquid YGMM for 18 h at 37°C. Then, different concentrations of sorbitol (0.25, 0.5, and 1 M final concentrations) (C) or fludioxonil (0.1, 1, and 10 μg/mL final concentrations) (D) were added. After 15 min, the mycelium was harvested, and the RNA was extracted. cDNAs was synthesized and used for real-time RT-PCR analysis. Relative expression ratios were calculated relative to WT. Error bars represent the standard deviations based on three replicates. *statistically significant difference relative to the Control in any strain ($P<0.05$). P-values were calculated by student's t-test followed by Bonferroni method.

the SakA MAPK cascade, none of the other HKs affect the HOG pathway functions.

Discussion

In this study, we found that the absence of NikA resulted in pleiotropic phenotypes such as growth retardation, reduction of conidia, aberrant hyphae, tolerance to cell wall-perturbing reagents and fungicides, and marked sensitivity to high osmolarity stress. However, NikA is likely to function independently of SskA and SakA in these phenotypes except for osmotic adaptation, which raised the question of what alternative component functions downstream of NikA. Because NikA is a component of the His-Asp phosphorelay circuitry in the TCS, NikA is thought to regulate

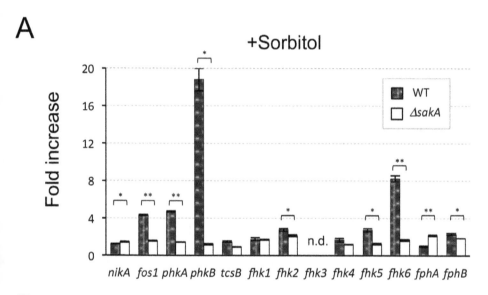

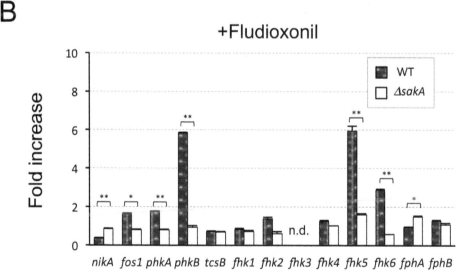

Figure 7. Transcriptional responses of _A. fumigatus_ histidine kinase (HK) genes to osmotic shock and fungicide. (A and B) WT and $\Delta sakA$ were grown in liquid YGMM for 18 h at 37°C. Then, sorbitol (A) or fludioxonil (B) was added (1 M and 10 μg/mL final concentrations, respectively). After 15 min, the mycelium was harvested, and the RNA was extracted. cDNAs was synthesized and used for real-time RT-PCR analysis. The relative expression ratios were compared between samples with and without osmotic shock or fludioxonil treatment; the fold increases of each HK gene in WT and $\Delta sakA$ were determined. Error bars represent the standard deviations based on three replicates. *$P<0.005$; **$P<0.001$ versus WT. P-values were calculated by the Student's t-test. (n.d.: not determined).

RRs via the YpdA HPt in a phosphorelay-dependent manner. Given that SskA is not involved in the phenotypes, a plausible candidate is AfSkn7, the other type of RR in *A. fumigatus*. In *A. nidulans*, the double gene deletion of SskA and SrrA, which is an ortholog of AfSkn7, resulted in a Δ*nikA*-level tolerance to the fungicides and a marked sensitivity to osmotic stress [10]. A similar result was also reported in *Cochliobolus heterostrophus*, suggesting that SskA and SrrA/Skn7 are redundantly or cooperatively involved in the adaptations to fungicides and osmotic environment downstream of NikA [13]. Importantly, *A. nidulans* single deletion mutants of *sskA* or *srrA* showed only slight resistance to fungicides and a moderate sensitivity to the osmotic stress [9,10]. Another research group characterized AfSkn7, where AfSkn7 was implicated in the oxidative stress response but not in morphology and pathogenicity in a murine model of aspergillosis [40]. In addition, the *afskn7* deletion mutant constructed by our group showed no significant resistance to cell wall-perturbing reagents and fungicides (unpublished data). Thus, to better understand how NikA regulates conidiation, morphology, and responses to chemicals, the generation and investigation of a double deletion mutant of *sskA* and *afskn7* is necessary. This will be the subject of our future study.

As in other fungi, the *A. fumigatus* SakA MAPK was transiently phosphorylated by osmotic shock and fungicide treatment, which was fully dependent on SskA. This clearly indicates that under the conditions tested here the SakA MAPK cascade is solely regulated by the TCS. Unexpectedly, we found that SakA was phosphorylated in response to osmotic shock in Δ*nikA* cells, suggesting that another mechanism modulates the SakA phosphorylation instead of NikA, possibly via SskA. On the one hand, NikA seemed to be indispensable for the accumulation of phosphorylated SakA in response to low concentrations of fludioxonil treatment (up to 1 μg/mL), whereas it was dispensable when treated with 10 μg/mL fludioxonil. Judged by the roughly estimated growth ICs (0.03 and 0.1 μg/mL as IC50 and IC90, respectively), 10 μg/mL of fludioxonil seems high and may cause a leaky effect at a specific interaction between the chemical and the target protein. In *C. heterostrophus* and *N. crassa*, the accumulation of phosphorylated Hog1-type MAPK was observed in the cells lacking the Nik1-type HK in the presence of high concentrations of fungicides or high osmolar reagents [41]. These results could support the idea that another HK or mechanism contributes to the phosphorylation of SakA via SskA in some filamentous fungi. Notably, in contrast to our results, in a previous study by McCormick et al. (2012), TcsC/NikA has been shown to be required for the phosphorylation of SakA in response to fludioxonil treatment (10 μg/mL) and osmotic shock (1.2 M sorbitol) in *A. fumigatus* [22]. We think that this discrepancy is derived from the differences in media (AMM vs. YGMM), cultivation time (9 h vs. 18 h), and treatment time (2 min vs. 15 min for fludioxonil, and 20 min vs. 15 min for osmotic shock). Considering these results, the regulation of the SakA MAPK cascade by NikA may be growth stage-specific.

We also found that the phosphorylation level of the SakA MAPK in Δ*nikA* was moderately high in the absence of stimuli. This increased phosphorylation level is reminiscent of a role of the *S. cerevisiae* Sln1p in modulating the Hog1 MAPK phosphorylation, where a defect in the *sln1* gene causes abnormal activation (hyperphosphorylation) of the Hog1 MAPK and leads to growth defects [6]. In fact, Δ*nikA* showed a slight defect in radial growth on YGMM plates and increased the phosphorylation of SakA. These findings could support the idea that similarly to Sln1p NikA is a negative regulator of the SakA MAPK, whereas it plays a positive role at least in response to fludioxonil treatment. Here, it should be noted that the expression level of *catA*, *dprA*, and *dprB* in Δ*nikA* in the absence of stimuli was comparable to that in WT, suggesting

that the moderate accumulation of phosphorylated SakA dose not necessarily lead to the high expression of the genes. There may be unknown regulatory mechanisms, for example a negative feedback system, concerning this. In any event, judged by the results of SakA phosphorylation and transcriptional responses to the stimuli, the phosphorylated form of SakA appears to be an active form. Also, we showed a straightforward signaling from SskA to the SakA MAPK in the *A. fumigatus* HOG pathway both at the transcriptional and post-translational levels.

In hyperosmotic stress conditions, the Δ*nikA*, Δ*sskA*, and Δ*sakA* strains showed growth sensitivity at distinct levels. As has been well studied in *S. cerevisiae* [42,43], the HOG pathway plays an important role in osmotic adaptation by regulating glycerol accumulation; thus, it is possible that glycerol accumulation occurred in response to hyperosmotic stress conditions in *A. fumigatus*, and the HOG pathway was responsible for this. In addition, DprB is a dehydrin-like protein responsible for growth under hyperosmotic conditions [39]. This seems to be one of the reasons for the growth retardation of Δ*sskA* and Δ*sakA*, in which not only the transcriptional response to osmotic shock but also the basal level of *dprB* expression were impaired. On the other hand, moderate levels of the transcriptional response and basal expression of *dprB* were observed in Δ*nikA*. These findings presented the possibility that the considerable growth impairment of Δ*nikA* under osmotic stress conditions resulted from not only a shut-off of the SskA-SakA-DprB function but also from a defect in another component. As stated above, the Skn7-type RR and SskA additively function in the osmotic adaptation of several fungi [9,10,13]. Thus, AfSkn7 may be a downstream component of NikA; together, they play an important role in osmotic adaptation.

In general, the fungal TCS is a target of fungicides such as fludioxonil, iprodione, and pyrrolnitrin, and the Nik1-type HK is supposedly the molecular target [10,12,15,25,31,44–48]. We revealed that the *A. fumigatus* NikA was indispensable for fungicide activity (growth inhibitory effects). This result is consistent with the results of studies on other fungi. As stated above, relatively high concentrations of fludioxonil may have a leaky effect on HOG pathway activation. With regard to this, we hypothesized that nonspecific interactions between fludioxonil and the other HKs occurred during the 10 μg/mL fludioxonil treatment. To examine this, we tried to clarify any implication of other HKs in this effect, but no obvious interplay was observed. These results allowed us to suggest another possibility: fludioxonil interacts directly with YpdA protein or the NikA-YpdA complex to inhibit its function. The reason for this assumption is twofold. First, as in most fungi including *S. cerevisiae*, *N. crassa*, *Cryptococcus neoformans*, and *A. nidulans* [6,48–50], *afypdA* is likely to be an essential gene (unpublished data). Second, when the expression of *afypdA* is downregulated, growth is inhibited and the SakA MAPK pathway is activated (unpublished data). These phenomena are reminiscent of how cells react to fludioxonil treatment. The clarification of this hypothesis is under investigation in our group.

A detailed molecular mechanism of the mode of action of fungicides such as fludioxonil is one of the long-sought topics concerning the biological control of plant pathogenic fungi [51,52]. The fungicides used here have a broad spectrum, and they affect animal pathogenic fungi such as *Candida*, *Cryptococcus*, *Aspergillus*, and *Fusarium* species [7]. Hence, this type of chemical seems to be a promising compound for antifungal drugs for life-threatening deep mycoses [53]. The elucidation of the molecular target, the signaling pathway, and the mode of action in each pathogen will aid the development of newly designed drugs with a broad spectrum.

In general, the TCS of filamentous fungi is composed of more than 10 HKs, two to four RRs, and one HPt factor [7,17]. In *A. fumigatus*, only a few HKs have been characterized so far [18–22]. To gain more comprehensive insight into the *A. fumigatus* TCS, we determined the expression profiles of all HKs. Interestingly, some HKs were upregulated in response to osmotic shock or fludioxonil treatment, and they were dependent on the SakA MAPK. We assumed that these HKs play certain roles in the signaling of the SakA MAPK cascade. From investigations of the deletion mutants, however, we had no clear evidence for this assumption. Notably, Fos1 has been reported to be involved in pathogenicity, sensitivity to fungicides, and cell wall construction [18,19]. Our *fos1* deletion mutant, however, showed no role in sensitivity to fungicide and cell wall stress reagents. We think that this discrepancy of phenotypes is derived from the different background of the strain. In addition, Δ*phkA*, Δ*phkB*, Δ*fhk5*, and Δ*fhk6* showed no visible phenotypes (colony size, surface color, and conidiation ability), including susceptibility to hydrogen peroxide, fungicides, and cell wall stressors, when compared with WT (data not shown). To elucidate the physiological role of these HKs, further functional analyses are necessary. Recently, Ji et al. (2012) have reported that TcsB plays an important role in growth in a high-temperature environment and likely modulates the SakA MAPK phosphorylation state in *A. fumigatus* [21]. Taken together with findings in this study, it appears that a robust and complex signal transduction system for adaptation to extracellular environments exists in *A. fumigatus*.

In conclusion, we found that NikA plays an important role in conidiation, morphology, and stress responses, and that the SakA MAPK cascade is regulated through SskA (TCS) in response to osmotic shock and fungicide treatment. We also characterized the other HKs including Fos1, PhkA, PhkB, Fhk5, and Fhk6. There were no interactions between these HKs and NikA or SakA, at least under the conditions tested here. Although NikA seems to be dispensable in *A. fumigatus* pathogenicity, molecular insights provided in this study may open possibilities in the development of new antifungal chemicals focusing on the TCS signaling of medically relevant fungal pathogens.

Materials and Methods

Strains and Growth Media

A. fumigatus strain Afs35 (FGSC A1159) (*akuA::loxP*) was used to generate the following gene deletion strains: Δ*nikA*, Δ*sakA*, Δ*sskA*, Δ*fos1*, Δ*phkB*, Δ*fhk5*, and Δ*fhk6* [35]. *A. fumigatus* strain Af293 was used to generate Δ*phkA*. These genes (*nikA*, Afu2g03560; *sakA*, Afu1g12940; *sskA*, Afu5g08390; *fos1*, Afu6g10240; *phkA*, Afu3g12550; *phkB*, Afu3g12530; *fhk5*, Afu4g00660; and *fhk6*, Afu4g00320) were replaced with the hygromycin resistance marker (*hph* or HygBʳ), sequences of which were derived from plasmids pSH75 (a generous gift from Dr. Takashi Tsuge, Nagoya University) or pCB1004, respectively [54]. An ectopic *nikA*-complemented strain (*Co-nikA*) was generated by transformation with a fragment containing *nikA* on pPTR I (Takara BIO, Ohtsu, Japan) (see below). Fungal strains used in this study are listed in Table 1. All strains were routinely cultivated in 0.1% yeast extract containing glucose minimal medium (YGMM) at 37°C. To collect conidia of each strain, potato dextrose agar (PDA) was used. For plates containing osmotic stress, 1.2 M sorbitol, 1.2 M mannitol, 1 M NaCl, or 1 M KCl was added. In liquid culture, 20 mL of 3 M sorbitol or 2.4 M NaCl was added to 40 mL of YGMM (final concentration, 1 M or 0.8 M, respectively). For antifungal chemicals, a 1000-fold concentrated stock solution was prepared and added to media in appropriate concentrations. Fludioxonil,

Table 1. Fungal strains used in this study.

Strain	Description	Reference
Afs35 (FGSC A1159)	Wild type (*akuA::loxP*)	From FGSC
Af293	Wild type	From FGSC
Δ*nikA*	*nikA::HygBʳ*	This study
Co-*nikA*	*nikA::HygBʳ+nikA*	This study
Δ*sskA*	*sskA::HygBʳ*	This study
Δ*sakA*	*sakA::hph*	This study
Δ*fos1*	*fos1::HygBʳ*	This study
Δ*phkA*	*phkA::HygBʳ*	This study
Δ*phkB*	*phkB::HygBʳ*	This study
Δ*fhk5*	*fhk5::HygBʳ*	This study
Δ*fhk6*	*fhk6::HygBʳ*	This study

iprodione, and MCFG were a generous gift from Kumiai-Chem. Co. Pyrrolnitrin (from *Pseudomonas cepacia*), CFW (fluorescent brightener 28), and CR were commercially obtained (Sigma-Aldrich Co., St. Louis, MO, USA).

Construction of the Gene Deletion and Reconstituted Strains

To construct the *A. fumigatus* deletion and the reconstituted strains, plasmids for each purpose were generated. DNA manipulation was performed according to standard laboratory procedures for recombinant DNA. To amplify DNA fragments from the genome and prepare gene replacement cassettes using one-step fusion PCR, Platinum Taq DNA Polymerase High Fidelity (Invitrogen, Carlsbad, CA, USA) or PrimeSTAR HS (Takara BIO) were used. Plasmids for transformation were constructed by the GeneArt system (Invitrogen) or one-step fusion PCR [55]. Primers used in this study are listed in Table S3.

For the generation of pSH75-sakA, 5′- and 3′-flanking regions of the *sakA* gene were obtained using the primers sakA-U(Bgl)-FW and sakA-U(Bgl)-RV (for the 5′-region) and sakA-D(Xho)-FW and sakA-D(Xho)-RV (for the 3′-region). The fragments were digested using the restriction enzymes *Bgl*II and *Xho*I and were ligated into pSH75 carrying the *hph* cassette. The resulting plasmid, pSH75-sakA, was used for transformation to construct the *sakA* deletion strain.

For the generation of plasmid pUC119-nikA, 5′- and 3′-flanking regions of the *nikA* gene were obtained using the primers nikA-U(Pst)-FW and nikA-U-RV-(HygBr) (for the 5′-region) and nikA-D-FW(HygBr) and nikA-D(Pst)-RV (for the 3′-region). The fragment of the HygBʳ cassette was obtained from the pCB1004 plasmid using the primers HygBr-FW and HygBr-RV. HygBʳ flanked by the *nikA*-flanking regions was obtained by one-step fusion PCR. The fragment was digested with the restriction enzyme *Pst*I and was ligated into *Pst*I-linearized pUC119. The resulting plasmid, pUC119-nikA, was used for transformation to construct the *nikA* deletion strain.

For the generation of plasmid pUC119-sskA, 5′- and 3′-flanking regions of the *sskA* gene were obtained using the primers sskA-U-F(BamHI) and sskA-U-RV(HygBr) (for the 5′-region) and sskA-D-FW(HygBr) and sskA-D-R(EcoRI) (for the 3′-region). HygBʳ flanked by the *sskA*-flanking regions was obtained by one-step fusion PCR. The fragment was digested with the restriction enzymes *Bam*HI and *Eco*RI and was ligated into pUC119 that was linearized with *Bam*HI and *Eco*RI. The resulting plasmid, pUC119-sskA, was used for transformation to construct the *sskA* deletion strain.

For the generation of plasmid pUC119-fos1, 5′- and 3′-flanking regions of the *fos1* gene were obtained using the primers fos1A-U(Pst) and fos1-U-R(HygBr) (for the 5′-region) and fos1-D-F(HygBr) and fos1-D(Pst) (for the 3′-region). HygB^r flanked by the *fos1*-flanking regions was obtained by one-step fusion PCR. The fragment was digested with the restriction enzyme *Pst*I and was ligated into pUC119 that was linearized with *Pst*I. The resulting plasmid, pUC119-fos1, was used for transformation to construct the *fos1* deletion strain.

For the generation of plasmids pUC119-phkA, pUC119-phkB, pUC119-fhk5, and pUC119-fhk6, 5′- and 3′-flanking regions of each gene were obtained using the primers phkA-U2-F(pUC119B) and phkA-U2-R(HygBr) (for the 5′-region of *phkA*), phkA-D-F(HygBr) and phkA-R(pUC119E) (for the 3′-region of *phkA*), phkB-F(pUC119B) and phkB-U-R(HygBr) (for the 5′-region of *phkB*), phkB-D-F(HygBr) and phkB-R(pUC119E) (for the 3′-region of *phkB*), fhk5-F(pUC119B) and fhk5-U-R(HygBr) (for the 5′-region of *fhk5*), and fhk5-D-F(HygBr) and fhk5-R(pUC119E) (for the 3′-region of *fhk5*), fhk6-F2(pUC119B) and fhk6-U-R(HygBr) (for the 5′-region of *fhk6*), and fhk6-D-F(HygBr) and fhk6-R2(pUC119E) (for the 3′-region of *fhk6*). These flanking regions and the HygB^r fragment were fused into pUC119 using the GeneArt system, resulting in pUC119-phkA, pUC119-phkB, pUC119-fhk5, and pUC119-fhk6, which were used for transformation to construct Δ*phkA*, Δ*phkB*, Δ*fhk5*, and Δ*fhk6*, respectively.

For the generation of plasmid pPTRI-nikA+, the *nikA* fragments containing 5′- and 3′-flanking regions were obtained using the primers nikA-N-F(pPTR-P) and nikA-N-R (for the fragment containing the 5′-region and N-terminus of *nikA*), and nikA-C-F and nikA-C-R(pPTR-K) (for the fragment containing the 3′-region and C-terminus of *nikA*). These two fragments were fused into pPTRI using the GeneArt system, resulting in pPTRI-nikA+, which was used for transformation of Δ*nikA* to construct the ectopically complemented strain *Co-nikA*.

A. fumigatus transformation was performed according to conventional methods for protoplast-polyethylene glycol (PEG) transformation for *Aspergillus*. Briefly, a mycelium that was cultured for one night was harvested from YGMM liquid medium and was treated with Lysing Enzymes (Sigma-Aldrich Co.) and Yatalase (Takara BIO) for more than 3 h to obtain a sufficient amount of protoplasts. The DNA was coincubated with the protoplasts, and osmotic treatment with PEG4000 led to the incorporation of the DNA. The protoplast was mixed into agar medium containing 1.2 M sorbitol and appropriate antibiotics for selection.

Homologous recombination and gene replacement were confirmed by PCR of the genomic DNA, and the absence of the mRNA of the target gene was verified using real-time RT-PCR analysis. We obtained at least three independent transformants for each gene deletion mutant, and the phenotypes of the multiple strains were clarified.

Conidia Preparation and Culture Conditions

Conidia of each strain were stored in a deep freezer (−80°C) to prevent unexpected mutations. To prepare fresh conidia suspensions, the stored conidia were inoculated on a PDA slant and incubated at 37°C for four to five days. Conidia were harvested with phosphate-buffered saline (PBS) containing 0.1% Tween 20 and kept at 4°C. Throughout this study, conidia were freshly harvested every two weeks. The number of conidia was counted using a hemocytometer (Watson, Kobe, Japan).

For the synchronized induction of asexual development, conidia (10^5 conidia/mL) were cultivated in liquid YGMM for 18 h, and mycelia were harvested using Miracloth (Merck, Frankfurt, Germany), washed with distilled water, and transferred onto YGMM plates. These plates were incubated at 37°C (this time point was set as 0 h), and the mycelia were harvested at the appropriate times.

Colony Diameter Measurement

About 10^4 conidia of each strain were point-inoculated on YGMM plates. After incubation for 72 h at 37°C, the diameter of the colony was measured (n = 3), and the mean values were calculated. The growth ratio under stressed conditions was calculated by comparing the radial diameter on the YGMM plate with and without antifungal chemicals or osmotic stress reagents. To judge sensitivity or resistance to the stresses, the calculated ratio of the diameter of each strain was compared with that of WT.

Calculation of Conidia Numbers

Before the agar solidified, conidia were mixed into YGMM agar medium (final concentration, 10^4 conidia/mL). The medium containing conidia was poured into a 6-well plate (3 mL per well), solidified, and then incubated at 37°C. After 48 h or 96 h of incubation, the agar from each well including mycelia and produced conidia was vortexed in 10 mL of PBS containing 0.1% Tween 20 for 1 min. The number of conidia in the suspensions was counted using a hemocytometer.

RNA and cDNA Preparation

The mycelia were harvested and frozen in liquid nitrogen, and total RNA was isolated using the FastRNA Pro Red Kit (MP Biomedicals, Santa Ana, CA, USA). The possible contaminating DNA was digested with Deoxyribonuclease for Heat Stop (Wako Pure Chemical Industries, Osaka, Japan). To obtain cDNA pools from the total RNA, reverse transcription was performed using the High Capacity cDNA Reverse Transcription Kit (Life Technologies Corporation, Carlsbad, CA, USA).

Quantitative Real-Time RT-PCR

Real-time RT-PCR was performed using the 7300 system (Life Technologies Corporation) with SYBR Green detection according to the manufacturer's instructions. For reaction mixture preparation, the THUNDERBIRD SYBR qPCR Mix was used (TOYOBO, Osaka, Japan). The primers listed in Table S3 were used to quantify the gene expression of interest. Primer specificity was verified by disassociation analysis. A cDNA sample that was obtained from reverse transcription reactions using 1 μg of total RNA was applied to each reaction mixture. The *actin* gene was used as the normalization reference (internal control) for target gene expression ratios, and WT without stress or at 0 h was set as the calibrator in each experiment. Relative expression ratios were calculated by first calculating the threshold cycle changes in sample and calibrator as $\Delta C_t^{sample} = C_{t(target)} - C_{t(actin)}$ and $\Delta C_t^{calibrator} = C_{t(target)} - C_{t(actin)}$, followed by calculating the expression rates in both sample and calibrator as Expression rates$_{sample} = 2^{\Delta C_t^{sample}}$ and Expression rates$_{calibrator} = 2^{\Delta C_t^{calibrator}}$, and finally, Relative expression rates $= \dfrac{\text{Expression rate}_{sample}}{\text{Expression rate}_{calibrator}}$. Each sample was tested in triplicate. For biological replicates, each test was repeated at least twice on different days.

Immunoblot Analysis

Conidia (final concentration, 10^5/mL) were inoculated into liquid YGMM and cultured for 18 h prior to addition of sorbitol or fungicides. Mycelia were collected with Miracloth, frozen in liquid nitrogen, ground to a powder, and immediately suspended in protein extraction buffer containing protease inhibitors [50 mM Tris-HCl, pH 7.5, 1% sodium deoxycholate, 1% Triton X-100, 0.1% SDS, 50 mM NaF, 5 mM sodium pyrophosphate, 0.1 mM sodium vanadate, and protease inhibitor cocktail (Wako)]. The suspension was immediately boiled for 10 min with an appropriate sample buffer for SDS-polyacrylamide gel electrophoresis (PAGE), and cell debris was then removed by centrifugation for 10 min at 14,000 rpm. The protein concentration of the supernatant was determined using a Pierce BCA Protein Assay Kit–Reducing Agent Compatible (PIERCE, Rockford, IL, USA). Total protein was loaded onto a NuPAGE Novex Bis-Tris 4–12% gel (Invitrogen) and blotted using the iBlot gel transfer system (Invitrogen). To detect total SakA and phospho-SakA proteins, a rabbit polyclonal IgG antibody against Hog1 y-215 (sc9079, Santa Cruz Biotechnology, Santa Cruz, CA, USA) and a rabbit polyclonal IgG antibody against dually phosphorylated p38 MAPK (#4631, Cell Signaling Technology, Beverly, MA, USA) were used, respectively. To detect these signals on blotted membranes, the ECL Prime Western Blotting Detection System (GE Healthcare, Little Chalfont, UK) and LAS1000 (FUJIFILM, Tokyo, Japan) were used.

Microscopy Experiments

For SEM analysis, strains were grown according to the same procedure as that in the calculation of conidia number. After 48 h of incubation at 37°C, appropriate-sized agar blocks were prefixed with 2% (w/v) glutaraldehyde containing 0.1 M phosphate buffer for 24 h, followed by 1% osmium tetraoxide in phosphate buffer for 1 h. Samples were dehydrated through a graded ethanol series to anhydrous 100% ethanol, substituted with isoamyl acetate, and then dried by the critical-point method (EM CPD030; Leica). After sputter coating with platinum (E102 Ion sputter, Hitachi, Tokyo, Japan), the samples were visualized under an S-3400 Scanning Electron Microscope (Hitachi).

For TEM analysis, hyphae that were cultured for 18 h were prefixed with 2% (w/v) glutaraldehyde containing 0.1 M phosphate buffer at 4°C for 10 h, followed by 1% osmium tetraoxide in phosphate buffer for 1 h. After dehydration, tissues were embedded in epoxy resin. Ultrathin sections were stained with uranyl acetate and lead citrate, and were visualized under a JEM 1400 electron microscope (JEOL, Tokyo, Japan). The hyphal diameter and cell wall thickness were measured using SMileView software (JEOL).

Determination of Cell Wall Carbohydrate Composition

The cell wall fractionation by alkali treatment and quantitative determination of the carbohydrate composition of the fractions were performed as previously described [56]. Briefly, WT and $\Delta nikA$ strains were cultured in 1 L of YGMM for 24 h, harvested, squeezed, and freeze-dried. One gram of the mycelial powder was used for fractionation of cell walls by alkali treatment. All of the fractions (HW: hot-water soluble, AS1 and AS2: alkali-soluble, AI: alkali-insoluble) were freeze-dried, and the weight was measured. Each freeze-dried cell wall fraction was hydrolyzed by H_2SO_4 at 100°C and neutralized with barium carbonate. Finally, the carbohydrate composition of the samples was determined using a high-performance anion-exchange chromatography (HPAC) system.

Supporting Information

Figure S1 Radial growth rate on plates containing fungicides. Growth inhibitory effects of fludioxonil (Flu), iprodione (Ipr), and pyrrolnitrin (PN) were examined. The WT strain was inoculated on YGMM agar with or without the indicated concentration of chemicals and grown at 37°C for 40 h. The radial growth rates were calculated by comparing the diameter of a colony to that of a colony grown on YGMM without chemicals. Each plot shows the mean based on three replicates.

Figure S2 Growth of the Co-nikA strain under high osmotic or fungicide stress conditions. Conidia of Co-nikA were inoculated onto YGMM containing sorbitol, NaCl, iprodione (Ipr), or fludioxonil (Flu) at the indicated concentrations and were incubated at 37°C for 44 h.

Figure S3 Resistance to cell wall-perturbing reagents. (A) Conidia of $\Delta sakA$, $\Delta sskA$, $\Delta nikA$, and WT were inoculated onto YGMM containing 1 ng/mL micafungin (MCFG), 30 μg/mL congo red (CR), or 15 μg/mL calcofluor white (CFW) and incubated at 37°C for 48 h.

Figure S4 Time-course expression of catA, dprA, and dprB in response to osmotic shock and fungicide. (A and B) The WT strain was grown for 18 h at 37°C; then, sorbitol (A) or fludioxonil (B) was added (1 M and 10 μg/mL final concentrations, respectively). The mycelia were harvested at the indicated time points. The expression ratios were investigated by real-time RT-PCR. Relative expression ratios were calculated relative to the 0 min sample. Error bars represent the standard deviations based on three independent replicates.

Figure S5 Comparison of growth of A. fumigatus HK gene deletion mutants under osmotic or fungicide stress conditions. (A) Growth on plates containing hyperosmotic stress. Conidia of WT (Afs35), $\Delta nikA$, $\Delta fos1$, $\Delta phkB$, $\Delta fhk5$, and $\Delta fhk6$ were inoculated onto YGMM containing 1.2 M sorbitol and 1 M NaCl and incubated at 37°C for 72 h. Conidia of WT (Af293) and $\Delta phkA$ were inoculated onto YGMM containing 1.2 M sorbitol and were incubated at 37°C for 72 h. (B) Growth on plates containing fungicides. Conidia of WT (Afs35), $\Delta nikA$, $\Delta fos1$, $\Delta phkB$, $\Delta fhk5$, $\Delta fhk6$, WT (Af293), and $\Delta phkA$ were inoculated onto YGMM containing 5 μg/mL iprodione (Ipr), 0.1 μg/mL fludioxonil (Flu), or 0.02 μg/mL pyrrolnitrin (PN), and were incubated at 37°C for 48 h.

Table S1 Radial growth of each strain on the plate containing fungicide or osmotic stress.

Table S2 Cell wall composition of each cell wall fraction.

Table S3 PCR primers used in this study.

Acknowledgments

We thank Dr. Masashi Yamaguchi for technical advice and Dr. Misako Ohkusu for technical assistance.

Author Contributions

Conceived and designed the experiments: DH. Performed the experiments: DH ATN TT AY. Analyzed the data: DH ATN TT AY KA. Wrote the paper: DH. Advisor for analysis of the data: KK TG SK.

References

1. Latgé JP (1999) *Aspergillus fumigatus* and *Aspergillosis*. Clin Microbiol Rev 12: 310–350.
2. Hoch JA, Silhavy TJ (1995) "Two-Component Signal Transduction," ASM Press, Washington DC, 1–473.
3. Mizuno T (1998) His-Asp phosphotransfer signal transduction. J Biochem 123: 555–563.
4. Appleby JL, Parkinson JS, Bourret RB (1996) Signal transduction via the multistep phosphorelay: Not necessarily a road less traveled. Cell 86: 845–848.
5. Maeda T, Wurgler-Murphy SM, Saito H (1994) A two-component system that regulates an osmosensing MAP kinase cascade in yeast. Nature 369: 242–245.
6. Posas F, Wurgler-Murphy SM, Maeda T, Witten EA, Thai TC, et al. (1996) Yeast HOG1 MAP kinase cascade is regulated by a multistep phosphorelay mechanism in the SLN1-YPD1-SSK1 'two-component' osmosensor. Cell 86: 865–875.
7. Bahn YS (2008) Master and commander in fungal pathogens: The two-component system and the HOG signaling pathway. Eukaryot Cell 7: 2017–2036.
8. Furukawa K, Hoshi Y, Maeda T, Nakajima T, Abe K (2005) *Aspergillus nidulans* HOG pathway is activated only by two-component signalling pathway in response to osmotic stress. Mol Microbiol 56: 1246–1261.
9. Hagiwara D, Asano Y, Marui J, Furukawa K, Kanamaru K, et al. (2007) The SskA and SrrA response regulators are implicated in oxidative stress responses of hyphae and asexual spores in the phosphorelay signaling network of *Aspergillus nidulans*. Biosci Biotechnol Biochem 71: 1003–1014.
10. Hagiwara D, Matsubayashi Y, Marui J, Furukawa K, Yamashino T, et al. (2007) Characterization of the NikA histidine kinase implicated in the phosphorelay signal transduction of *Aspergillus nidulans*, with special reference to fungicide responses. Biosci Biotechnol Biochem 71: 844–847.
11. Jones CA, Greer-Phillips SE, Borkovich KA (2007) The response regulator RRG-1 functions upstream of a mitogen-activated protein kinase pathway impacting asexual development, female fertility, osmotic stress, and fungicide resistance in *Neurospora crassa*. Mol Biol Cell 18: 2123–36.
12. Ochiai N, Fujimura M, Motoyama T, Ichiishi A, Usami R, et al. (2001) Characterization of mutations in the two-component histidine kinase gene that confer fludioxonil resistance and osmotic sensitivity in the os-1 mutants of *Neurospora crassa*. Pest Manag Sci 57: 437–442.
13. Izumitsu K, Yoshimi A, Tanaka C (2007) Two-component response regulators Ssk1p and Skn7p additively regulate high-osmolarity adaptation and fungicide sensitivity in *Cochliobolus heterostrophus*. Eukaryot Cell 6: 171–181.
14. Rispail N, Di Pietro A (2010) The two-component histidine kinase Fhk1 controls stress adaptation and virulence of *Fusarium oxysporum*. Mol Plant Pathol 11: 395–407.
15. Viaud M, Fillinger S, Liu W, Polepalli JS, Le Pêcheur P, et al. (2006) A class III histidine kinase acts as a novel virulence factor in *Botrytis cinerea*. Mol Plant Microbe Interact 19: 1042–1050.
16. Chauhan N, Calderone R (2008) Two-component signal transduction proteins as potential drug targets in medically important fungi. Infect Immun 76: 4795–4803.
17. Catlett NL, Yoder OC, Turgeon BG (2003) Whole-Genome Analysis of Two-Component Signal Transduction Genes in Fungal Pathogens. Eukaryot Cell 2: 1151–1161.
18. Pott GB, Miller TK, Bartlett JA, Palas JS, Selitrennikoff CP (2000) The isolation of FOS-1, a gene encoding a putative two-component histidine kinase from *Aspergillus fumigatus*. Fungal Genet Biol 31: 55–67.
19. Clemons KV, Miller TK, Selitrennikoff CP, Stevens DA (2002) *fos-1*, a putative histidine kinase as a virulence factor for systemic aspergillosis. Med Mycol 40: 259–62.
20. Du C, Sarfati J, Latge JP, Calderone R (2006) The role of the *sakA* (*Hog1*) and *tcsB* (*sln1*) genes in the oxidant adaptation of *Aspergillus fumigatus*. Med Mycol 44: 211–218.
21. Ji Y, Yang F, Ma D, Zhang J, Wan Z, et al. (2012) HOG-MAPK signaling regulates the adaptive responses of *Aspergillus fumigatus* to thermal stress and other related stress. Mycopathologia 174: 273–282.
22. McCormick A, Jacobsen ID, Broniszewska M, Beck J, Heesemann J, et al. (2012) The two-component sensor kinase TcsC and its role in stress resistance of the human-pathogenic mold *Aspergillus fumigatus*. PLoS One 7: e38262.
23. Schumacher MM, Enderlin CS, Selitrennikoff CP (1997) The osmotic-1 locus of *Neurospora crassa* encodes a putative histidine kinase similar to osmosensors of bacteria and yeast. Curr Microbiol 34: 340–347.
24. Cui W, Beever RE, Parkes SL, Weeds PL, Templeton MD (2002) An osmosensing histidine kinase mediates dicarboximide fungicide resistance in *Botryotinia fuckeliana* (*Botrytis cinerea*). Fungal Genet Biol 36: 187–198.
25. Yoshimi A, Tsuda M, Tanaka C (2004) Cloning and characterization of the histidine kinase gene Dic1 from *Cochliobolus heterostrophus* that confers dicarboximide resistance and osmotic adaptation. Mol Genet Gennomics 271: 228–236.
26. Motoyama T, Kadokura K, Ohira T, Ichiishi A, Fujimura M, et al. (2005a) A two-component histidine kinase of the rice blast fungus is involved in osmotic stress response and fungicide action. Fungal Genet Biol 42: 200–212.
27. Avenot H, Simoneau P, Iacomi-Vasilescu B, Bataillé-Simoneau N (2006) Characterization of mutations in the two-component histidine kinase gene AbNIK1 from *Alternaria brassicicola* that confer high dicarboximide and phenylpyrrole resistance. Curr Genet 47: 234–243.
28. Boyce KJ, Schreider L, Kirszenblat L, Andrianopoulos A (2011) The two-component histidine kinases DrkA and SlnA are required for in vivo growth in the human pathogen *Penicillium marneffei*. Mol Microbiol 82: 1164–1184.
29. Motoyama T, Ohira T, Kadokura K, Ichiishi A, Fujimura M, et al. (2005b) An Os-1 family histidine kinase from a filamentous fungus confers fungicide-sensitivity to yeast. Curr Genet 47: 298–306.
30. Bahn YS, Kojima K, Cox GM, Heitman J (2006) A unique fungal two-component system regulates stress responses, drug sensitivity, sexual development, and virulence of *Cryptococcus neoformans*. Mol Bio Cell 17: 3122–3135.
31. Dongo A, Bataillé-Simoneau N, Campion C, Guillemette T, Hamon B, et al. (2009) The group III two-component histidine kinase of filamentous fungi is involved in the fungicidal activity of the bacterial polyketide ambruticin. Appl Environ Microbiol 75: 127–134.
32. Meena N, Kaur H, Mondal AK (2010) Interactions among HAMP domain repeats act as an osmosensing molecular switch in group III hybrid histidine kinases from fungi. J Biol Chem 285: 12121–12132.
33. Buschart A, Gremmer K, El-Mowafy M, van den Heuvel J, Mueller PP, et al. (2012) A novel functional assay for fungal histidine kinases group III reveals the role of HAMP domains for fungicide sensitivity. J Biotechnol 157: 268–277.
34. Nemecek JC, Wüthrich M, Klein BS (2006) Global control of dimorphism and virulence in fungi. Science 312: 583–588.
35. Krappmann S, Sasse C, Braus GH (2006) Gene targeting in *Aspergillus fumigatus* by homologous recombination is facilitated in a nonhomologous end-joining-deficient genetic background. Eukaryot Cell 5: 212–215.
36. Mah JH, Yu JH (2006) Upstream and downstream regulation of asexual development in *Aspergillus fumigatus*. Eukaryotic Cell 5: 1585–1595.
37. Thau N, Monod M, Crestani B, Rolland C, Tronchin G, et al. (1994) Rodletless mutants of *Aspergillus fumigatus*. Infect Immun. 62: 4380–4388.
38. Latge JP, Mouyna I, Tekaia F, Beauvais A, Nierman W (2005) Specific molecular features in the organization and biosynthesis of the cell wall of *Aspergillus fumigatus*. Med Mycol 43: S1: 15–22.
39. Wong Sak Hoi J, Lamarre C, Beau R, Meneau I, Berepiki A, et al. (2011) A novel family of dehydrin-like proteins is involved in stress response in the human fungal pathogen *Aspergillus fumigatus*. Mol Biol Cell 22: 1896–1906.
40. Lamarre C, Ibrahim-Granet O, Du C, Calderone R, Latgé JP (2007) Characterization of the SKN7 ortholog of *Aspergillus fumigatus*. Fungal Genet Biol 44: 682–690.
41. Yoshimi A, Kojima K, Takano Y, Tanaka C (2005) GroupIII histidine kinase is a positive regulator of Hog1-type mitogen-activated protein kinase in filamentous fungi. Eukaryot Cell 4: 1820–1828.
42. Beever RE, Laracy EP (1986) Osmotic adjustment in the filamentous fungus *Aspergillus nidulans*. J Bacteriol 168: 1358–65.
43. Hohmann S, Krantz M, Nordlander B (2007) Yeast osmoregulation. Methods Enzymol 428: 29–45.
44. Ochiai N, Fujimura M, Oshima M, Motoyama T, Ichiishi A, et al. (2002) Effects of iprodione and fludioxonil on glycerol synthesis and hyphal development in *Candida albicans*. Biosci Biotechnol Biochem 66: 2209–2215.
45. Lin CH, Chung KR (2010) Specialized and shared functions of the histidine kinase- and HOG1 MAP kinase-mediated signaling pathways in *Alternaria alternata*, a filamentous fungal pathogen of citrus. Fungal Genet Biol 47: 818–827.
46. Luo YY, Yang JK, Zhu ML, Liu CJ, Li HY, et al. (2012) The group III two-component histidine kinase AlHK1 is involved in fungicides resistance, osmosensitivity, spore production and impacts negatively pathogenicity in *Alternaria longipes*. Curr Microbiol 64: 449–456.
47. Furukawa K, Katsuno Y, Urao T, Yabe T, Yamada-Okabe T, et al. (2002) Isolation and functional analysis of a gene, *tcsB*, encoding a transmembrane hybrid-type histidine kinase from *Aspergillus nidulans*. Appl Environ Microbiol 68: 5304–5310.
48. Banno S, Noguchi R, Yamashita K, Fukumori F, Kimura M, et al. (2007) Roles of putative His-to-Asp signaling modules HPT-1 and RRG-2, on viability and sensitivity to osmotic and oxidative stresses in *Neurospora crassa*. Curr Genet 51: 197–208.
49. Vargas-Pérez I, Sánchez O, Kawasaki L, Georgellis D, Aguirre J (2007) Response regulators SrrA and SskA are central components of a phosphorelay system involved in stress signal transduction and asexual sporulation in *Aspergillus nidulans*. Eukaryot Cell 6: 1570–1583.

50. Lee JW, Ko YJ, Kim SY, Bahn YS (2011) Multiple roles of Ypd1 phosphotransfer protein in viability, stress response, and virulence factor regulation in *Cryptococcus neoformans*. Eukaryot Cell 10: 998–1002.

51. Leroux P, Fritz R, Debieu D, Albertini C, Lanen C, et al. (2002) Mechanisms of resistance to fungicides in field strains of *Botrytis cinerea*. Pest Manag Sci 58: 876–888.

52. Islas-Flores I, Sanchez-Rodriguez Y, Brito-Argaez L, Peraza-Echeverria L, Rodriguez-Garcia C, et al. (2011) The amazing role of the group III of histidine kinases in plant pathogenic fungi, an insight to fungicide resistance. Asian Journal of Biochemistry 6: 1–14.

53. Li D, Agrellos OA, Calderone R (2010) Histidine kinases keep fungi safe and vigorous. Curr Opin Microbiol 13: 424–430.

54. Carroll AM, Sweigard JA, Valent B (1994) Improved vectors for selecting resistance to hygromycin. Fungal Genet Newsl 41: 20–21.

55. Yu JH, Hamari Z, Han KH, Seo JA, Reyes-Domínguez Y, et al. (2004) Double-joint PCR: a PCR-based molecular tool for gene manipulations in filamentous fungi. Fungal Genet Biol 41: 973–981.

56. Yoshimi A, Sano M, Inaba A, Kokubun Y, Fujioka T, et al. (2013) Functional Analysis of the α-1,3-Glucan Synthase Genes *agsA* and *agsB* in *Aspergillus nidulans*: AgsB Is the Major α-1,3-Glucan Synthase in This Fungus. PLoS ONE 8: e54893.

14

Identification of ABC Transporter Genes of *Fusarium graminearum* with Roles in Azole Tolerance and/or Virulence

Ghada Abou Ammar[1◐], **Reno Tryono**[1◐], **Katharina Döll**[2], **Petr Karlovsky**[2], **Holger B. Deising**[1,3], **Stefan G. R. Wirsel**[1,3]*

1 Institute of Agricultural and Nutritional Sciences, Faculty of Natural Sciences III, Martin-Luther-University Halle-Wittenberg, Halle (Saale), Germany, **2** Molecular Phytopathology and Mycotoxin Research Section, Georg-August-Universität Göttingen, Göttingen, Germany, **3** Interdisziplinäres Zentrum für Nutzpflanzenforschung, Martin-Luther-Universität Halle-Wittenberg, Halle (Saale), Germany

Abstract

Fusarium graminearum is a plant pathogen infecting several important cereals, resulting in substantial yield losses and mycotoxin contamination of the grain. Triazole fungicides are used to control diseases caused by this fungus on a worldwide scale. Our previous microarray study indicated that 15 ABC transporter genes were transcriptionally upregulated in response to tebuconazole treatment. Here, we deleted four ABC transporter genes in two genetic backgrounds of *F. graminearum* representing the DON (deoxynivalenol) and the NIV (nivalenol) trichothecene chemotypes. Deletion of *FgABC3* and *FgABC4* belonging to group I of ABC-G and to group V of ABC-C subfamilies of ABC transporters, respectively, considerably increased the sensitivity to the class I sterol biosynthesis inhibitors triazoles and fenarimol. Such effects were specific since they did not occur with any other fungicide class tested. Assessing the contribution of the four ABC transporters to virulence of *F. graminearum* revealed that, irrespective of their chemotypes, deletion mutants of *FgABC1* (ABC-C subfamily group V) and *FgABC3* were impeded in virulence on wheat, barley and maize. Phylogenetic context and analyses of mycotoxin production suggests that *FgABC3* may encode a transporter protecting the fungus from host-derived antifungal molecules. In contrast, *FgABC1* may encode a transporter responsible for the secretion of fungal secondary metabolites alleviating defence of the host. Our results show that ABC transporters play important and diverse roles in both fungicide resistance and pathogenesis of *F. graminearum*.

Editor: Jae-Hyuk Yu, University of Wisconsin - Madison, United States of America

Funding: This work was supported by stipends provided by the Islamic Development Bank (GAA, 57/SYR/P29) and the Deutscher Akademischer Austauschdienst (RT, A/09/75891). The work received additional support from the Interdisciplinary Center for Crop Plant Research (Interdisziplinäres Zentrum für Nutzpflanzenforschung, IZN). The funders had no role in study design, data collection and analysis, decision to publish, or preparation of the manuscript.

Competing Interests: The authors have declared that no competing interests exist.

* E-mail: stefan.wirsel@landw.uni-halle.de

◐ These authors contributed equally to this work.

Introduction

Fusarium head blight (FHB), caused by a number of closely related species including *Fusarium graminearum* Schwabe (teleomorph *Gibberella zeae* (Schwein.) Petch), is a major disease of wheat and other small-grain cereals. These fungi can cause considerable economic losses not only due to diminishing yield and quality of the harvest but also because of the production of mycotoxins in infected grains [1]. In *F. graminearum*, the most important mycotoxins are B-trichothecenes such as deoxynivalenol (DON) and nivalenol (NIV), but also zearalenone (ZEN) [1,2]. Infection of cereals leading to contamination of food and feed with these mycotoxins poses a health risk to consumers. The major sources of inoculum in FHB are ascospores produced by *F. graminearum* growing saprophytically on cereal debris. After expulsion from the perithecium, airborne ascospores infect wheat heads. Infection occurs most effectively at the stage of anthesis. Some FHB-causing fungi including *F. graminearum* may infect cereals at other developmental stages resulting in seedling blight, foot, crown or root rots [1].

Control of FHB includes agronomic practices such as appropriate crop rotation, tilling and fungicide application, and the utilisation of resistant cultivars. Management practices integrating several control measures performed better than the application of measures separately [3,4]. In North America and Europe, the preferred fungicides to control FHB are triazoles such as tebuconazole, prothioconazole and metconazole, all of which are sterol biosynthesis inhibitors (SBI) class I [5]. Recently, declining efficacies of these fungicides was reported [6,7].

In our previous work, we investigated the capability of *F. graminearum* to develop resistance to azoles and the molecular mechanisms underlying this process. Cultivation of strain NRRL 13383 in the presence of a sublethal concentration of tebuconazole allowed to recover isolates with enhanced tolerance to that fungicide [8]. Transcriptome analysis of *F. graminearum* challenged with tebuconazole *in vitro* [9] showed strong responses for some genes of the sterol biosynthesis pathway, noticeably *FgCyp51A* to

FgCyp51C encoding cytochrome P450 sterol 14α-demethylase, which is the molecular target of azoles. Furthermore, 15 out of 54 genes encoding ABC transporters were more than twofold upregulated by tebuconazole treatment. Functional proof for a contribution of CYP51 to azole resistance in *F. graminearum* was provided by deletion analyses [10,11]. It is however uncertain whether mutations in any of the three *Cyp51* genes or changes in their regulation cause increased azole tolerance in field strains.

In addition to CYP51, membrane-bound transporters affect the sensitivity of fungal pathogens to azoles [12,13,14]. Contribution of these proteins to azole resistance in *F. graminearum* has not been shown before. Taking advantage of our previous transcriptome analysis, we have chosen in this study four genes encoding ABC transporters for functional analyses. We deleted these genes to determine their contribution to fungicide resistance, virulence and mycotoxin production.

Materials and Methods

Fungal Cultivation

The strains *F. graminearum* PH1 and NRRL 13383 used in this study, as well as the procedures used for their growth, sporulation and storage were described previously [8,9].

Vegetative growth rates were determined on PDA plates (Ø 90 mm) at 15°C, 23°C and 30°C. Mycelial plugs (Ø 5 mm) taken from margins of colonies grown on PDA at 23°C for five days were used for inoculation. Two perpendicular measurements of colony diameters were taken during seven days and averaged. Each variant was replicated four times.

The capacity of fungal strains to produce macroconidia was determined in 50 ml Mung Bean Broth (MBB) [15] in 250 ml Erlenmeyer flasks inoculated with five mycelial plugs per flask as above. Cultures were incubated at 23°C with 100 rpm for 7 days. Conidia were harvested by filtering through Miracloth (Merck, Darmstadt, Germany) and collected by centrifugation at 3000×g for 10 min. Conidial density was determined using a haemacytometer (Brand, Wertheim, Germany). Each strain was grown in four cultures and conidia were counted twice. Statistical significances were determined by T-test ($p<0.05$) as implemented in the Microsoft Excel 2010 software.

Germination efficiency of macroconidia was determined on glass slides inoculated with 20 μl of a conidial suspension (10^4 ml^{-1}), covered with a cover glass and incubated on three layers of moistened paper towels inside a $120 \times 120 \times 17$ mm plastic dish (Greiner Bio-One, Solingen, Germany) at 23°C for 24 h under illumination with near-UV light (L18W/73, Osram, Munich, Germany). Germinated and ungerminated conidia were counted twice in four replicates per strain. Statistical analysis was carried out as described above.

Procedures for Nucleic Acid Isolation Manipulations

The isolation of fungal genomic DNA and total RNA followed published methods [9]. Cultivation and treatment of mycelia with tebuconazole, preparation and validation of RNA, RT-qPCR and data analysis were performed as outlined previously [9]. Each variant was represented by four biological replicas (RNAs from independent cultures), each of which was analysed twice by RT-qPCR. Expression data of candidate genes were normalised with those from three reference genes (FGSG_01244, FGSG_06245, FGSG_10791) as reported [9].

DNA fragments used for fungal transformation were generated by the DJ-PCR method [16]. Marker genes employed for selection encoded hygromycin phosphotransferase (*hph*), nourseothricin acetyltransferase (*nat1*) and neomycin phosphotransferase (*npt*).

Using plasmids pAN7-1 [17], pNR1 [18] and pII99 [19] as templates, DNA fragments comprising these markers genes including heterologous constitutive promoters were generated by PCR and then fused by DJ-PCR with the left and right flanks of the respective target gene. The deletion constructs generated included *hph* for *FgABC1* (FGSG_10995), *nat1* for *FgABC2* (FGSG_17046) and *FgABC4* (FGSG_17058) and *npt* for *FgABC3* (FGSG_04580) (see Fig. S1). Oligonucleotides used in this study are listed in Table S1.

Generation and Validation of *F. graminearum* Transformants

For the preparation of protoplasts, 5×10^6 macroconidia were incubated for 12 h in 100 mL of YEPD at 28°C and 175 rpm. The mycelium was recovered on a sterile paper filter and then incubated for 4 h at 30°C and 90 rpm in 20 ml of protoplasting mix (500 mg driselase, 1 mg chitinase, 100 mg lysing enzyme of *Trichoderma harzianum* (all from Sigma-Aldrich, Schnelldorf, Germany) in 1.2 M KCl). Protoplasts were harvested at R.T. by centrifugation at 1000×g and suspended in 1 mL STC buffer (1.2 M sorbitol, 50 mM CaCl$_2$, 10 mM Tris-HCl, pH 7.5). A transformation reaction contained 10^7 protoplasts in 100 μL STC buffer, 50 μL 30% PEG 8000 and 8 μg DNA of the deletion construct in 50 μL water. After incubation for 20 min at RT and 50 rpm, 2 mL 30% PEG 8000 and 5 min later 4 mL STC buffer were added. Aliquots of 600 μL were mixed with 15 mL molten regeneration medium (275 g of sucrose, 0.5 g yeast extract, 0.5 g casein hydrolysate, 5 g of agar per litre) and poured into a Petri dish. After incubation for 12 h at 26°C, 15 mL of molten regeneration medium containing, depending on the marker gene used for selection, either hygromycin B, nourseothricin or G418 at concentrations of 200 μg/mL, were poured onto the surface of the agar. Colonies that started emerging after 4 d were harvested to obtain single spore isolates as described before [8].

Transformants were analysed by PCR and Southern hybridisation. Probes for the latter were generated with the PCR DIG Labeling Mix^{PLUS} kit (Roche Diagnostics, Mannheim, Germany) as recommended by the manufacturer. Bioluminescence was visualized by exposing Nylon membranes to Hyperfilm ECL X-ray film (Amersham Pharmacia Biotech, Piscataway, USA).

Determination of Sensitivity to Fungicides and Plant Metabolites

The sensitivity of transformants to fungicides and plant compounds was tested on PDA plates (12×12 cm, Greiner Bio-One) amended with appropriate concentrations of a given substance. For each compound, we used three concentrations that were optimized in preliminary experiments. The following fungicides were obtained as commercial formulations: azoxystrobin (Amistra, BASF), fenpropimorph (Corbel, BASF), metconazole (Caramba, BASF), prochloraz (Sportak, BASF) and tebuconazole (Folicur, Bayer). Pure active compounds epoxyconazole, fenarimol, spiroxamine, boscalid and dithianon were obtained from Sigma-Aldrich (Schnelldorf, Germany), except for prothioconazole, which was kindly provided by Bayer. Tolnaftat (Sigma-Aldrich) was included as a control xenobiotic that has never been used in agriculture. Sensitivity of fungal strains against plant secondary metabolites 2-benzoxazolinone (= BOA), 3-(dimethylaminomethyl)indole (= gramine), 2,3-dihydro-5,7-dihydroxy-2-(4-hydroxyphenyl)-4H-1-benzopyran-4-one, 4′,5,7-trihydroxyflavanone (= naringenin) and 3,3′,4′,5,7-pentahydroxyflavone dihydrate (= quercetin) purchased from Sigma-Aldrich was tested in the same way. For all substances, stock solutions were prepared in

DMSO; the final concentration of the solvent in culture media was at most 0.3%. Two µL of a suspension of macroconidia $(1 \times 10^5 \text{ mL}^{-1})$ were used as inoculum to assess the effect of the compounds on germination. Mycelial plugs on unamended PDA were used as inoculum to determine the effect of the compounds on vegetative growth. At least three plates were used for each compound and concentration. Incubation was carried out in the dark at 23°C for strains derived from NRRL 13383 and at 30°C for strains with PH-1 background. Colony area was determined on digital photographs using ImageJ software version 1.46 (http://rsbweb.nih.gov/ij/index.html). Statistical analysis was done as above.

Wheat Ear Infection Assay

Cultivation of wheat in the greenhouse and environmentally controlled growth chambers and ear inoculation was described earlier [8]. Shortly, *F. graminearum* strains were point-inoculated into the 9th spikelet of wheat cultivar Kadrilj (SW Seed Hadmersleben, Hadmersleben, Germany) when it reached anthesis. The inoculum consisted of 300 macroconidia suspended in 10 µL of 0.02% Tween 20. For each strain tested, at least ten wheat heads were inoculated and covered with plastic bags misted with water to maintain high humidity. The bags were removed after 2 days post inoculation (dpi) and the incubation continued at 25°C, 70% relative humidity until 14 dpi. The development of bleached spikelets in the heads was recorded daily.

Maize Stem Infection Assay

Maize plants cultivar Golden Jubilee (Territorial Seed Company, Cottage Grove, OR, USA) were cultivated for six weeks in a greenhouse at 24°C with 50% relative humidity and a 14 h photoperiod, which employed lamps (Plantstar 600 Watt E40, Osram, Munich, Germany) providing 4.2×10^{17} photons $\text{sec}^{-1} \text{ m}^{-2}$ at the surface of the bench. For each strain tested, at least five plants were inoculated by punching a hole into the stem at the first internode using a sterile tootpick, followed by injection of 1000 macroconidia in 10 µL of 0.02% Tween 20. The control plants were inoculated with 0.02% Tween 20. The hole was covered with Parafilm for 7d to maintain high humidity and exclude other organisms. At 14 dpi, the stalks were split longitudinally and the symptoms were documented by photography. The extent of the necrotic area was quantified using ImageJ software version 1.46 as above. Statistical analysis was performed as above.

Barley Ear Infection Assay

Barley cultivar Barke (Saatzucht Josef Breun, Herzogenaurach, Germany) was cultivated for ten weeks in a greenhouse using the same conditions as described for maize. For each strain tested, sixteen mature ears were inoculated employing a glass flacon to spray 2000 macroconidia in 2 ml of 0.02% Tween 20 onto each ear. The inoculated ears were enclosed in a misted plastic bag for 2d. After an additional incubation of 12d, the number of bleached spikelets was recorded for each head. Statistical analysis was carried out as above.

Analysis of Mycotoxin Production *in vitro*

Fusarium isolates were grown in rice media, culture material was extracted with acetonitrile/water and the extracts were defatted as described [20]. Mycotoxins were separated by HPLC on an RP column (Polaris C18 ether, 100×2 mm, 3 µm particle size; Agilent, Darmstadt, Germany) at 40°C at a flow rate of 0.2 ml/min. The solvent system consisted of (A) water with 5%

acetonitrile and (B) methanol, both containing 7 mM acetic acid. The elution gradient rose linearly from 10% to 98% of solvent B followed by washing and equilibration steps. The detection was performed by tandem mass spectrometry using triple quadrupole 1200L (Varian, Darmstadt, Germany) after electrospray ionization in negative mode as described before [21,22]. For each isolate, at least five independent cultures were analysed. Statistical analysis was done as above.

Results

Targeted Deletion of ABC Transporter Genes *FgABC1*, *FgABC2*, *FgABC3* and *FgABC4*

We chose four genes encoding ABC transporters, which were previously found upregulated in strain PH-1 after tebuconazole treatment, for targeted deletion mutagenesis. These genes clustered in three ABC subfamilies [9]. ABC transporters have been classified according to several schemes. Using one that was based on the yeast nomenclature classified FgABC1 (FGSG_10995) and FgABC4 (FGSG_17058) into the MRP (multidrug resistance-related protein) subfamily, FgABC3 (FGSG_04580) into the PDR (pleiotropic drug resistance) subfamily and FgABC2 (FGSG_17046) into a distinct unnamed clade [9]. Applying an alternative classification scheme, which was originally used to classify human ABC transporters, FgABC1 and FgABC4 were assigned to subfamily ABC-C group V, FgABC3 to subfamily ABC-G group I and FgABC2 to subfamily ABC-A group I [23]. Since expression of these four genes was previously only analysed in strain PH-1, we determined their transcript levels by RT-qPCR in strain NRRL 13383 after a 12 h treatment with 5 ppm tebuconazole. In comparison to untreated controls, the transcript levels of *FgABC1*, *FgABC2*, *FgABC3* and *FgABC4* were increased 3.0-, 3.1-, 3.4- and 3.9-fold, respectively. The transcriptional responses of strain NRRL 13383 were thus similar to those previously observed in strain PH-1 [9], except for gene *FgABC4*, which responded stronger in PH-1 (10.6-fold). We introduced deletions of these four genes into both strains, i.e. NRRL 13383 and PH-1, which have the NIV and the 15ADON chemotypes, to examine whether any of the resulting phenotypes may occur independently from the genetic background of the host used for transformation including its chemotype.

DNA cassettes, which comprised a dominant resistance marker gene for an antibiotic controlled by a heterologous promoter and the left and right flanks of the targeted genes, were transformed into both strains, i.e. PH-1 and NRRL 13383 (Fig. S1). Single spore isolates were analysed by PCR and Southern hybridisation for the mode of DNA integration (Fig. S1). Several transformants containing the resistance cassette integrated by a double-cross-over event at the targeted locus (type III integration according to [24]) were obtained for each gene and recipient strain.

Vegetative Fitness and Fungicide Sensitivities of the Deletion Mutants

We assessed whether the deletion of a given ABC transporter gene would impair the resulting transformants with respect to vegetative growth and asexual reproduction. For each deletion, we examined growth rates on PDA at three temperatures (Fig. S2) and quantified the formation (Fig. S3) and the germination of macroconidia *in vitro* in two transformants (Fig. S4). None of the deletions led to any significant change in any of the three attributes. This was true for the transformants in the PH-1 and the NRRL 13383 backgrounds.

We determined the impact of 11 fungicides belonging to the chemical groups of anthraquinones, imidazoles, methoxy-acrylates, morpholines, pyridine-carboxamides, pyrimidines, spiroketal-amines and triazoles on germination and vegetative growth of the transformants. In the background of NRRL 13383, we observed significantly reduced tolerance in ΔFgABC3 strains for the triazoles tebuconazole, prothioconazole and epoxyconazole (Fig. 1). Similarly, ΔFgABC4 mutants were significantly less tolerant for the latter two. In addition, both of these deletions led to significantly reduced tolerance against fenarimol, which has the same target as the triazoles (SBI class I) but is a pyrimidine. There existed no significantly changed sensitivities against fungicides grouped into SBI class II, QoI, SDHI, anthraquinone and N-phenyl carbamate, as well as tolnaftat. The deletion of the same four ABC transporter genes in the PH-1 background affected the resulting mutants in a similar way as in NRRL 13383. However, the reductions in tolerance to the above-mentioned fungicides were less severe so that in some cases these trends were not statistically significant (Fig. 1). Deletion of the genes FgABC1 and FgABC2 did not significantly reduce the tolerance levels for any fungicide in any of the two backgrounds. The impact of the SBI class I fungicides prothioconazole and fenarimol on vegetative hyphae was observed by microscopy (Fig. S5). In untreated control cultures, hyphal morphology of all mutants resembled that of the wild type strains. In contrast, treatment with 3 ppm of either fungicide induced aberrant hyphal morphology in ΔFgABC3 and ΔFgABC4 mutants, but not in the WT and in ΔFgABC1 and ΔFgABC2 mutants. Such hyphae appeared thicker and had swellings that emerged throughout the mycelium, but most often apically. Occasionally, such structures collapsed. These effects resembled those reported previously for tebuconazole treatment of Fusarium culmorum [25].

We examined whether the deletion of the four genes encoding ABC transporters might have affected the sensitivity of the transformants to four commercially available secondary metabolites with antifungal properties produced by cereals. However, at none of the concentrations tested neither BOA, gramine, naringenin nor quercetin impaired the growth of any mutant in any background significantly different from the respective wild type strain (not shown).

Transcript Abundances of FgABC1 to FgABC4 in the Deletion Mutants

We determined by RT-qPCR the transcript levels of the four ABC transporter genes in untreated and tebuconazole-treated wild type NRRL 13383 and a single mutant for each deletion. As expected, no transcripts were detected for the deleted gene in the mutant that was deleted for that gene (Fig. 2). In the wild type, the transcript levels of all genes increased significantly after tebuconazole treatment (Fig. 2, #). Interestingly, this transcriptional response was lost for FgABC1 in the deletion mutants of genes FgABC2, FgABC3 and FgABC4. Likewise, transcriptional responses to the fungicide treatment were also lost for FgABC4 in the deletion mutants of genes FgABC1, FgABC2 and FgABC3. The same two genes, i.e. FgABC1 and FgABC4, were the only showing significant differences when comparing the mutants to the wild type (Fig. 2, asterisks). Transcript levels of FgABC1 were significantly lower in the deletion mutants of genes FgABC2, FgABC3 and FgABC4, but only in cultures treated with tebuconazole. The corresponding effects were observed for gene FgABC4, although this proved only significant for the deletion mutants of FgABC1 and FgABC3.

Virulence of the Deletion Mutants

For each background and each deleted gene, two transformants were point-inoculated into central spikelets of wheat ears. Over the entire period monitored, the percentage of bleached spikelets per head was higher in heads inoculated with strain PH-1 than with strain NRRL 13383 (Figs. 3, 4). Deletion of FgABC1 caused a strong reduction of virulence in both backgrounds. Although the mutants were still able to cause local infections, they spread very slowly, as compared to the corresponding wild type strains. At the end of the scoring period, in the PH-1 background the ΔFgABC1 mutants had caused disease in only about one third of the spikelets, as compared to wild type strain (Fig. 3a). In the NRRL 13383 background, symptom development was even slower (Fig. 3b). Deletion of FgABC3 also resulted in strongly reduced symptoms in both genetic backgrounds, comparable to ΔFgABC1 deletion strains. In contrast, deletion mutants of FgABC2 and FgABC4 caused symptom developments resembling those of the corresponding wild type references (Fig. 3).

F. graminearum has a rather broad host range encompassing several cultivated and wild grasses, allowing to determine whether virulence factors discovered to be essential for infection of wheat are also essential for the infection of other host species. Interestingly, the same ABC transporter genes required for full virulence in wheat were also required for full virulence in maize (Figs. 4, 5) and barley (Figs. 4, 6). Compared to the respective wild type strains, deletion mutants of FgABC1 and FgABC3 were significantly reduced for virulence on maize stems, which was true in both genetic backgrounds, i.e. PH-1 (Fig. 5a) and NRRL 13383 (Fig. 5b). Virulence defects were more severe for ΔFgABC1 than for ΔFgABC3 strains. In barley, the reduction in virulence of the ΔFgABC1 mutants was more evident in the NRRL 13383 (Fig. 6b) than in the PH-1 background (Fig. 6a). The ΔFgABC3 mutants showed similar effects in both backgrounds.

Production of Mycotoxins by the Deletion Mutants

We analysed whether the deletion of the ABC transporter genes affected the levels of B-trichothecenes and zearalenone, and whether possible alterations might explain the results seen in the virulence assays. In the PH-1 background, the levels of DON (Fig. 7a) and 15ADON (Fig. 7b) produced in vitro were increased in all four deletion mutants when compared to the wild type strain. The ZEN levels produced by the mutants were similar to PH-1, except for the deletion mutant of FgABC4 that showed higher levels (Fig. 7c). In the NRRL 13383 background, deletion of FgABC1 led to higher and deletion of FgABC3 led to lower NIV levels, whereas the other two mutants resembled the wild type (Fig. 7d). None of the deletion mutants produced ZEN (Fig. 7f) at levels that differed significantly from the wild type. This experiment indicated that the strongly reduced virulence observed in the deletion mutants of FgABC1 and FgABC3 was likely not caused by a reduction of trichothecenes that represent virulence factors for the infection of wheat and maize, but not barley [26].

Discussion

We study mechanisms mediating azole resistance in F. graminearum. Exploiting transcriptomic and phylogenetic data [9], we chose four genes for functional analyses encoding full-size ABC transporters belonging to three subfamilies. We found that the deletion mutants ΔFgABC3 and ΔFgABC4 had acquired a higher sensitivity to several fungicides belonging to the SBI class I. Remarkably, the deletion of FgABC1 and FgABC3 caused a strong reduction of virulence on three economically important crops, wheat, barley and maize.

Effects of Deletions on Fungicide Sensitivity

Our study shows that the deletion of FgABC3 and of FgABC4 caused enhanced sensitivity to several triazoles and to fenarimol

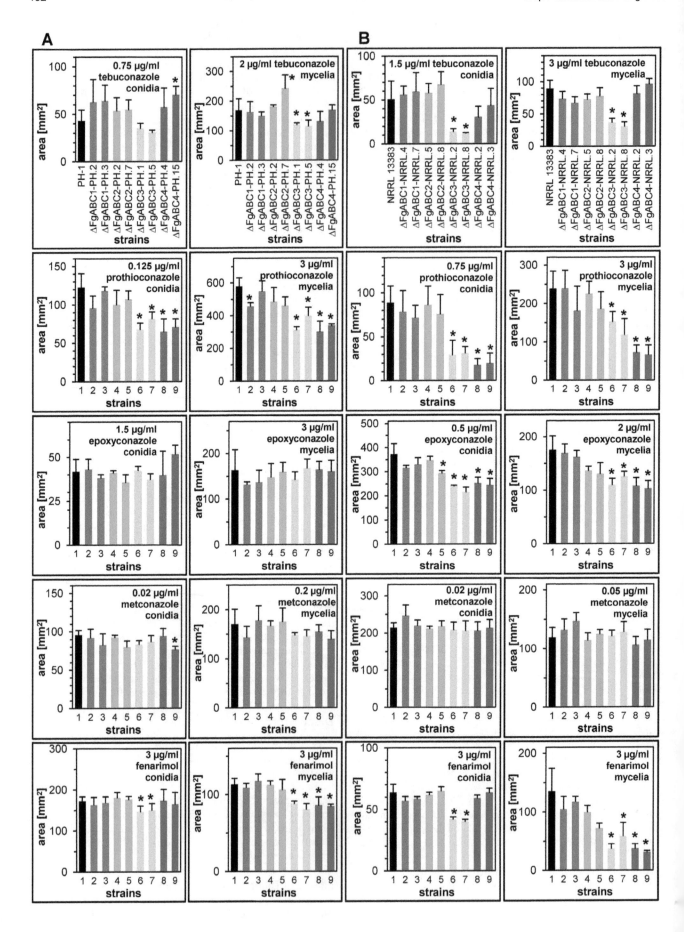

Figure 1. Sensitivity to SBI class I fungicides. For each deletion, colonial areas of two transformants of each genetic background were assessed on PDA amended with the indicated concentration of fungicides. Error bars represent SD. An asterisk indicates a significant difference between a mutant and the wild type. (A) PH-1 background, strains tested: 1) PH-1, 2) ΔFgABC1-PH.2, 3) ΔFgABC1-PH.3, 4) ΔFgABC2-PH.2, 5) ΔFgABC2-PH.7,6) ΔFgABC3-PH.1, 7) ΔFgABC3-PH.5, 8) ΔFgABC4-PH.4, 9) ΔFgABC4-PH.15; (B) NRRL 13383 background, strains tested: 1) NRRL 13383, 2) ΔFgABC1-NRRL.4, 3) ΔFgABC1-NRRL.7, 4) ΔFgABC2-NRRL.5, 5) ΔFgABC2-NRRL.8, 6) ΔFgABC3-NRRL.2, 7) ΔFgABC3-NRRL.8, 8) ΔFgABC4-NRRL.2, 9) ΔFgABC4-NRRL.3.

that are classified as SBI class I fungicides. These effects are rather specific since such deletion mutants do neither suffer from general fitness impairment nor do they show increased sensitivity against other fungicide classes. Considering that the genome of *F. graminearum* comprises 54 genes putatively encoding ABC transporters [9], it is notable that single deletions already yielded this phenotype. In addition to *FgABC3* and *FgABC4*, the PDR and the MRP subfamilies harbour additional genes that also responded to tebuconazole [9]. This could suggest that distinct transcriptional responses only occurring for *FgABC3* and *FgABC4* may not explain

why similar paralogs were not able to complement the deletions. Nonetheless, our RT-qPCR analyses revealed for two of the four genes, *FgABC1* and *FgABC4*, a loss of transcriptional upregulation, which occurs in the wild type, in response to triazole, if any of the other three genes studied was deleted. In triazole-treated Δ*FgABC3* mutants, transcript levels of *FgABC4* were significantly lower than those of the fungicide-treated wild type. As shown above, *FgABC4* is needed to maintain wild type levels of fungicide tolerance. Thus, the reduced sensitivity in the Δ*FgABC3* mutants might have been indirectly caused by the decrease of *FgABC4* transcript levels.

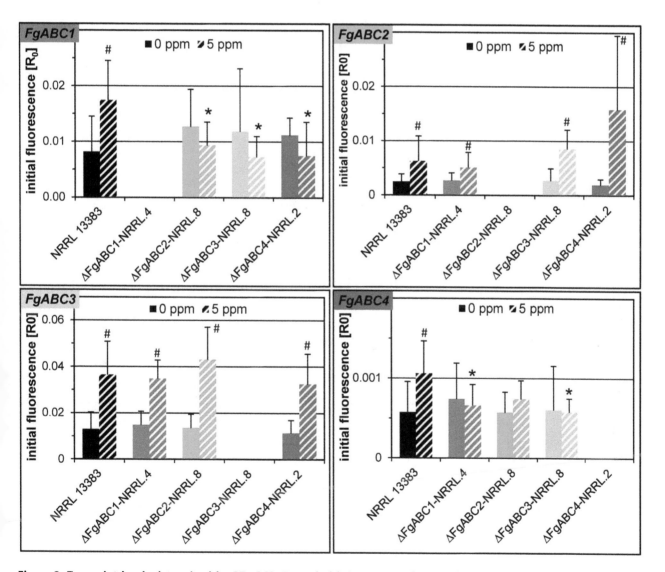

Figure 2. Transcript levels determined by RT-qPCR. For each deletion, one transformant of the NRRL 13383 background was analysed. Columns show calculated initial fluorescence after normalisation with three reference genes. The analysed gene is indicated in the upper left corner of a given box. For each strain, RNA preparations were assayed originating from cultures amended with 5 ppm tebuconazole or not. Error bars represent SD. For each strain, # indicate significant differences between the fungicide treatment and the control. Within each treatment, an asterisk indicates a significant difference between a mutant and the wild type.

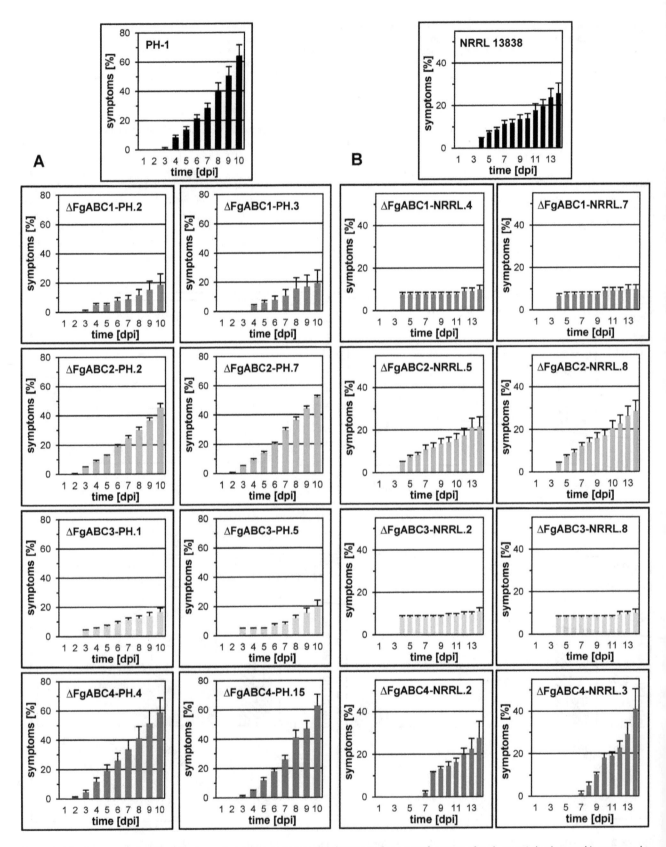

Figure 3. Virulence on wheat heads. For each deletion, symptom development of two transformants of each genetic background is compared to the respective wild type strain for up to 14 dpi. Columns show the percentage of symptomatic spikelets in point-inoculated wheat heads. Error bars represent SE. A) PH-1 background, B) NRRL 13383 background.

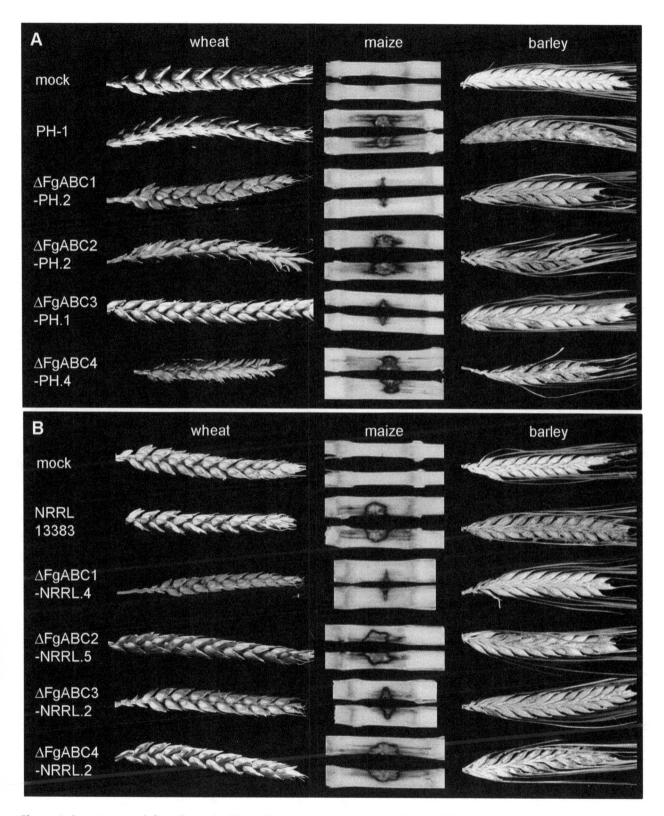

Figure 4. Symptoms on infected cereals. Photos show symptoms occurring at the end of the monitoring. One representative example is provided for each genotype. Mocks were treated with 0.02% Tween 20. A) PH-1 background, B) NRRL 13383 background.

However, this is unlikely because also the deletion of *FgABC1* analogously affected the transcript levels of *FgABC4*. Conversely, such mutants were similar to the wild type with respect to

fungicide sensitivity. Therefore, the reduction in triazole tolerance seen in the mutants of *FgABC3* is mainly resulting from the deletion of that gene. Another explanation why similar paralogs

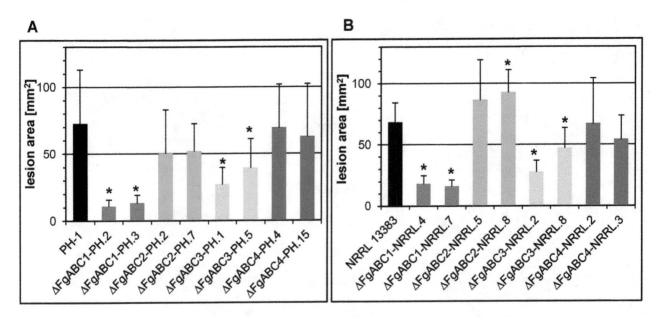

Figure 5. Virulence on maize stems. For each deletion, two transformants of each genetic background are compared to the respective wild type strain. Columns give symptomatic areas in maize stems that were harvested at 14 dpi and then split longitudinally. Error bars represent SD. An asterisk indicates a significant difference between a mutant and the wild type. A) PH-1 background, B) NRRL 13383 background.

were not able to complement adequately the deletions of *FgABC3* and *FgABC4* could be the existence of distinct post-transcriptional regulation. Alternatively, the transporters missing in the deletion mutants could have distinct substrate specificities that only poorly matched those of other transporters.

Typically, fungal ABC transporters known to mediate fungicide resistance belong to the PDR (ABC-G) and the MDR (ABC-B) and to a lesser extent to the MRP (ABC-C) subfamilies. PDR transporters, whose contribution to azole resistance had been analysed in detail, are for example CDR1 and CDR2 in *Candida albicans* and PDR5 in *Saccharomyces cerevisiae* [27]. Like FgABC3,

these proteins belong to group I in the ABC-G subfamily [23]. In contrast to human pathogens, far fewer functional genetic analyses have been performed for PDR transporters potentially mediating azole resistance in plant pathogenic fungi. Disruption mutants of *PMR1* (ABC-G group I) in *Penicillium digitatum*, a pathogen of citrus, exhibited increased sensitivity to azoles [28]. By employing gene replacement and overexpression in *Botrytis cinerea*, a necrotroph with a wide host range, BcAtrD (ABC-G group I) was shown to mediate azole resistance [29]. *ShAtrD*, a homolog of BcAtrD, was found to be overexpressed in azole resistant field isolates of *Sclerotinia homoeocarpa* causing dollar spot disease of turf

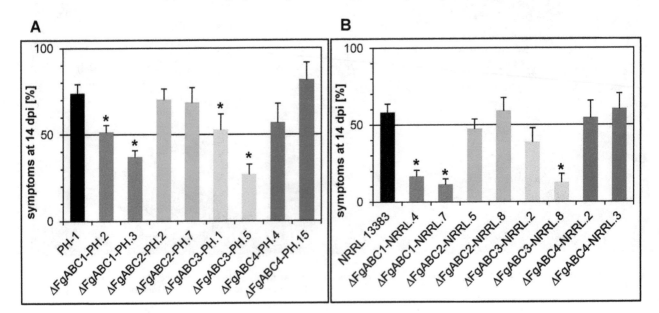

Figure 6. Virulence on barley heads. For each deletion, two transformants of each genetic background are compared to the respective wild type strain. Columns show the percentage of symptomatic spikelets in spray-inoculated barley heads at 14 dpi. Error bars represent SE. A) PH-1 background, B) NRRL 13383 background.

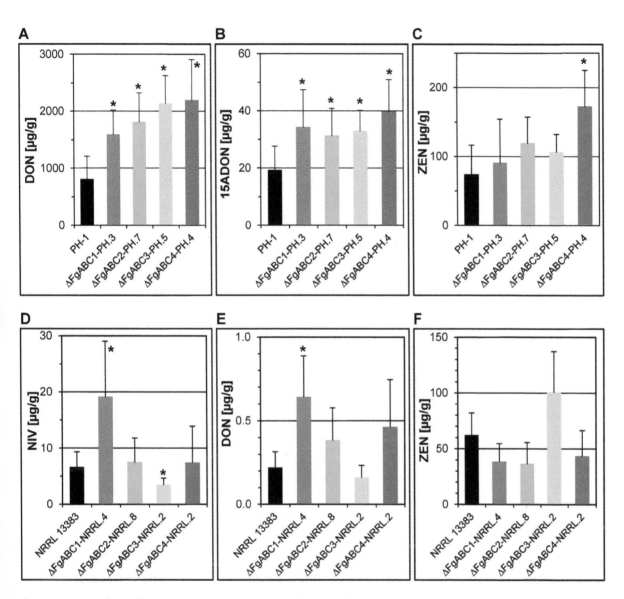

Figure 7. Mycotoxin production. For each deletion, one transformant of each genetic background is compared to the respective wild type strain. Columns show the concentration of a given substance in rice cultures determined by HPLC-MS/MS. Error bars represent SD. An asterisk indicates a significant difference between a mutant and the wild type. A) to C) PH-1 background, D) to F) NRRL 13383 background. A), E) DON; B) 15ADON; C), F) ZEN; D) NIV.

grasses [30]. In *Mycosphaerella graminicola*, a pathogen infecting leaves of wheat, several ABC transporters were functionally analysed [31]. Heterologous expression of *MgAtr1* (ABC-G group III), *MgAtr2* (ABC-G group I) and *MgAtr4* (ABC-G group I) in a *S. cerevisiae* mutant with deletions of six ABC transporter genes resulted in increased tolerance against azoles. However, when the genes *MgAtr1* to *MgAtr5* were deleted individually in *M. graminicola*, no change in azole sensitivity was observed, possibly because of redundant substrate specificities [31]. At the moment, it is uncertain whether FgABC3 and FgABC4 possess very distinct substrate specificities that do not overlap sufficiently with those of other ABC transporters to allow for a compensation of the observed fungicide phenotype.

Effects of Deletions on Virulence and Mycotoxin Production

Interestingly, in addition to fungicide tolerance, the deletion of *FgABC3* reduced the virulence on wheat and barley heads but also maize stems, suggesting an important role of the encoded protein, and thus of the molecules transported by it, during pathogenesis. The biological roles of several transporters in the PDR subfamily have been studied earlier. This applies for *FgABC3* that was previously identified in a microarray analysis as a down-regulated gene (*FgZRA1*) in a deletion mutant of *FgZEB2* (FGSG_02398) [32]. *FgZEB2* encodes a transcription factor regulating the gene cluster for zearalenone biosynthesis [33]. Deletion mutants of *FgZRA1* (= *FgABC3* = FGSG_04580) accumulated less ZEN in liquid medium as well as in the mycelium [32]. The authors discussed that FgZRA1 is unlikely to export ZEN. The effect of *FgZRA1* deletion on ZEN production could not be explained, its role in fungicide sensitivity and virulence was not investigated.

Other reports indicated that deletion mutants for the genes involved in ZEN biosynthesis were causing the same levels of FHB on wheat and barley as the wild type strains, suggesting that ZEN is dispensable for virulence on these hosts [33,34]. In contrast to that earlier study on *FgZRA1* [32], our analysis did not show reduced ZEN levels in the deletion mutants of *FgABC3* in any of the two backgrounds studied. Differences in the genetic backgrounds and/or culture conditions between the two studies may account for this discrepancy. On the other hand, an involvement of PDR subfamily transporters in pathogenesis was demonstrated in several cases. MgAtr4 of *M. graminicola* is needed to attain full virulence on wheat and it was proposed that it may protect the pathogen against host defence molecules [35]. Similarly, BcAtrB (ABC-G group V), was described to protect *B. cinerea* against the phytoalexins resveratrol in grapevine [36] and camalexin in *Arabidopsis thaliana* [37]. In *Magnaporthe oryzae*, a hemibiotrophic pathogen of rice, the most similar protein to FgABC3 is MoABC1 (ABC-G group I). The deletion of *MoABC1* yielded mutants that were severely reduced in virulence [38]. Again, it was suggested that MoABC1 might protect the invading fungus from plant defence molecules. Later research detected a subclade within the ABC-G subfamily group I, which is distinctive to *Fusarium* spp. [39]. Functional characterisation of three members of this subclade, FcABC1 in *F. culmorum* [40], NhABC1 in *Nectria haematococca* (anamorph: *F. solani*) [39] and GpABC1 in *Gibberella pulicaris* (anamorph: *F. sambucinum*) [41] demonstrated in all cases that the encoded proteins are essential for full virulence. It was shown for the latter two transporters that they are needed to protect the pathogen from phytoalexins of their hosts, i.e. pisatin and rishitin. In conclusion, considering the literature and the results of our ZEN measurements, we propose that the biological function of FgABC3 may rather be to export a host-derived defence compound than to export the fungal secondary metabolite ZEN. Our rationale is supported by the considerably decreased levels of virulence caused by Δ*FgABC3* mutants on all three hosts tested. A virulence defect is not expected if the function of FgABC3 would be to export ZEN, because as outlined above, ZEN does not contribute to virulence. Currently, the exported molecule remains unknown, since none of the cereal metabolites that we have tested showed noteworthy variation in their effect on deletion mutants and wild type strains. Published microarray data comparing the transcriptome of *F. graminearum* during FHB on wheat and barley [42] show that *FgABC3* has the highest transcript levels among the four genes studied here (Fig. S6). In wheat, *FgABC3* transcripts peaked at 4 dpi, in barley they continuously increased until to the end of the experiment. This may indicate that FgABC3 is more important during late than early stages of infection.

Deletion mutants of *FgABC1* were impeded in infections of wheat, barley and maize irrespective of their trichothecene chemotype. The phylogenetically most similar protein to FgABC1 is FgABC4 [9,23]; both of which are members of the MRP subfamily (ABC-C group V). Despite their similarity, deletion of *FgABC4* did not significantly affect virulence on any host tested, regardless of the chemotype. As outlined above, the opposite phenomenon was observed with respect to fungicide tolerance since the deletion of *FgABC1*, in contrast to *FgABC4*, did not cause changes compared to wild type strains, whereas deletion of *FgABC4* did so. Interrogation of published microarray data [42] indicates that transcriptional patterns of the two genes during the infection of wheat and barley were quite similar and thus inappropriate to explain why *FgABC4* cannot functionally compensate the virulence defects of the *FgABC1* deletion mutants (Fig. S6). Other explanations for this failure are contrasting post-

transcriptional regulation, contrasting substrate specificities leading to discrimination between the native substrates of the two transporters or different subcellular localisations. The latter appears however questionable since the softwares Euk-mPLoc 2.0 (http://www.csbio.sjtu.edu.cn/bioinf/euk-multi-2/) and YLoc Fungi (http://abi.inf.uni-tuebingen.de/Services/YLoc/webloc.cgi) predicted for both transporters the plasma membrane as their most likely localisation (not shown).

Transporters of the MRP subfamily (ABC-C) have been less studied than those in the PDR subfamily. Recently, in *M. oryzae* three members of this subfamily have been functionally characterised [43]. Only the deletion of *MoABC5* resulted in reduced virulence on rice. The encoded protein belongs to ABC-C group V, which comprises only two members in *M. oryzae* [23]. In *Aspergillus*, several members in ABC-C group V seem to transport fungal secondary metabolites produced by nonribosomal peptide synthetase (NRPS) and polyketide synthase (PKS) enzymes [23]. *FgABC1* (FGSG_10995) resides in a supposed NRPS gene cluster [44]. In a recent microarray study, this gene and the other members of the cluster were found upregulated during infection of wounded wheat coleoptiles. Individual deletions of three genes of the cluster that encoded the transporter, an NRPS and a putative peptidoglycan deacetylase yielded mutants that exhibited reduced virulence on wheat [44]. Our Δ*FgABC1* mutants also showed strongly reduced virulence on wheat and furthermore on barley and maize. The effect on FHB in wheat was reminiscent, even though less severe, than that seen in Δ*FgTri5* mutants, which are unable to produce B-trichothecenes [45]. The latter were reported to remain restricted just to the initially infected spikelet. We observed often a similar effect in the background of NRRL 13383, which, however, is less aggressive on wheat than PH-1. In NRRL 13383, the Δ*FgABC1* mutants spread at most to two additional spikelets. Our mycotoxin analyses show that the production of trichothecenes is not impeded in the Δ*FgABC1* mutants. Therefore, the hitherto unknown secondary metabolites synthesized by the NRPS cluster, to which *FgABC1* belongs, are likely required for infection of wheat, barley and maize.

We have functionally analysed the four ABC transporter genes in two genetic backgrounds, i.e. NRRL 13383 and PH-1, to assess whether the respective genomic context may influence the effect of gene deletion. Whereas deletion of *FgABC3* and *FgABC4* caused in NRRL 13383 significantly reduced tolerance to certain class I sterol biosynthesis inhibitors, this effect was somewhat less prominent in PH-1. Due to the lack of the genome sequence of NRRL 13383 it is unknown whether this strain has exactly the same set of ABC transporters as PH-1. Variations in their numbers, sequences and regulation could cause putative compensatory effects, although other reasons may apply. Nevertheless, our results show that alterations in fungicide sensitivities resulting from gene deletions may vary in their extents in different genomic contexts. In contrast, in the virulence tests we observed rather similar consequences of the deletions in NRRL 13383 and PH-1 indicating that the virulence defects observed in the Δ*FgABC1* and Δ*FgABC3* mutants do occur independently from the trichothecene chemotype, highlighting the importance of these genes for achieving full virulence on cereals.

Supporting Information

Figure S1 Gene deletions. Strategy employed to generate deletion mutants of *FgABC1* to *FgABC4* (A). Results from PCR (B) and Southern hybridisations (C) document the respective genotypes.

Figure S2   Growth kinetics *in vitro*. For each deletion, two transformants of each genetic background are compared to the respective wild type strain at three temperatures on PDA medium. Boxes on the left side show results for the PH-1 and those on the right side for the NRRL 13383 background. Each data point represents the mean of four replicated cultures.

Figure S3   Formation of macroconidia *in vitro*. For each deletion, two transformants of each genetic background are compared to the respective wild type strain. Data shown give the average conidial densities formed in MBB medium in four replicated cultures after incubation for 7 d at 23°C. Error bars represent SE. None of the variations between the mutants and the wild type is significant. A) PH-1 background, B) NRRL 13383 background.

Figure S4   Germination of macroconidia *in vitro*. For each deletion, two transformants of each genetic background are compared to the respective wild type strain. Data shown give the average frequencies of germinated macroconidia on glass slides in four replicated cultures after incubation for 24 h at 23°C. Error bars represent SD. Variations between mutants and wild types are not significant. A) PH-1 background, B) NRRL 13383 background.

Figure S5   Impact of SBI class I fungicides on hyphal morphology. For each strain, cultures containing 3 ppm of prothioconazole or fenarimol or no fungicide were grown for 4 d in liquid PDA. Only Δ*FgABC3* and Δ*FgABC4* mutants are shown, since Δ*FgABC1* and Δ*FgABC2* mutants were like the wild type references. Observation by bright field microscopy at 400x magnification.

Figure S6   Transcript levels during FHB. Data for *FgABC1* to *FgABC4* transcript levels were taken from published work (Lysoe et al., 2011). A) Time course of infection of wheat, B) of barley.

Table S1   Oligonucleotides. Used for the generation of deletion constructs, Southern blots, analytical PCR and RT-qPCR.

Acknowledgments

We thank Rayko Becher for stimulating discussions, Elke Vollmer for greenhouse services and Andrea Beutel for technical support.

Author Contributions

Conceived and designed the experiments: SGRW. Performed the experiments: GAA RT KD. Analyzed the data: GAA RT KD SGRW. Contributed reagents/materials/analysis tools: PK. Wrote the paper: GAA RT KD HBD SGRW.

References

1. Becher R, Miedaner T, Wirsel SGR (2013) Biology, diversity and management of FHB-causing *Fusarium* species in small-grain cereals. In: Kempken F, editor. The Mycota XI, Agricultural Applications. 2 ed. Berlin Heidelberg: Springer-Verlag. pp. in press, DOI 10.1007/1978–1003–1642–36821–36829_36828.

2. Stepien L, Chelkowski J (2010) *Fusarium* head blight of wheat: pathogenic species and their mycotoxins. World Mycotoxin J 3: 107–119.

3. Willyerd KT, Li C, Madden LV, Bradley CA, Bergstrom GC, et al. (2012) Efficacy and stability of integrating fungicide and cultivar resistance to manage *Fusarium* head blight and deoxynivalenol in wheat. Plant Disease 96: 957–967.

4. Blandino M, Haidukowski M, Pascale M, Plizzari L, Scudellari D, et al. (2012) Integrated strategies for the control of *Fusarium* head blight and deoxynivalenol contamination in winter wheat. Field Crop Res 133: 139–149.

5. Paul PA, McMullen MP, Hershman DE, Madden LV (2010) Meta-analysis of the effects of triazole-based fungicides on wheat yield and test weight as influenced by *Fusarium* head blight intensity. Phytopathology 100: 160–171.

6. Klix MB, Verreet JA, Beyer M (2007) Comparison of the declining triazole sensitivity of *Gibberella zeae* and increased sensitivity achieved by advances in triazole fungicide development. Crop Prot 26: 683–690.

7. Yin Y, Liu X, Li B, Ma Z (2009) Characterization of sterol demethylation inhibitor-resistant isolates of *Fusarium asiaticum* and *F. graminearum* collected from wheat in China. Phytopathology 99: 487–497.

8. Becher R, Hettwer U, Karlovsky P, Deising HB, Wirsel SGR (2010) Adaptation of *Fusarium graminearum* to tebuconazole yielded descendants diverging for levels of fitness, fungicide resistance, virulence, and mycotoxin production. Phytopathology 100: 444–453.

9. Becher R, Weihmann F, Deising HB, Wirsel SGR (2011) Development of a novel multiplex DNA microarray for *Fusarium graminearum* and analysis of azole fungicide responses. BMC Genomics 12: 52.

10. Liu X, Yu F, Schnabel G, Wu J, Wang Z, et al. (2011) Paralogous *Cyp51* genes in *Fusarium graminearum* mediate differential sensitivity to sterol demethylation inhibitors. Fungal Genet Biol 48: 113–123.

11. Fan J, Urban M, Parker JE, Brewer HC, Kelly SL, et al. (2013) Characterization of the sterol 14alpha-demethylases of *Fusarium graminearum* identifies a novel genus-specific CYP51 function. New Phytol 198: 821–835.

12. Cannon RD, Lamping E, Holmes AR, Niimi K, Baret PV, et al. (2009) Efflux-mediated antifungal drug resistance. Clin Microbiol Rev 22: 291–321.

13. Becher R, Wirsel SG (2012) Fungal cytochrome P450 sterol 14alpha-demethylase (CYP51) and azole resistance in plant and human pathogens. Appl Microbiol Biotechnol 95: 825–840.

14. De Waard MA, Andrade AC, Hayashi K, Schoonbeek HJ, Stergiopoulos I, et al. (2006) Impact of fungal drug transporters on fungicide sensitivity, multidrug resistance and virulence. Pest Manag Sci 62: 195–207.

15. Bai GH, Shaner G (1996) Variation in *Fusarium graminearum* and cultivar resistance to wheat scab. Plant Dis 80: 975–979.

16. Yu JH, Hamari Z, Han KH, Seo JA, Reyes-Dominguez Y, et al. (2004) Double-joint PCR: A PCR-based molecular tool for gene manipulations in filamentous fungi. Fungal Genet Biol 41: 973–981.

17. Punt PJ, Oliver RP, Dingemanse MA, Pouwels PH, van den Hondel CA (1987) Transformation of *Aspergillus* based on the hygromycin B resistance marker from *Escherichia coli*. Gene 56: 117–124.

18. Malonek S, Rojas MC, Hedden P, Gaskin P, Hopkins P, et al. (2004) The NADPH-cytochrome P450 reductase gene from *Gibberella fujikuroi* is essential for gibberellin biosynthesis. J Biol Chem 279: 25075–25084.

19. Namiki F, Matsunaga M, Okuda M, Inoue I, Nishi K, et al. (2001) Mutation of an arginine biosynthesis gene causes reduced pathogenicity in *Fusarium oxysporum* f. sp. *melonis*. Mol Plant Microbe Interact 14: 580–584.

20. Nutz S, Döll K, Karlovsky P (2011) Determination of the LOQ in real-time PCR by receiver operating characteristic curve analysis: Application to qPCR assays for *Fusarium verticillioides* and *F. proliferatum*. Anal Bioanal Chem 401: 717–726.

21. Adejumo TO, Hettwer U, Karlovsky P (2007) Occurrence of *Fusarium* species and trichothecenes in Nigerian maize. Int J Food Microbiol 116: 350–357.

22. Adejumo TO, Hettwer U, Karlovsky P (2007) Survey of maize from south-western Nigeria for zearalenone, alpha- and beta-zearalenols, fumonisin B-1 and enniatins produced by *Fusarium* species. Food Addit Contam 24: 993–1000.

23. Kovalchuk A, Driessen AJ (2010) Phylogenetic analysis of fungal ABC transporters. BMC Genomics 11: 177.

24. Fincham JR (1989) Transformation in fungi. Microbiol Rev 53: 148–170.

25. Kang ZS, Huang LL, Krieg U, Mauler-Machnik A, Buchenauer H (2001) Effects of tebuconazole on morphology, structure, cell wall components and trichothecene production of *Fusarium culmorum* in vitro. Pest Manag Sci 57: 491–500.

26. Maier FJ, Miedaner T, Hadeler B, Felk A, Salomon S, et al. (2006) Involvement of trichothecenes in fusarioses of wheat, barley and maize evaluated by gene disruption of the trichodiene synthase (*Tri5*) gene in three field isolates of different chemotype and virulence. Mol Plant Pathol 7: 449–461.

27. Prasad R, Goffeau A (2012) Yeast ATP-binding cassette transporters conferring multidrug resistance. Annu Rev Microbiol 66: 39–63.

28. Nakaune R, Adachi K, Nawata O, Tomiyama M, Akutsu K, et al. (1998) A novel ATP-binding cassette transporter involved in multidrug resistance in the phytopathogenic fungus *Penicillium digitatum*. Appl Environ Microbiol 64: 3983–3988.

29. Hayashi K, Schoonbeek HJ, De Waard MA (2002) Expression of the ABC transporter BcatrD from *Botrytis cinerea* reduces sensitivity to sterol demethylation inhibitor fungicides. Pestic Biochem Phys 73: 110–121.

30. Hulvey J, Popko JT, Sang H, Berg A, Jung G (2012) Overexpression of *ShCYP51B* and *ShatrD* in *Sclerotinia homoeocarpa* isolates exhibiting practical field resistance to a demethylation inhibitor fungicide. Appl Environ Microb 78: 6674–6682.

31. Zwiers LH, Stergiopoulos I, Gielkens MM, Goodall SD, De Waard MA (2003) ABC transporters of the wheat pathogen *Mycosphaerella graminicola* function as protectants against biotic and xenobiotic toxic compounds. Mol Genet Genomics 269: 499–507.

32. Lee S, Son H, Lee J, Lee YR, Lee YW (2011) A putative ABC transporter gene, *ZRA1*, is required for zearalenone production in *Gibberella zeae*. Curr Genet 57: 343–351.

33. Kim YT, Lee YR, Jin JM, Han KH, Kim H, et al. (2005) Two different polyketide synthase genes are required for synthesis of zearalenone in *Gibberella zeae*. Mol Microbiol 58: 1102–1113.

34. Gaffoor I, Brown DW, Plattner R, Proctor RH, Qi WH, et al. (2005) Functional analysis of the polyketide synthase genes in the filamentous fungus *Gibberella zeae* (anamorph *Fusarium graminearum*). Eukaryot Cell 4: 1926–1933.

35. Stergiopoulos I, Zwiers LH, De Waard MA (2003) The ABC transporter MgAtr4 is a virulence factor of *Mycosphaerella graminicola* that affects colonization of substomatal cavities in wheat leaves. Mol Plant Microbe Interact 16: 689–698.

36. Schoonbeek H, Del Sorbo G, De Waard MA (2001) The ABC transporter BcatrB affects the sensitivity of *Botrytis cinerea* to the phytoalexin resveratrol and the fungicide fenpiclonil. Mol Plant Microbe Interact 14: 562–571.

37. Stefanato FL, Abou-Mansour E, Buchala A, Kretschmer M, Mosbach A, et al. (2009) The ABC transporter BcatrB from *Botrytis cinerea* exports camalexin and is a virulence factor on *Arabidopsis thaliana*. Plant J 58: 499–510.

38. Urban M, Bhargava T, Hamer JE (1999) An ATP-driven efflux pump is a novel pathogenicity factor in rice blast disease. EMBO J 18: 512–521.

39. Coleman JJ, White GJ, Rodriguez-Carres M, VanEtten HD (2011) An ABC transporter and a cytochrome P450 of *Nectria haematococca* MPVI are virulence factors on pea and are the major tolerance mechanisms to the phytoalexin pisatin. Mol Plant Microbe Interact 24: 368–376.

40. Skov J, Lemmens M, Giese H (2004) Role of a *Fusarium culmorum* ABC transporter (FcABC1) during infection of wheat and barley. Physiol Mol Plant Pathol 64: 245–254.

41. Fleissner A, Sopalla C, Weltring KM (2002) An ATP-binding cassette multidrug-resistance transporter is necessary for tolerance of *Gibberella pulicaris* to phytoalexins and virulence on potato tubers. Mol Plant Microbe Interact 15: 102–108.

42. Lysoe E, Seong KY, Kistler HC (2011) The transcriptome of *Fusarium graminearum* during the infection of wheat. Mol Plant Microbe Interact 24: 995–1000.

43. Kim Y, Park SY, Kim D, Choi J, Lee YH, et al. (2013) Genome-scale analysis of ABC transporter genes and characterization of the ABCC type transporter genes in *Magnaporthe oryzae*. Genomics 101: 354–361.

44. Zhang XW, Jia LJ, Zhang Y, Jiang G, Li X, et al. (2012) In planta stage-specific fungal gene profiling elucidates the molecular strategies of *Fusarium graminearum* growing inside wheat coleoptiles. Plant Cell 24: 5159–5176.

45. Jansen C, von Wettstein D, Schäfer W, Kogel KH, Felk A, et al. (2005) Infection patterns in barley and wheat spikes inoculated with wild-type and trichodiene synthase gene disrupted *Fusarium graminearum*. Proc Natl Acad Sci USA 102: 16892–16897.

15

Studies on Inhibition of Respiratory Cytochrome bc_1 Complex by the Fungicide Pyrimorph Suggest a Novel Inhibitory Mechanism

Yu-Mei Xiao[1,2], Lothar Esser[2], Fei Zhou[2], Chang Li[1], Yi-Hui Zhou[1,2], Chang-An Yu[3], Zhao-Hai Qin[1]*, Di Xia[2]*

1 Department of Applied Chemistry, China Agricultural University, Beijing, China, 2 Laboratory of Cell Biology, Center for Cancer Research, National Cancer Institute, NIH, Bethesda, Maryland, United States of America, 3 Department of Biochemistry and Molecular Biology, Oklahoma State University, Stillwater, Oklahoma, United States of America

Abstract

The respiratory chain cytochrome bc_1 complex (cyt bc_1) is a major target of numerous antibiotics and fungicides. All cyt bc_1 inhibitors act on either the ubiquinol oxidation (Q_P) or ubiquinone reduction (Q_N) site. The primary cause of resistance to bc_1 inhibitors is target site mutations, creating a need for novel agents that act on alternative sites within the cyt bc_1 to overcome resistance. Pyrimorph, a synthetic fungicide, inhibits the growth of a broad range of plant pathogenic fungi, though little is known concerning its mechanism of action. In this study, using isolated mitochondria from pathogenic fungus *Phytophthora capsici,* we show that pyrimorph blocks mitochondrial electron transport by affecting the function of cyt bc_1. Indeed, pyrimorph inhibits the activities of both purified 11-subunit mitochondrial and 4-subunit bacterial bc_1 with IC_{50} values of 85.0 μM and 69.2 μM, respectively, indicating that it targets the essential subunits of cyt bc_1 complexes. Using an array of biochemical and spectral methods, we show that pyrimorph acts on an area near the Q_P site and falls into the category of a mixed-type, noncompetitive inhibitor with respect to the substrate ubiquinol. *In silico* molecular docking of pyrimorph to cyt b from mammalian and bacterial sources also suggests that pyrimorph binds in the vicinity of the quinol oxidation site.

Editor: Lijun Rong, University of Illinois at Chicago, United States of America

Funding: This research was supported in part by the Intramural Research Program of the NIH, National Cancer Institute, Center for Cancer Research, and by a grant from the National Basic Research Science Foundation of China (2010CB126100) to ZQ. The funders had no role in study design, data collection and analysis, decision to publish, or preparation of the manuscript.

Competing Interests: The authors have the following interests. Dimethomorph was a gift from Jiangsu Frey Chemical Co. Ltd. (Jiangsu Province, China).

* E-mail: qinzhaohai@263.net (ZQ); xiad@mail.nih.gov (DX)

Introduction

The cytochrome bc_1 complex (cyt bc_1, also known as ubiquinone:cyt c oxidoreductase, Complex III or bc_1) is a central component of the cellular respiratory chain of mitochondria. It catalyzes the reaction of electron transfer (ET) from ubiquinol to cyt c and couples this reaction to proton translocation across the mitochondrial inner membrane, contributing to the cross-membrane proton motive force essential for cellular functions such as ATP synthesis [1,2]. The indispensible function of cyt bc_1 in cellular energy metabolism makes it a prime target for numerous natural and synthetic antibiotics. More than 20 synthetic fungicides targeting cyt bc_1 are in widespread use in agriculture with an annual sale exceeding $2.7 billion [3].

All cyt bc_1 inhibitors target either the ubiquinol oxidation site (Q_P or Q_o) or the ubiquinone reduction site (Q_N or Q_i), which are defined by the Q-cycle mechanism of cyt bc_1 function [4,5]. Despite variations in subunit compositions of bc_1 from various organisms, only three subunits are essential for ET-coupled proton translocation function: they are cyt b, cyt c_1 and the iron-sulfur protein (ISP). The cyt b subunit contains two b-type hemes (b_L and b_H), the cyt c_1 subunit has a c-type heme, and the ISP possesses a

2Fe-2S cluster. Both active sites are located within the cyt b subunit, as demonstrated by crystallographic studies of mitochondrial and bacterial bc_1 complexes [6–12]. Resistance to known cyt bc_1 fungicides has been reported at an alarming rate, rendering many of these reagents ineffective. Most common mechanisms of resistance involve target site mutations and corresponding strategies to overcome drug resistance have been proposed [13]. Developing new agents targeting areas outside the Q_P and Q_N sites of cyt bc_1 is most attractive primarily because the new compounds presumably are able to circumvent existing fungal resistance.

Pyrimorph, (Z)-3-[(2-chloropyridine-4-yl)-3-(4-*tert*-butylphenyl)-acryloyl] morpholine, is a novel systemic antifungal agent that belongs to the family of carboxylic acid amide (CAA) fungicides [14], whose members include mandipropamid, dimethomorph, flumorph, and valinine derivatives. Pyrimorph exhibits excellent activity inhibiting mycelial growth of the fungal species *Phytophthora infestans*, *Phytophthora capsici*, and *Rhizoctonia solani* and is able to suppress zoosporangia germination of *Pseudoperonospora cubensis* with EC_{50} values in the range between 1.3 and 13.5 μM [15]. The *in vitro* sensitivities of various asexual stages of *Peronophythora litchii* to pyrimorph were studied with four single-sporangium isolates, showing high sensitivity at the stage of mycelial growth with an EC_{50} of 0.3 μM [16].

Although pyrimorph is currently in use to control various fungal pathogens [15–17], its functional mechanism has remained unclear. The presence of a common CAA moiety has led to the suggestion that pyrimorph may work in a fashion similar to that of other CAA-type fungicides [18]. One CAA member, mandipropamid, was shown to target the pathway of cell wall synthesis by inhibiting the CesA3 cellulose synthases [19]. However, treatment of fungal pathogens with pyrimorph appeared to affect multiple cellular pathways, including, but not limited to, those of cellular energy metabolism and cell wall biosynthesis, either directly or indirectly [20]. Indeed, a recent report has correlated the pyrimorph resistance phenotype in *P. capsici* with mutations in the CesA3 gene [21].

Other mechanisms of pyrimorph action have yet to be investigated. In particular, its potential interference with cellular respiratory chain components leading to reduced ATP synthesis appears to be a reasonable hypothesis for the observed inhibitory effects on energy demanding processes such as mycelial growth and cytospore germination of fungi. Here, we report the effects of pyrimorph on electron flow through the isolated fungal mitochondrial respiratory chain and the identification of the cyt bc_1 complex as pyrimorph's primary target. Kinetic experiments suggest that the mode of pyrimorph inhibition is to interfere with substrate access to the ubiquinol oxidation site but in a way that differs from other bc_1 inhibitors, suggesting a novel mode of inhibitory mechanism.

Materials and Methods

The pyrimorph used in all experiments was synthesized in our laboratory. Dimethomorph was a gift from Jiangsu Frey Chemical Co. Ltd. (Jiangsu Province, China). Cyt c (from horse heart, type III) was purchased from Sigma-Aldrich (St. Louis, MI). 2,3-dimethoxy-5-methyl-6-(10-bromodecyl)-1,4-benzoquinol ($Q_0C_{10}BrH_2$) was prepared as previously reported [22]. N-dodecyl-β-D-maltoside (β-DDM) and N-octyl-β-D-glucoside (β-OG) were purchased from Affymetrix (Santa Clara, CA). All other chemicals were purchased and are of the highest grade possible.

Preparation of Light Mitochondria from *Phytophthora capsici*

Light mitochondrial fraction were prepared from cultured mycelia from laboratory strain *Phytophthora capsici Leonia* (*P. capsici*), which was grown in CA liquid medium (8% carrot juice and 2% glucose) for 5 days in the dark at 25°C [23]. 10 g mycelia (fresh weight) were washed with 0.6 M mannitol solution and ground up for 5 minutes with an ice-cold mortar and pestle in 100 ml buffer A containing 10 mM MOPS•KOH, pH 7.1, 0.3 M mannitol, 1 mM EDTA and 0.1% (w/v) bovine serum albumin (BSA) and 30 g of sea sand. The homogenate was centrifuged at 3,200×g for 10 min at 4°C and the supernatant was further centrifuged at 12,000×g for 30 min. The precipitate, light mitochondrial fraction, was resuspended and washed with 20 ml buffer B containing 10 mM MOPS•KOH, pH 7.1, 0.25 M sucrose and 1 mM EDTA and pelleted again by centrifugation at 12,000×g for 20 min at 4°C. The mitochondrial preparation was resuspended in buffer A and the protein concentration was adjusted to 0.1 mg/ml.

Table 1. Inhibition of respiratory complexes I, II, and III by pyrimorph.

| Complex | Concentration of pyrimorph (μM) | | | | | | | | | |
| | 0 | | 4 | | 8 | | 12 | | 16 | |
	Activity*	% Inh	Activity	% Inh	Activity	% Inh	Activity	% Inh	Activity	% Inh
I (Inhibition of NADH oxidation)	51.8±2	0	52.7±3	-1.7	48.3±6	6.8	52.1±7	-0.5	49.8±1	3.8
II (Inhibition of DCPIP reduction)	112.7	0	83.6±4	25.8	58.3±6	48.3	25.1±7	77.7	6.8±1	94.6
III (Inhibition of cyt c reduction)	27.24	0	1.29±1	95.3	—	—	—	—	—	—

*Millimolar extinction coefficients for oxidation NADH is 6.2 mM^{-1} cm^{-1} [49], for DCPIP is 21 mM^{-1} cm^{-1} [50], and for cyt c is 18.5 mM^{-1} cm^{-1} [24].

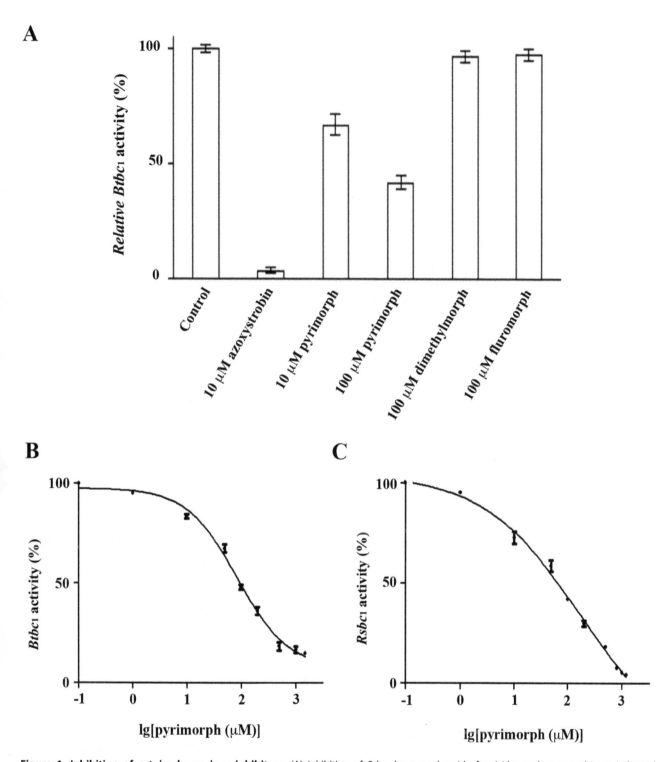

Figure 1. Inhibition of cyt bc_1 by various inhibitors. (A) Inhibition of $Btbc_1$ by several amide fungicides and azoxystrobin at indicated concentrations. The control is the activity of bc_1 in the absence of inhibitor, which is set to 100%. (B) Concentration-dependent inhibition of $Btbc_1$ by pyrimorph. (C) Concentration-dependent inhibition of $Rsbc_1$ by pyrimorph.

Inhibition of the ET Activity of *P. capsici* Mitochondria by Pyrimorph

The activities of mitochondrial respiratory chain components were assayed using the Mitochondria Complex Activity Assay Kit (Genmed Scientifics, Inc. USA, Wilmington, DE) following manufacturer's instruction. Briefly, Complex I activity was measured by following the oxidation of NADH by monitoring the decrease in absorbance difference between 340 nm and 380 nm. The reaction mixture (1 ml) consisted of 50 mM potassium phosphate buffer, pH 7.6, 0.25 mM NADH and 50 mM decylubiquinone as the electron acceptor. Crude mitochondria (200 µg protein) were added to start the reaction.

Complex II activity was estimated as the rate of reduction of ubiquinone to ubiquinol by succinate, which can be followed by the secondary reduction of 2,6-dichlorophenolindophenol (DCPIP) as the ubiquinol forms. The reaction mixture (1 ml) contained 50 mM potassium phosphate buffer, pH 7.6, 20 mM succinate, 1.0 mM EDTA, 0.05 mM DCPIP and 3 mM NaN_3, and 50 mM decylubiquinone. Crude mitochondria (65 µg) were added to initiate the reaction and the decrease in absorbance at 600 nm was followed as DCPIP becomes reduced. Complex III activity was assayed by following the increase in absorbance at 550 nm as cyt c becomes reduced using decylubiquinol as an electron donor. Here, the reaction mixture (1 ml) consisted of 50 mM potassium phosphate buffer, pH 7.6, 0.1% BSA, 0.1 mM EDTA, 60 mM oxidized cyt c, and 150 µM decylubiquinol. Crude mitochondria (10 µg protein) were then added to initiate the reaction.

Purification of Cyt bc_1 Complexes from Beef Heart and Photosynthetic Bacterium *Rhodobacter sphaeroides*

Bovine heart mitochondrial bc_1 ($Btbc_1$) complex was prepared starting from highly purified succinate-cyt c reductase, as previously reported [24]. The bc_1 particles were solubilized by deoxycholate and contaminants were removed by a 15-step ammonium acetate fractionation. The purified bc_1 complex was recovered in the oxidized state from the precipitates formed between 18.5% and 33.5% ammonium acetate saturation. The final product was dissolved in 50 mM Tris•HCl buffer, pH 7.8, containing 0.66 M sucrose resulting in a stock solution with a protein concentration of 30 mg/ml, which was stored at −80°C. The concentrations of cyt b and c_1 were determined spectroscopically using millimolar extinction coefficients of 28.5 and 17.5 $mM^{-1} cm^{-1}$ for cyt b and c_1, respectively.

To prepare cyt bc_1 complex from the photosynthetic bacterium *R. sphaeroides* ($Rsbc_1$), *R. sphaeroides* strain BC17 cells bearing the pRKD418-*fbc*FBC6HQ plasmid [25] were grown photosynthetically at 30°C in an enriched Sistrom medium containing 5 mM glutamate and 0.2% casamino acids [26]. The growth was monitored by measuring the OD_{600} value every 3–5 h. Cells were transferred to a larger batch or harvested when OD_{600} reached 1.8–2.0. Chromatophore membranes were prepared from BC17 cells as described previously [27] and stored at a very high concentration in the presence of 20% glycerol at −80°C. To purify the hexahistidine-tagged $Rsbc_1$ complex, the freshly prepared chromatophores or frozen chromatophores thawed on ice were adjusted to a cyt b concentration of 25 µM with a solubilization buffer containing 50 mM Tris•HCl, pH 8.0 at 4°C, and 1 mM $MgSO_4$. 10% (w/v) β-DDM was added to the chromatophore suspension to a final concentration of 0.56 mg detergent/nmole of cyt b followed by addition of 4M NaCl solution to a final concentration of 0.1 M. After stirring on ice for 1 hour, the admixture was centrifuged at 220,000×g for 90 minutes; the supernatant was collected and diluted with equal volume of the solubilization buffer followed by passing through a Ni-NTA agarose column (100 nmole of cyt b/ml of resin) pre-equilibrated with two volumes of the solubilization buffer. After loading, the column was washed sequentially with the following buffers until the absence of greenish color in effluent was reached: washing buffer (50 mM Tris•HCl, pH 8.0 at 4°C, containing 100 mM NaCl) +0.01% β-DDM; washing buffer +0.01% β-DDM and 5 mM histidine; washing buffer +0.5% β-OG; washing buffer +0.5% β-OG and 5 mM histidine. The cyt bc_1 complex was eluted with the washing buffer +0.5% β-OG and 200 mM histidine. Pure fractions were combined and concentrated by Centriprep-30

concentrator. Glycerol was added to a final concentration of 10% before storage at −80°C.

Measurement of bc_1 Activity and its Inhibition (IC_{50}) by Various Inhibitors

The activities of isolated cyt bc_1 complexes were assayed following the reduction of substrate cyt c. The purified bc_1 complexes were diluted to a final concentration of 0.1 µM and 1 µM based on the concentration of cyt b for $Btbc_1$ and $Rsbc_1$, respectively, in the B200 buffer (50mM Tris•HCl, pH 8.0, 0.01% β-DDM, 200 mM NaCl). The assay mixture contains 100 mM phosphate buffer, pH 7.4, 0.3 mM EDTA, and 80 µM cyt c, and $Q_0C_{10}BrH_2$ at a final concentration of 5 µM. The addition of 3 µl of diluted bc_1 solution initiates the reaction, which is recorded immediately following the cyt c reduction at 550 nm wavelength for 100 seconds in a two-beam Shimadzu UV-2250 PC spectrophotometer at 23°C. The amount of cyt c reduced over a given period of time was calculated using a millimolar extinction coefficient of 18.5 $mM^{-1} cm^{-1}$.

To measure the effect of bc_1 inhibitors, bc_1 was pre-incubated at various concentrations of an inhibitor for 15 minutes prior to the measurement of its activity. The IC_{50} value was calculated by a least-squares procedure fitting the equation ($Y = A_{min} + (A_{max} - A_{min})/(1+10^{(X-\log IC50)})$ implemented in the commercial package Prism, where A_{max} and A_{min} are maximal and minimal activities, respectively. Although the chemical properties of $Q_0C_{10}BrH_2$ are comparable to those of $Q_0C_{10}H_2$, the former is a better substrate for the cyt bc_1 complex isolated in detergent solution [22].

Reaction Kinetics of bc_1 in the Presence of Inhibitors

To measure the enzyme kinetics of cyt bc_1 complex under inhibitory conditions, purified cyt bc_1, either $Btbc_1$ or $Rsbc_1$, was assayed at different concentrations of substrates. When the $Q_0C_{10}BrH_2$ concentration was varied (1 µM, 2 µM, 5 µM, 10 µM, 20 µM) the cyt c concentration was kept constant at 80 µM, whereas when the concentration of cyt c varied (1 µM, 2 µM, 4 µM, 8 µM, 12 µM, 16 µM) the $Q_0C_{10}BrH_2$ concentration was kept constant at 50 µM. The reactions were initiated by adding 3 µl of diluted bc_1 solution (0.1 µM for $Btbc_1$ or 1.0 µM for $Rsbc_1$) pre-incubated with various concentrations of inhibitors for 15 minutes. The time course of the absorbance change due to cyt c reduction was recorded continuously at 550 nm. Initial rates were determined from the slopes in the linear portion of cyt c_1 reduction time course.

Analysis of Cyt bc_1 Spectra in the Presence of Inhibitors

For each run a solution of 1 ml bovine cyt bc_1 at a cyt b concentration of 5 µM was fully reduced with addition of a tiny amount of sodium dithionite and its spectrum was obtained in the range of 520–600 nm. A specific inhibitor was added at various concentrations to the reduced bc_1 complex and was scanned repeatedly until no changes were observed. All scans were stored digitally and difference spectra were produced by subtracting the corresponding spectrum of the inhibitor-free, fully reduced bc_1 complex.

Measurement of Cyt b and c_1 Reduction Time Course in a Single Turnover Reaction

The enzyme was diluted to a final concentration of about 4 µM of cyt c_1 in 1 ml of B200 buffer (50 mM Tris•HCl, pH 8.0, 0.01% β-DDM, 200 mM NaCl) and oxidized fully by adding a tiny amount of potassium ferricyanide. The spectrum of the fully oxidized enzyme in the range of 520–590 nm was stored.

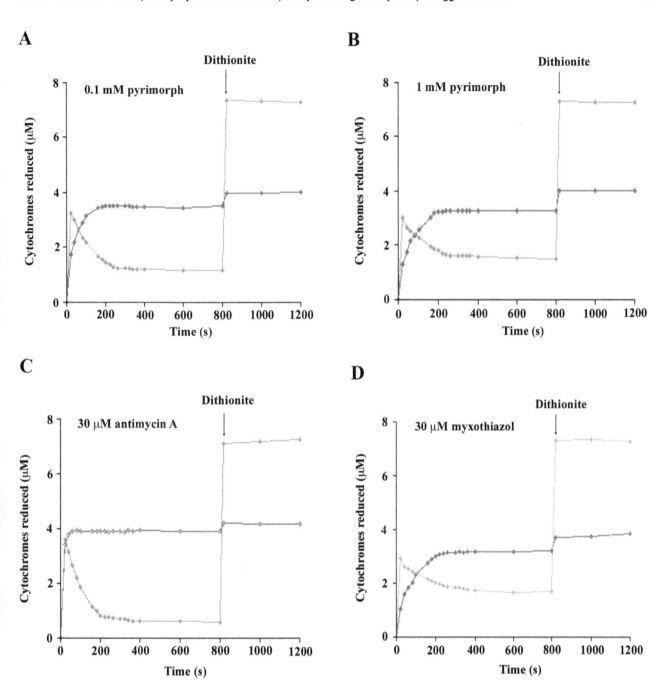

Figure 2. Influences of cyt *b* and *c*$_1$ reduction by pyrimorph. Isolated *Btbc*$_1$ was incubated with indicated inhibitors followed by single turn-over reaction initiated by addition of 10 µM Q$_0$C$_{10}$BrH$_2$. The spectra were recorded immediately following the mixing and every 20 seconds thereafter. At the 800 second time point, a tiny amount of sodium dithionite was added to reduce both cyt *b* and *c*$_1$. The amounts of reduced cyt *b* and *c*$_1$ were calculated and plotted as a function of time. The green trace is the amount of cyt *b* reduced over time and the red one is the amount of cyt *c*$_1$ reduced. (A) 100 µM pyrimorph, (B), 1000 µM pyrimorph, (C) 30 µM myxothiazol, and (D) 30 µM antimycin A.

Inhibitors at various concentrations were introduced and incubated for 2 min followed by addition of the ubiquinol analog Q$_0$C$_{10}$BrH$_2$ to a final concentration of 10 µM to start the reaction. Spectra were recorded at 20-second intervals starting immediately after mixing. After 800 seconds the enzyme was fully reduced by dithionite. The spectrum of the fully oxidized complex was subtracted from that at each time point and the amounts of reduced cyt *c*$_1$ and *b* at a given time were calculated from the difference spectra at 552–540 nm and 560–576 nm, respectively.

Molecular Docking of Pyrimorph to Cyt *b*

Coordinates for the receptor molecule were taken from the protein data bank (pdb) entry 1SQX, for the stigmatellin inhibited complex of *Btbc*$_1$. The side chains of residues E271 and F274 were modeled as standard rotamers consistent with their positions in apo *Btbc*$_1$. The ligand molecule, pyrimorph, was drawn and converted to a SMILES string using tools from the CADD Group [28]. The SMILES string was converted by the program Elbow [29] to 3D coordinates and energy minimized in GAMESS [30].

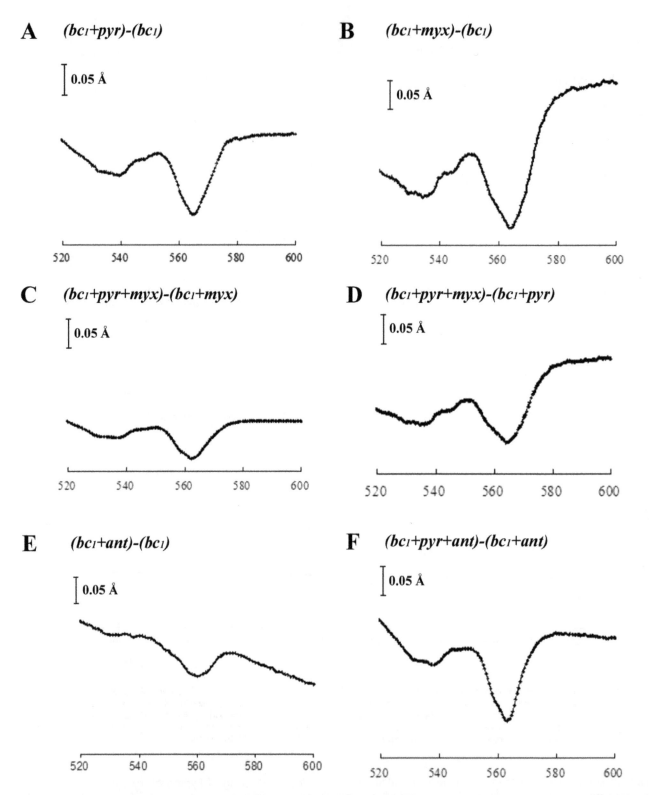

Figure 3. Difference spectra of inhibitors and inhibitors combinations to reduced Btbc1. All spectra were recorded with purified $Btbc_1$ at a concentration of 5 µM of cyt b with the concentrations of inhibitor as indicated. Prior to spectral scan, the bc_1 complex was reduced by addition of dithionite. (A) Spectrum of reduced $Btbc_1$ in the presence of 1 mM pyrimorph (pyr) minus that of reduced $Btbc_1$ alone. (B) Spectrum of reduced $Btbc_1$ in the presence of 10 µM myxothiazol (myx) minus that of reduced $Btbc_1$ alone. (C) and (D) The spectrum of reduced $Btbc_1$ in equilibration with 1 mM pyrimorph followed by addition of 10 µM myxothiazol minus spectrum of reduced $Btbc_1$ in the presence of 10 µM myxothiazol or 1 mM pyrimorph, respectively. (E) Spectrum of reduced $Btbc_1$ in the presence of 10 µM antimycin A (ant) minus spectrum of reduced $Btbc_1$ alone. (F) Spectrum of reduced $Btbc_1$ after equilibration with 1 mM pyrimorph and 10 µM antimycin A in sequence minus spectrum of reduced $Btbc_1$ in the presence of antimycin A.

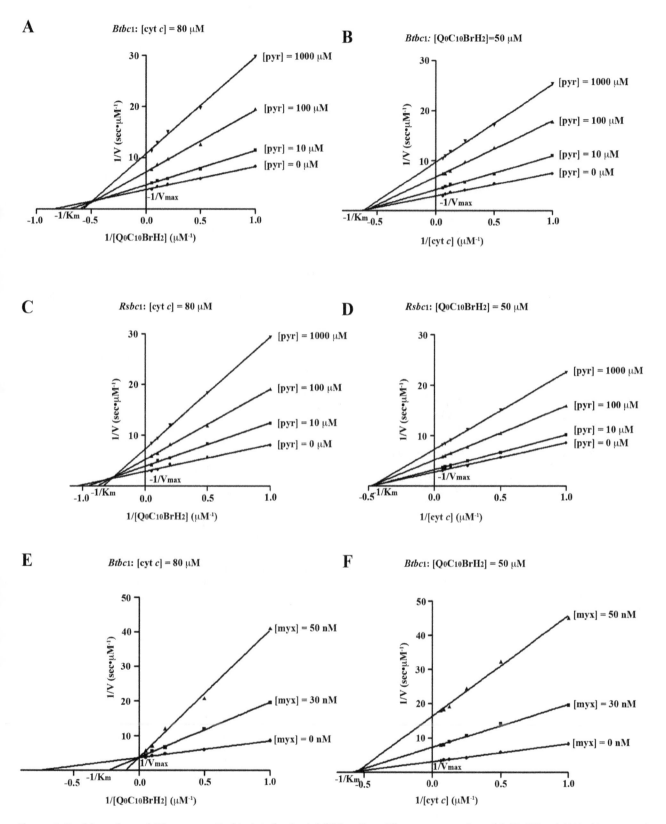

Figure 4. Double-reciprocal (Lineweaver-Burk) plots for bc_1 inhibition. Four different concentrations of 0, 10, 100 and 1000 μM were used for pyrimorph and three, 0, 30 and 50 nM were used for myxothiazol. Each point represents a mean value of at least 3 independent experimental measurements. (A) Inhibition of $Btbc_1$ by pyrimorph with variations in concentration of $Q_0C_{10}BrH_2$. (B) Inhibition of $Btbc_1$ by pyrimorph with variations in concentration of cyt c. (C) Inhibition of $Rsbc_1$ by pyrimorph with variations in concentration of $Q_0C_{10}BrH_2$. (D) Inhibition of $Rsbc_1$ by pyrimorph with variations in concentration of cyt c. (E) Inhibition of $Btbc_1$ by myxothiazol with variations in concentration of $Q_0C_{10}BrH_2$. (F) Inhibition of $Btbc_1$ by myxothiazol with variations in concentration of cyt c.

A

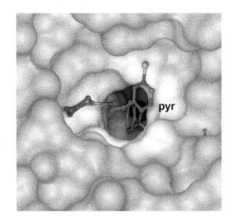

B

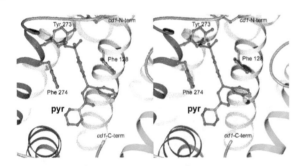

Figure 5. Molecular docking showing the possible binding site and interaction of pyrimorph in the cyt *b* subunit. (A) Molecular surface of the cyt *b* subunit is given, showing the access portal leading to the Q_P site, which is blocked by the docked pyrimorph in a ball-and-stick model with atoms of carbon in magenta, nitrogen in blue, oxygen in red and chlorine in green. (B) Stereoscopic pair showing the detailed interactions between residues in the cyt *b* subunit with the docked pyrimorph molecule.

Both receptor and ligand molecules were converted to standard pdbqt files using mgltools [31]. For all docking runs, the acrylamide moiety (C7 = C8–C9 = O16) of pyrimorph was fixed in the *syn-periplanar* conformation because the alternative *anti*-conformation would bring the larger morpholino and pyridyl groups into close contact.

Two approaches to docking were taken: (1) as the most likely binding sites for inhibitors are the Q_P and the Q_N sites, docking attempts were made first at those known sites. (2) In a second set of runs, no prior knowledge of sites was imposed but the program Q-site-finder [32] was used to locate all potential binding sites and docking was carried out at all sensible locations. Docking pyrimorph to known and unknown inhibitor binding sites was performed using the program Autodock Vina [33].

Results

Pyrimorph Blocks the Mitochondrial Respiratory Chain by Targeting cyt bc_1 Complex

To test whether pyrimorph inhibits fungal growth by interfering with the cellular energy metabolism pathway, in particular the mitochondrial respiratory chain, we isolated light mitochondrial fraction from the pathogenic fungus *P. capsici* and examined the ability of pyrimorph to inhibit various segments of the respiratory

chain (Table 1). It is quite clear that pyrimorph has no effect on the activity of Complex I, as a concentration of pyrimorph as high as 16 μM was unable to inhibit NADH oxidation catalyzed by Complex I. By contrast, under the same conditions, pyrimorph inhibits 94.6% of Complex II activity. Most importantly, Complex III, the cyt bc_1 complex, shows the highest sensitivity toward pyrimorph, with 95.3% inhibition even at 4 μM concentration.

Although the effects of pyrimorph on the mitochondrial respiratory chain of *P. capsici* were clearly demonstrated, such effects could be indirect. To ascertain that the target of pyrimorph is indeed the cyt bc_1 complex, we used highly purified cyt bc_1 from beef heart (*Bos taurus* bc_1 or $Btbc_1$) and assayed inhibition of its cyt c reductase activity by pyrimorph. The result was compared to the well-known anti-bc_1 fungicide azoxystrobin and two other CAA-type fungicides, dimethomorph and flumorph. As shown in Fig. 1A, $Btbc_1$ activities are reduced to 67% and 42% of the control, respectively, in the presence of 10 μM and 100 μM of pyrimorph. Azoxystrobin is able to inhibit bc_1 activity by more than 95% at 10 μM concentration. However, the two CAA-type fungicides, dimethomorph and flumorph, displayed no activity at all against $Btbc_1$. These results indicated that pyrimorph is different from other members of CAA-type fungicides and more importantly established that pyrimorph is indeed an inhibitor of the cyt bc_1 complex, albeit a weak one.

We subsequently determined 50% inhibitory concentration (IC_{50}) for pyrimorph against isolated cyt bc_1 complexes from both bovine mitochondria and photosynthetic bacterium *R. sphaeroides* ($Rsbc_1$). Pyrimorph is slightly more potent against $Rsbc_1$, giving an IC_{50} value of 69.2 μM; it gives an IC_{50} of 85.0 μM for $Btbc_1$ (Figs. 1B and 1C). As a comparison, the well-known bc_1 inhibitor stigmatellin and azoxystrobin give IC_{50} values of 2.8 nM and 47.7 nM, respectively, for isolated $Btbc_1$ by our measurement (data not shown) under the same assay conditions. Since $Rsbc_1$ has only four subunits, it is therefore certain that pyrimorph targets the essential subunits of the bc_1 complex.

Effect of Pyrimorph Binding on Reduction of the Cyt b and c_1 by Ubiquinol

Nearly all cyt bc_1 inhibitors bind to the Q_N site, Q_P site or both [34]. It is known that binding of inhibitors produces various effects on spectra of cyt b and c_1 heme groups, as well as on redox potential and conformation of the iron-sulfur protein [34–38]. These effects ultimately determine the rate and amount of cyt b or c_1 reduced under equilibrium conditions and can be exploited to compare modes of action of different inhibitors [39], distinguishing for example a Q_N site inhibitor from a Q_P site one. Starting with a completely oxidized enzyme, $Btbc_1$ was mixed with substrate $Q_0C_{10}BrH_2$ in the presence of pyrimorph at two different concentrations (0.1 mM and 1.0 mM); no cyt c was used as the terminal electron acceptor. The amount of cyt b including b_L and b_H hemes and cyt c_1 reduced as a function of time was recorded (Figs. 2A and 2B). The results were compared with those produced by Q_N site inhibitor antimycin A (Fig. 2C) and by Q_P site inhibitor myxothiazol (Fig. 2D).

In the absence of the high-potential electron acceptor cyt c, only a single enzymatic turnover at the Q_P site is possible when the Q_N site inhibitor antimycin A is bound. Under such conditions, both cyt b and c_1 were rapidly reduced reaching maximal reduction of nearly half of the b-type hemes and all of the c-type heme perhaps even before the first measurement was recorded (Fig. 2C). Once reaching maximal reduction, cyt b began non-enzymatic oxidation rather rapidly, whereas the redox state of cyt c_1 remained unchanged. When the Q_P site is occupied by an inhibitor such as myxothiazol, not a single turnover is possible (Fig. 2D). The initial rapid reduction of cyt b was most likely via the revere reaction at the Q_N site and cyt c_1 reduction was entirely non-enzymatic. Thus, the rate of cyt c_1 reduction and that of cyt b re-oxidation can be employed to determine which site an inhibitor targets.

Clearly, the reduction behavior of cytochromes b and c_1 induced by pyrimorph distinguishes it from that induced by antimycin A (Figs. 2A, 2B and 2C), demonstrating that pyrimorph does not target the Q_N site. By contrast, the time courses of cyt b and c_1 reduction in the presence of pyrimorph or myxothiazol resemble each other (Figs. 2A, 2B, and 2D), except that the latter inhibitor takes a longer time for cyt c_1 to be maximally reduced than pyrimorph does. This result puts pyrimorph into the category of a Q_P-site inhibitor by either directly or indirectly competing with ubiquinol for the Q_P site.

Spectral Analyses Suggest the Binding of Pyrimorph near Q_P Site

Single turnover experiments suggested that pyrimorph acts in a fashion similar to that of Q_P site inhibitors. It was thus necessary to determine how pyrimorph interferes with bc_1 function at the Q_P site. Since spectral changes, especially red shifts in the α- and β-band caused by binding of inhibitors to reduced bc_1, were successfully used to deduce information about their binding interactions [37,40], a similar approach was taken for pyrimorph. By comparing the spectra of reduced bc_1 bound with pyrimorph to those obtained with known Q_P or Q_N site inhibitors such as myxothiazol or antimycin A, we hoped to gain insight into the binding interactions and location.

Binding of pyrimorph causes the spectrum to red shift, as the difference spectrum $[(bc_1+pyr)-bc_1]$ shows a trough centered around 565 nm (Fig. 3A), which is an indication that binding of pyrimorph affects cyt b hemes. This spectrum was compared to spectra with bound Q_P site inhibitor myxothiazol $[(bc_1+myx)-bc_1]$ and Q_N site inhibitor antimycin A $[(bc_1+ant)-bc_1]$, respectively (Figs. 3B and 3E). At a first glance, it seems that the spectral change due to pyrimorph binding resembles that caused by myxothiazol binding, despite considerable differences (see below), indicating that pyrimorph binds closer to the b_L heme or the Q_P site. Indeed, binding pyrimorph to bc_1 does not seem to interfere with subsequent binding of antimycin A, as the difference spectrum of $[(bc_1+pyr+ant) - (bc_1+ant)]$ (Fig. 3F) looks almost identical to $[(bc_1+pyr)-(bc_1)]$ (Fig. 3A). This experiment confirms that pyrimorph does not target the Q_N site.

However, binding of pyrimorph to bc_1 does affect subsequent binding of myxothiazol and *vise versa*, because the difference spectrum $[(bc_1+pyr+myx) - (bc_1+myx)]$ (Fig. 3C) does not look like that of $[(bc_1+pyr)-(bc_1)]$ (Fig. 3A), nor does the difference spectrum $[(bc_1+pyr+myx) - (bc_1+pyr)]$ (Fig. 3D) resemble that of $[(bc_1+myx)-(bc_1)]$ (Fig. 3B). These spectra indicate the possibility that both inhibitors can co-exist near the Q_P pocket and influence each other. Since the binding of myxothiazol to the Q_P pocket is well established, the experiment further suggests that pyrimorph may have a different binding mode from that of myxothiazol.

Inhibitory Kinetics of Cyt bc_1 Suggests the Mode of Pyrimorph Action

The possibility that pyrimorph has a different mode of action is of particular interest in light of our extensive knowledge on the development of resistance to existing inhibitors. To further probe the mechanism of pyrimorph's action, we investigated the kinetic properties of bc_1 function under pseudo first-order reaction conditions by measuring its activity with respect to changes in concentration of either substrates $Q_0C_{10}BrH_2$ or cyt c in the presence of different amount of pyrimorph, allowing double reciprocal or Lineweaver-Burk plots to reveal the relationship between $1/V$ and $1/[S]$. The measurements were done for both $Btbc_1$ and $Rsbc_1$, revealing nearly identical kinetic behavior (Fig. 4). As shown in Figs. 4A and 4C, in the presence of a constant 80 μM cyt c and with increasing concentrations of $Q_0C_{10}BrH_2$, both K_m and V_{max} are altered as is the K_m/V_{max} ratio. Expectedly, as the concentration of pyrimorph increases, the V_{max} decreases; at a constant pyrimorph concentration, the reciprocal enzyme activity $1/V$ has a positive slope with respect to $1/[S]$. However, the K_m value for the substrate quinol changed, falling between competitive (Fig. 4E) and non-competitive (Fig. 4F) inhibitions. Thus, pyrimorph falls into the category of a mixed-type, noncompetitive inhibitor with respect to the substrate ubiquinol, suggesting that it competes, both directly and indirectly, with ubiquinol to occupy the Q_P site.

At a constant 50 μM $Q_0C_{10}BrH_2$ concentration and with varying concentrations of cyt c, double-reciprocal plots show that x-intercepts remain the same with or without pyrimorph (Figs. 4B and 4D), suggesting that the apparent K_m for substrate cyt c remains unchanged. Thus, pyrimorph is a noncompetitive inhibitor with respect to cyt c. As a control, we performed the same experiments with $Btbc_1$ using the classic Q_P-site inhibitor myxothiazol, showing that myxothiazol is a competitive inhibitor for the substrate quinol but a non-competitive inhibitor for cyt c (Figs. 4E and 4F).

Docking of Pyrimorph to cyt b Subunit

Docking of pyrimorph to known inhibitor-binding sites in the cyt b subunit of $Btbc_1$ were performed with Autodock Vina and resulted in top solutions at the Q_P site with a binding free energy of -9.7 kcal/mol and -9.2 kcal/mol at the Q_N site, representing a 2.3-fold difference in binding affinity between the two sites. These energy values can be compared with binding of other known bc_1 inhibitors such as stigmatellin, giving rise to a binding free energy of -10.5 kcal/mol. Potential inhibitor binding sites outside the known active sites were searched by Q-site-finder and the top 20 sites suggested (which included the Q_P and Q_N site) were subjected to extensive docking trials using Autodock Vina but no new locations showing improved affinity over the classic sites were identified. The Q_P site showed the highest binding affinity to pyrimorph. Unlike traditional inhibitors, pyrimorph does not enter the Q_P site, but rather blocks the entrance or portal to the quinol oxidation site (Fig. 5A). While its morpholino and 4-(2-chloro pyridyl) moieties stay in the central cavity of the bc_1 dimer, its 4-(*tert*-butyl) phenyl group enters the access portal, where it is stabilized by the aromatic side chain of F274 and partially by F128. The latter primarily interacts with the pyridyl moiety via aromatic-aromatic (Ar-Ar) interactions. The *tert*-butyl group that penetrates into the Q_P site is flanked by the residues Y273, Y131 and P270, establishing beneficial van-der-Waals contacts.

Discussion

Resistance to cyt bc_1 inhibitors has been extensively investigated, revealing a wide variety of underlying mechanisms including target site mutations [41,42], activation of alternative oxidase pathways [43,44], altered metabolic degradation [45], reduced uptake and increased efflux [46,47]. By far, target site mutation is the most prevalent form of resistance that develops against bc_1 inhibitors. Thus, extensive research has been focusing on how to overcome resistance caused by target site mutations. Finding inhibitors that target alternative sites seems to be an attractive strategy.

Pyrimorph is a Multi-target Fungicide Displaying Inhibitory Activity against Cyt bc_1

Pyrimorph is a fungicide containing a carboxylic acid amide (CAA) moiety and was shown to be cross-resistant with other CAA fungicides such as mandipropamid, dimethomorph and flumorph [21], suggesting the possibility that pyrimorph may function in a manner similar to that of other CAA-type fungicides such as mandipropamid for which the mode of action was established by inhibiting cellulose synthase 3 or CesA3 [19]. In a recent publication [21], pyrimorph-resistant isolates of *P. capsici* were selected in the presence of the inhibitor and the three most resistant strains share a common mutation (Q1077K) in the CesA3 gene, which is different from the one (G1104V) selected for mandipropamid resistance [19]. However, it remains to be seen whether transfer of the resistant allele to the sensitive parental strain would make the latter pyrimorph-resistant. So far there is no direct evidence from *in vitro* biochemical experiments that shows at the protein level the inhibition of CesA3 by pyrimorph.

In the current study, we followed up on the previous observation that pyrimorph may act on the cellular respiratory chain of pathogenic fungi [20]. We showed that pyrimorph is able to suppress the respiratory chain function at 4 μM concentration by inhibiting the activity of Complex III in isolated mitochondria of *P. capsici* mycelia (Table 1). We further showed conclusively that pyrimorph inhibits purified mitochondrial as well as bacterial bc_1 complexes with IC_{50} values at sub-millimolar range (Fig. 1). By contrast, two other CAA-type inhibitors, dimethomorph and flumorph, displayed no inhibitory activity against bc_1 complex (Fig. 1A).

It did not escape our notice that cyt bc_1 in light mitochondrial fraction isolated from *P. capsici* mycelia appears to be more sensitive to pyrimorph than purified bovine or bacterial bc_1 complexes, suggesting the following possibilities: (1) Direct comparison between results of two very different assays is not a fair comparison, because in isolated light mitochondrial fraction the estimation of cyt bc_1 concentration is difficult in the presence of many different proteins. However, the inhibitory concentrations or IC values are directly related to the amount of enzyme in the assay solution. So the lower IC value could be due to a lower concentration of bc_1 in the assay conditions. (2) Being a hydrophobic compound, pyrimorph may preferentially partition into the lipid bilayer of mitochondrial membranes, leading to a higher local concentration and in turn to the apparent 95% inhibition at 4 μM concentration (Table 1, Fig. 1). (3) Conversely, the presence of detergent (micelles) in the solution of purified bc_1 complex might lower the effective concentration of pyrimorph. This scenario is less likely, as the concentration of β-DDM in our assay buffer is barely above one critical micelle concentration (CMC). (4) The cyt bc_1 of *P. capcisi* is more sensitive to pyrimorph than either bovine or bacterial bc_1. However, we note that the purified bacterial complex is more sensitive to pyrimorph than bovine bc_1 but only by a factor of 1.2. The difference might simply be a reflection of changes in the sequences and we do observe that bacterial bc_1 exhibits slightly higher similarity to fungal than bovine mitochondrial cyt b.

Pyrimorph Likely Acts Near but not at the Q_P Site

The fact that pyrimorph inhibits both 11-subunit $Btbc_1$ and 4-subunit $Rsbc_1$ demonstrates that the inhibitor acts on cyt b, cyt c_1 or ISP subunits of the complex; it does not inhibit mitochondrial bc_1 function through binding to the so-called supernumerary subunits

(Figs. 1B and 1C). The two potential sites for pyrimorph binding are Q_N and Q_P in the cyt b subunit and so far all experimental evidence suggests a binding site near the Q_P site: (1) Single turnover experiments show the reduction rate of cyt c_1 and re-oxidation rate of cyt b in the presence of two different concentrations of pyrimorph (Figs. 2A and 2B) are very similar to those in the presence of the Q_P site inhibitor myxothiazol (Fig. 2D), but are drastically different from those in the presence of the Q_N site inhibitor antimycin A (Fig. 2C). (2) Difference spectra of reduced cyt bc_1 also provided strong evidence that pyrimorph targets the Q_P site (Fig. 3), because the difference spectrum of $[(bc_1+pyr)-(bc_1)]$ (Fig. 3A) resembles that of $[(bc_1+myx)-(bc_1)]$ (Fig. 3B) but not that of $[(bc_1+ant)-(bc_1)]$ (Fig. 3E).

While the analysis of the difference spectra points to the Q_P pocket as the target site, it also suggests that pyrimorph has a non-overlapping binding site with the classic Q_P site inhibitor myxothiazol (Fig. 3). Indeed, double-reciprocal or Lineweaver-Burk plots of bc_1 activity showed that pyrimorph acts as a mixed-type, non-competitive inhibitor with respect to the substrate ubiquinol (Figs. 4A and 4C), suggesting that pyrimorph may act both competitively and non-competitively for the substrate ubiquinol (Fig. 4E). Mechanistically, it means that pyrimorph is capable of modulating the binding of the substrate ubiquinol without directly competing with it at the active site, which categorizes it as a mixed-type, non-competitive inhibitor [48].

Unlike classic Q_P site inhibitors that compete directly with substrate ubiquinol for interactions with the same set of residues in the Q_P site, pyrimorph rather seems to block the portal to the Q_P site, through which the substrate ubiquinol has to pass to contact ISP. Simultaneously, a good portion of pyrimorph is held outside the substrate-binding pocket by hydrophobic forces. Consequently, ubiquinol has to actively displace pyrimorph from the entrance in order to gain access to the Q_P site as in the case of a competitive inhibitor. On the other hand, as pyrimorph has the ability to adhere well to the lipophilic sides of the portal that leads to the Q_P site of cyt b, it may stay close and possibly interfere with the necessary motion of the cd1/cd2 helix and with the release of ubiquinone, displaying inhibitory activities more characteristic of non-competitive inhibitors. This picture is entirely consistent with biochemical and spectral characterizations of pyrimorph, qualifying it as a mixed-type, non-competitive inhibitor.

Molecular modeling of other CAA-type fungicides such as dimethomorph indicate a very similar binding position and orientation to the Q_P site but with a significantly lower free energy for binding, consistent with the observation that both dimethomorph and flumorph are not bc_1 inhibitors (Fig. 1A). Since both dimethomorph and flumorph structurally resemble pyrimorph, it is clear that the shape of pyrimorph is not a dominant factor for its ability to bind bc_1. At the very least the binding of pyrimorph has sparked ideas for a dual approach to inhibitor design: (1) on the side that binds to the Q_P entrance, modifications could significantly increase the affinity, making it a better competitive inhibitor. (2) Improvements in hydrophobicity or geometric factors on the side that stays outside the Q_P pocket, the inhibitor could enhance its non-competitive properties. Should it be possible to design a dual type inhibitor, fungal resistance may be stalled for an extended period. Suggestions for improvements might include modifications of the *tert*-butyl group to include polar groups (hydroxy methyl, methoxy methyl, etc.) that are within the reach of E271 (both sidechain and backbone amide) as well as sterically demanding aromatic groups that may take advantage of the large, aromatic cavity of the Q_P site. On the side that stays outside of the Q_P pocket, variations of saturated and aromatic ring systems seem likely to improve binding properties, as it appears that the morpholino group and the chloro-pyridyl group can change places with minimal change in binding energy.

Acknowledgments

We thank George Leiman for editorial assistance during the preparation of this manuscript. This study utilized the high-performance computational capabilities of the Biowulf Linux cluster at the National Institutes of Health, Bethesda, MD (http://biowulf.nih.gov).

Author Contributions

Conceived and designed the experiments: YMX ZHQ DX. Performed the experiments: YMX LE FZ CL YHZ DX. Analyzed the data: YMX LE DX. Contributed reagents/materials/analysis tools: CAY. Wrote the paper: YMX LE DX.

References

1. Trumpower BL (1990) Cytochrome bc1 complexes of microorganisms. Microbiological Reviews 54: 101–129.
2. Keilin D (1925) On cytochrome, a respiratory pigment, common to animals, yeast, and higher plants. Proc R Soc Lond B98: 312–399.
3. Leadbeater A (2012) Resistance risk to QoI fungicides and anti-resistance strategies. In: Thind TS, editor. Fungicide resistance in crop protection: risk and mangement: CAB eBooks. 141–152.
4. Trumpower BL (1990) The protonmotive Q cycle. Energy transduction by coupling of proton translocation to electron transfer by the cytochrome bc1 complex. Journal of Biological Chemistry 265: 11409–11412.
5. Mitchell P (1975) Protonmotive redox mechanism of the cytochrome b-c1 complex in the respiratory chain: protonmotive uniquinone cycle. FEBS Lett 56: 1–6.
6. Xia D, Yu CA, Kim H, Xia JZ, Kachurin AM, et al. (1997) Crystal structure of the cytochrome bc1 complex from bovine heart mitochondria. Science 277: 60–66.
7. Iwata S, Lee JW, Okada K, Lee JK, Iwata M, et al. (1998) Complete structure of the 11-subunit bovine mitochondrial cytochrome bc1 complex [see comments]. Science 281: 64–71.
8. Zhang Z, Huang L, Shulmeister VM, Chi YI, Kim KK, et al. (1998) Electron transfer by domain movement in cytochrome bc1. Nature 392: 677–684.
9. Hunte C, Koepke J, Lange C, Rossmanith T, Michel H (2000) Structure at 2.3 A resolution of the cytochrome bc(1) complex from the yeast Saccharomyces cerevisiae co-crystallized with an antibody Fv fragment. Structure 15: 669–684.
10. Esser L, Elberry M, Zhou F, Yu CA, Yu L, et al. (2008) Inhibitor complexed structures of the cytochrome bc1 from the photosynthetic bacterium Rhodobacter sphaeroides at 2.40 Å resolution. J Biol Chem 283: 2846–2857.
11. Kleinschroth T, Castellani M, Trinh CH, Morgner N, Brutschy B, et al. (2011) X-ray structure of the dimeric cytochrome bc(1) complex from the soil bacterium Paracoccus denitrificans at 2.7-A resolution. Biochim Biophys Acta 1807: 1606–1615.
12. Berry EA, Huang L, Saechao LK, Pon NG, Valkova-Valchanova M, et al. (2004) X-ray structure of Rhodobacter capsulatus cytochrome bc1: comparison with its mitochondrial and chloroplast counterparts. Photosynthesis Research 81: 251–275.
13. Esser L, Yu CA, Xia D (2013) Structural Basis of Resistance to Anti-Cytochrome bc1 Complex Inhibitors: Implication for Drug Improvement. Curr Pharm Des.
14. Mu CW, Yuan HZ, Li N, Fu B, Xiao YM, et al. (2007) Synthesis and fungicidal activities of a novel seris of 4-[3-(pyrid-4-yl)-3-substituted phenyl acryloyl] morpholine. Chem J Chinese U 28: 1902–1906.
15. Chen XX, Yuan HZ, Qin ZH, Qi SH, Sun LP (2007) Preliminary studies on antifungal activity of pyrimorph. Chinese Journal of Pesticide Science 9: 229–234.
16. Wang HC, Sun HY, Stammler G, Ma JX, Zhou MG (2009) Baseline and differential sensitivity of Peronophythora litchii (lychee downy blight) to three carboxylic acid amide fungicides. Plant Pathology.
17. Du YN, Wang GZ, Li G (2008) Preventive and therapeutic experiment of 20% Bimalin on phytophthora blight of Capsicum. Modern Agrochemicals 7: 44–46.
18. Sun H, Wang H, Stammler G, Ma J, Zhou MG (2010) Baseline Sensitivity of Populations of Phytophthora capsici from China to Three Carboxylic Acid Amide (CAA) Fungicides and Sequence Analysis of Cholinephosphotranferases from a CAA-sensitive Isolate and CAA-resistant Laboratory Mutants. Journal of Phytopathology 158: 244–252.
19. Blum M, Boehler M, Randall E, Young V, Csukai M, et al. (2010) Mandipropamid targets the cellulose synthase-like PiCesA3 to inhibit cell wall biosynthesis in the oomycete plant pathogen, Phytophthora infestans. Mol Plant Pathol 11: 227–243.
20. Yan X, Qin W, Sun L, Qi S, Yang D, et al. (2010) Study of inhibitory effects and action mechanism of the novel fungicide pyrimorph against Phytophthora capsici. J Agric Food Chem 58: 2720–2725.
21. Pang Z, Shao J, Chen L, Lu X, Hu J, et al. (2013) Resistance to the novel fungicide pyrimorph in Phytophthora capsici: risk assessment and detection of point mutations in CesA3 that confer resistance. PLoS One 8: e56513.
22. Yu CA, Yu L (1982) Syntheses of biologically active ubiquinone derivatives. Biochemistry 21: 4096–4101.
23. Mitani S, Araki S, Takii Y, Ohshima T, Matsuo N, et al. (2001) The biochemical mode of action of the novel selective fungicide cyazofamid: specific inhibition of mitochondrial complex III in Phythium spinosum. Pesticide Biochemistry and Physiology 71: 107–115.
24. Yu L, Yang S, Yin Y, Cen X, Zhou F, et al. (2009) Chapter 25 Analysis of electron transfer and superoxide generation in the cytochrome bc1 complex. Methods Enzymol 456: 459–473.
25. Mather MW, Yu L, Yu CA (1995) The involvement of threonine 160 of cytochrome b of Rhodobacter sphaeroides cytochrome bc1 complex in quinone binding and interaction with subunit IV. Journal of Biological Chemistry 270: 28668–28675.
26. Tian H, Yu L, Mather MW, Yu CA (1997) The Involvement of Serine 175 and Alanine 185 of Cytochrome b of Rhodobacter sphaeroides Cytochrome bc1 Complex in Interaction with Iron-Sulfur Protein. Journal of Biological Chemistry 272: 23722–23728.
27. Yu L, Yu CA (1991) Essentiality of the molecular weight 15,000 protein (subunit IV) in the cytochrome b-c1 complex of rhodobacter sphaeroides. Biochemistry 30: 4934–4939.
28. Nicklaus M.C SM (2012) CADD Group Chemoinformatics Tools and User Services.
29. Adams PD, Afonine PV, Bunkoczi G, Chen VB, Davis IW, et al. (2010) PHENIX: a comprehensive Python-based system for macromolecular structure solution. Acta Crystallographica Section D 66: 213–221.
30. Schmidt MW, Baldridge KK, Boatz JA, Elbert ST, Gordon MS, et al. (1993) General Atomic and Molecular Electronic-Structure System. Journal of Computational Chemistry 14: 1347–1363.
31. Sanner MF (1999) Python: A programming language for software integration and development. Journal of Molecular Graphics and Modelling 17: 57–61.
32. Laurie AT, Jackson RM (2005) Q-SiteFinder: an energy-based method for the prediction of protein-ligand binding sites. Bioinformatics 21: 1908–1916.
33. Trott O, Olson AJ (2010) AutoDock Vina: Improving the speed and accuracy of docking with a new scoring function, efficient optimization, and multithreading. Journal of Computational Chemistry 31: 455–461.
34. Esser L, Quinn B, Li Y, Zhang M, Elberry M, et al. (2004) Crystallographic studies of quinol oxidation site inhibitors: A modified classification of inhibitors for the cytochrome bc1 complex. Journal of Molecular Biology 341: 281–302.
35. von Jagow G, Link TA (1986) Use of specific inhibitors on the mitochondrial bc1 complex. Methods in Enzymology 126: 253–271.
36. Link TA, Haase U, Brandt U, von Jagow G (1993) What information do inhibitors provide about the structure of the hydroquinone oxidation site of ubihydroquinone: cytochrome c oxidoreductase? Journal of Bioenergetics and Biomembrane 25: 221–232.
37. von Jagow G, Liungdahl PO, Craf P, Ohnishi T, Trumpower BL (1984) An Inhibitor of Mitochondrial Respiration Which Binds to Cytochrome b and Displaces Quinone from the Iron-Sulfur Protein of the Cytochrome bc1 Complex. Journal of Biological Chemistry 259: 6318–6326.
38. von Jagow G, Engel WD (1981) Complete inhibition of electron transfer from ubiquinol to cytochrome b by teh combined action of antimycin and myxothiazol. FEBS Letters 136: 19–24.
39. Berry EA, Huang LS, Lee DW, Daldal F, Nagai K, et al. (2010) Ascochlorin is a novel, specific inhibitor of the mitochondrial cytochrome bc1 complex. Biochim Biophys Acta 1797: 360–370.
40. Jordan DB, Livingston RS, Bisaha JJ, Duncan KE, Pember SO, et al. (1999) Mode of action of famoxadone. Pesticide Science 55: 105–118.
41. Bolgunas S, Clark DA, Hanna WS, Mauvais PA, Pember SO (2006) Potent inhibitors of the Qi site of the mitochondrial respiration complex III. J Med Chem 49: 4762–4766.
42. Brasseur G, Saribas AS, Daldal F (1996) A compilation of mutations located in the cytochrome b subunit of the bacterial and mitochondrial bc1 complex. BiochimBiophysActa 1275: 61–69.
43. Wood PM, Hollomon DW (2003) A critical evaluation of the role of alternative oxidase in the performance of strobilurin and related fungicides acting at the Qo site of complex III. Pest Manag Sci 59: 499–511.
44. Steinfeld U, Sierotzki H, Parisi S, Poirey S, Gisi U (2001) Sensitivity of mitochondrial respiration to different inhibitors in Venturia inaequalis. Pesticide Mangement Science 57: 787–796.
45. Dosnon-Olette R, Schroder P, Bartha B, Aziz A, Couderchet M, et al. (2011) Enzymatic basis for fungicide removal by Elodea canadensis. Environ Sci Pollut Res Int 18: 1015–1021.
46. Gaur M, Choudhury D, Prasad R (2005) Complete inventory of ABC proteins in human pathogenic yeast, Candida albicans. J Mol Microbiol Biotechnol 9: 3–15.
47. Hill P, Kessl J, Fisher N, Meshnick S, Trumpower BL, et al. (2003) Recapitulation in Saccharomyces cerevisiae of cytochrome b mutations

conferring resistance to atovaquone in Pneumocystis jiroveci. Antimicrob Agents Chemother 47: 2725–2731.

48. Garrett RH, Grisham CM (1999) Biochemistry. Fort Worth, Philadelphia, San Diego, New York, Orlando, Austin, San Antonio, Toronto, Montreal, London, Sydney, Tokyo: Saunders College Publishing, Harcourt Brace College Publishers.

49. Yoshida T, Murai M, Abe M, Ichimaru N, Harada T, et al. (2007) Crucial structural factors and mode of action of polyene amides as inhibitors for mitochondrial NADH-ubiquinone oxidoreductase (complex I). Biochemistry 46: 10365–10372.

50. Hatefi Y, Stiggall DL (1978) Preparation and properties of succinate: ubiquinone oxidoreductase (complex II). Methods Enzymol 53: 21–27.

16

Microcolony Imaging of *Aspergillus fumigatus* Treated with Echinocandins Reveals Both Fungistatic and Fungicidal Activities

Colin J. Ingham[1,2,3]*, Peter M. Schneeberger[1]

1 Department of Medical Microbiology and Infection Control, Jeroen Bosch Hospital, 's-Hertogenbosch, The Netherlands, **2** Laboratory of Microbiology, Wageningen University, Wageningen, The Netherlands, **3** MicroDish BV, Wageningen, The Netherlands

Abstract

Background: The echinocandins are lipopeptides that can be employed as antifungal drugs that inhibit the synthesis of 1,3-β-glucans within the fungal cell wall. Anidulafungin and caspofungin are echinocandins used in the treatment of *Candida* infections and have activity against other fungi including *Aspergillus fumigatus*. The echinocandins are generally considered fungistatic against *Aspergillus* species.

Methods: Culture of *A. fumigatus* from conidia to microcolonies on a support of porous aluminium oxide (PAO), combined with fluorescence microscopy and scanning electron microscopy, was used to investigate the effects of anidulafungin and caspofungin. The PAO was an effective matrix for conidial germination and microcolony growth. Additionally, PAO supports could be moved between agar plates containing different concentrations of echinocandins to change dosage and to investigate the recovery of fungal microcolonies from these drugs. Culture on PAO combined with microscopy and image analysis permits quantitative studies on microcolony growth with the flexibility of adding or removing antifungal agents, dyes, fixatives or osmotic stresses during growth with minimal disturbance of fungal microcolonies.

Significance: Anidulafungin and caspofungin reduced but did not halt growth at the micrcony level; additionally both drugs killed individual cells, particularly at concentrations around the MIC. Intact but not lysed cells showed rapid recovery when the drugs were removed. The classification of these drugs as either fungistatic or fungicidal is simplistic. Microcolony analysis on PAO appears to be a valuable tool to investigate the action of antifungal agents.

Editor: Gustavo Henrique Goldman, Universidade de Sao Paulo, Brazil

Funding: This work was funded by the JBZ Hospital. The funders had no role in study design, data collection and analysis, decision to publish, or preparation of the manuscript.

Competing Interests: The authors have declared that no competing interests exist.

* E-mail: colinutrecht@gmail.com

Introduction

The echinocandins are an important class of lipopeptide antifungal compounds consisting of cyclic hexapeptides linked to a long chain fatty acid. The echinocandins inhibit cell wall synthesis, by non-competitive inhibition of the 1,3-β-glucan synthase. This is thought to occur via interaction with the Fks1 subunit of this enzyme [1]. Echinocandins are commonly used to treat infections by many *Candida* species, against which they are fungicidal. Unlike an earlier class of 1,3-beta-glucan synthase inhibitors, the liposaccharide papulocandins, the activity of echinocandins extends beyond *Candida* spp., to include *Aspergillus* spp. and *Pneumocystis carinii* [2].

Aspergillus fumigatus is a widespread filamentous fungus, which is both highly allergenic and an opportunistic pathogen. Systemic infections by *A. fumigatus*, particularly in immunocompromised individuals, present a significant risk of mortality. Treatment is commonly with amphotericin B or triazole drugs such as itraconazole or voriconazole. Triazole resistant strains are known [3]. Caspofungin has been reported to be effective in salvage therapy for patients refractory to standard antifungal agents for invasive aspergillosis [4].

Despite 1,3-β-D-glucan being the dominant form of glucan in the cell wall of *Aspergillus* spp., echinocandins are generally considered fungistatic against these moulds rather than fungicidal [5,6]. Chitin synthesis appears to be able to reinforce cell walls and compensate, in some instances, for echinocandin-mediated damage [2]. Caspofungin and micafungin have been suggested to have a degree of fungicidal activity based on liquid culture. This has been investigated using 5,(6)-carboxyfluorescein as a fluorogenic indicator of esterase activity in live cells and bis-(1,3-dibutylbarbituric acid)trimethine oxonol to identify dead cells [7,8]. These studies suggest that the classification of effects of caspofungin on growth and viability in *A. fumigatus* may not be entirely straightforward. Work on this subject to date has generally been qualitative or semi-quantitative and has not looked at anidulafungin, an important drug, as a fungicidal agent.

Acquired resistance to echinocandins in pathogenic yeasts and moulds is mediated via mutations within the Fks1 subunit of the 1,3-β-D-glucan synthase [2,5,6,7,8]. Determining precise MIC values from susceptibility testing is therefore necessary for some clinical isolates, but when inhibition of growth is not necessarily complete, deriving exact MIC values can be problematic. An alternative, minimal effective concentration (MEC) based around echinocandin-induced morphology changes, has been proposed but determination of this value is rather subjective.

Culture on a porous aluminium oxide (PAO) support offers advantages in studying the effects of drugs on microorganisms over direct growth on agar or liquid culture. This ceramic material permits effective imaging of large numbers of microcolonies cultured on its upper surface – with imaging by fluorescence and scanning electron microscopy. Strips of PAO are highly porous (pore size 20 to 200 nm, 40% porosity) but only 60 μm thick. Therefore, it is possible to add or remove compounds (such as drugs or fluorogenic dyes or fixatives) rapidly but with minimal disturbance by moving the material between agars of different compositions [9]. This method has been previously used to osmotically stress bacteria and to dose microcolonies of bacteria and *Candida* spp. during rapid drug susceptibility testing [10,11]. Microcolony-based methods combined with fluorogenic dyes and imaging may offer particular advantages with respect to studying filamentous fungi, a group of organisms for which dispersed growth in liquid culture is not necessarily the most relevant [7]. In this study, PAO culture was used to explore the lethality of caspofungin and anidulafungin to *A. fumigatus*, and also the recovery from these drugs, by imaging individual microcolonies and by quantifying the effects.

Results

Germination and growth of *A. fumigatus* and *A. terreus* on PAO

Dispersal and germination of conidia on PAO. Conidia purified from clinical isolates and a reference strains (ATCC204305) of *A. fumigatus* and *A. terreus* (Tables 1 and 2) were inoculated onto 36×8 mm strips of PAO placed on Sabouraud agar to a density of from 5 to 50 cfu/mm². After incubation for 4 to 10 h at 37°C the fungi on PAO strips were stained with Fun-1/calcofluor white and the percentage germination assessed. Both germinated and ungerminated conidia of both species could be imaged and distinguished upon PAO by this method. The distribution of conidia immediately after inoculation was predominantly (>93%) of single spores. The median swelling and outgrowth times for all strains tested (Table 1) were similar on PAO placed on Sabouraud agar compared to growth directly on the same agar. This suggests conidia can be inoculated to single cfu on PAO and germination occurs with the same efficiency to that seen on agar.

Growth of mycelia on PAO. Hyphal extension rates were calculated from transmission light microscopy, and were found to be similar on PAO compared to culture directly on the same agar medium (Table 1). Scanning electron microscopy suggested that the tips of vegetative mycelia of *A. fumigatus* were lifted off the surface of the PAO during growth (Figure 1A and 1B); observation by light microscopy supported this conclusion. Visible colonies of all strains tested were smaller on PAO compared to agar, when observed after 24 and 36 h. The formation of conidiophores on PAO was delayed by 12 to 24 h. Taken together, these data suggest that PAO is suitable for the culture of fungi to microcolonies with no nutrient limitation caused by the interposition of a 40% porous filter between agar and the fungus, the changes in surface properties of the growth surface (such as wetness or texture) having a major impact. Macroscopic growth was somewhat restricted, presumably by nutrient

limitation, but this was not relevant to the microcolony studies presented in this work.

Growth of *A. fumigatus* and *A. terreus* with anidulafungin and caspofungin

Effect of echinocandins on microcolony growth. The strains listed in Table 1 were grown from conidia for 9 h and 14 h on 0.125 to 32 μg/ml anidulafungin or caspofungin, on PAO strips. After 7 to 9 h inhibition of conidial outgrowth by both drugs could be seen but a reproducible determination of a MIC was not possible. After 14 h culture, a clear dose-dependent effect of both echinocandins on microcolony diameter was observed, allowing MIC determinations that agreed well with an established method, the E-test using 36 h culture (Table 2). Using the PAO method, even high concentrations of anidulafungin and caspofungin allowed significant, if limited, outgrowth when judged at the microcolony level (Figure 2A). In order to test if hyphal extension occurred continuously with high levels anidulafungin and caspofungin, the average microcolony diameter (using >100 microcolonies per data point) was measured at different time points over 24 h (Figure 2). These data suggested that widespread, if very slow, growth was occurring for sensitive strains in the presence of anidulafungin and caspofungin, even at a concentration >10-fold above the MIC. In contrast, amphotericin B was completely inhibitory to microcolony growth, at concentrations equal to the MIC, and above (Figure 2) Reanalysis of the same images, calculating average microcolony area instead of diameter, led to the same conclusions (data not shown). Caspofungin resistance was apparent from the higher growth rate of a resistant strain of *A. fumigatus* on PAO when dosed with this drug (Figure 2B).

Cell lysis by anidulafungin and caspofungin. During culture of sensitive strains with both anidulafungin and caspofungin, lysed hyphal tips were seen (Figure 1B to 1E; Figure 3C and 3D). Lysis could be observed by fluorescence microscopy or by staining microcolonies with Fun-1/calcofluor white (Figure 3) or with Syto9/propidium iodide (Figure 4). Scanning electron microscopy (Figure 1 and Figure 3) confirmed that lysis of hyphal tips was induced by these drugs. Both liquid and vapour fixation methods for SEM preparation gave the same results, with a good correspondence with fluorescence microscopy. In all cases the debris resulting from tip lysis could be observed in an arc pattern radiating outwards from the lysed tip. The material liberated from lysed cells adhered to the PAO, providing a convenient marker for scoring the location of a lytic event. Using fluorescence microscopy (with microcolonies stained with Fun-1 and calcofluor white) the staining pattern with each dye individually indicated that the calcofluor white was responsible for detecting the damage. Therefore, it appears likely that cell wall debris was being stained. With combined Syto9 and propidium iodide, staining was primarily with the latter dye. When Syto9 was used alone it could also facilitate detection of the debris. It is likely, therefore, that nucleic acids liberated from lysed cells were the primary target for propidium iodide staining, which outcompeted the Syto9 in a mixture due to a higher affinity for DNA binding. When caspofungin resistant strains of *A. fumigatus* were tested (CWZ1243 shown, CWZ93 was similar), tip lysis was only rarely observed (Figure 4D–F) at a concentration sufficient to cause widespread lysis in sensitive strains. Taking these results together, we conclude that both anidulafungin and caspofungin caused apical lysis of hyphae in sensitive strains of *A. fumigatus*. A similar level of lysis with both echinocandins was observed for *A. terreus* (e.g. Figure 4G, Figure 4H and Figure 4I).

Table 1. Comparison of germination and growth of strains of *Aspergillus* spp. studied on Sabouraud agar and on PAO placed on Sabouraud plates.

Strain		Conidial Swelling[a]		Outgrowth[a]		Hyphal Extension (μm/h)[b]	
		Agar	PAO	Agar	PAO	Agar	PAO
JBZ11	A. fumigatus	4.5 h	4 h	6.5 h	7 h	24.9+/−4	19.1+/−4
JBZ17	A. fumigatus	4 h	5 h	6 h	6 h	18.0+/−2	16.1+/−3
JBZ32	A. fumigatus	4.5 h	4.5 h	7 h	7 h	19.0+/−5	16.8+/−4
CWZ93	A. fumigatus	5 h	5 h	7 h	6.5 h	22.5+/−6	20.6+/−47
CWZ855	A. fumigatus	4.5 h	5 h	6.5 h	6.5 h	16+/−3	17+/−4
ATCC204305	A. fumigatus	4 h	4.5 h	7 h	7 h	20.0+/−4	16+/−3
CWZ59	A. terreus	6 h	6 h	8 h	8.5 h	11.0	8.9

[a]Calculated from the area of >100 conidia or microcolonies per data points; taking measurements every 30 min after inoculation. ANOVA with Tukey post hoc test was to determine first time point with a significant increase in area (P<0.05) comparing unswollen conidia at T=0 with subsequent time points to determine swelling time. Outgrowth times were determined in a similar way, comparing the microcolony area of swollen conidia with subsequent time points .
[b]Mean of 20 determinations +/− S.D.

The concentrations of anidulafungin and caspofungin that triggered the greatest degree of cellular lysis for strain JBZ11 of *A. fumigatus* (0.125 μg/ml and 0.5 μg/ml respectively) were used in Sabouraud agar plates in viable counts. Viability was scored by counting the number of microcolonies that germinated after 14 h. Anidulafungin reduced viability to 86% and caspofungin to 88% relative to the untreated control (mean of 3 determinations). Therefore, the ability of these drugs to prevent microcolony formation was limited. Using higher concentrations (2 μg/ml for both) viable counts were 81% (anidulafungin) and 90% (caspofungin) of the control (n = 3). Therefore, again, at the microcolony level the fungicidal activity of these drugs was marginal. These trends (79 to 89% viability) were similar for other echinocandin sensitive isolates, i.e. strains ATCC204305, JBZ17 and JBZ32 of *A. fumigatus* and strain CWZ59 of *A. terreus*. In contrast, amphotericin B reduced viability <86% when used at the MICs of JBZ11, JBZ13 and JBZ32 (0.05 to 0.1 μg/ml).

Quantification of cell lysis by anidulafungin and caspofungin. Cell lysis was scored for a range of concentrations of caspofungin and anidulafungin after 14 h (Figure 5). Lysis was particularly common at intermediate concentrations around the MIC. In contrast, whilst higher concentrations of echinocandins were more effective at limiting

growth lysis was seen less often (Figure 5). Anidulafungin appeared somewhat more effective than caspofungin, with >50% cell lysis at the most effective concentrations (0.06 to 0.25 μg/ml). Fungal microcolonies growing on echinocandins appeared heterogeneous in their response. Subpopulations of lysed cells and intact ones coexisted within the same microcolony, ones that were derived, in most cases, from a single conidium. In both cases, a total count of microcolonies after 14 h suggested that neither drug reduced the number of microcolonies, despite having a significant impact on the number of intact hyphal tips within a microcolony. Caspofungin resistance (Figure 5D) increased the concentration of this drug required for optimal tip lysis commensurate with the increased MIC compared to sensitive strains.

A series of control experiments were performed, using different staining or fixing and imaging methods, to check that the observed tip lysis was not affected by sample preparation (Figure 6) or by dye choice. The latter point was considered, in part, as calcofluor white is known to destabilise the cell wall in *A. nidulans* [12]. Fun-1/calcofluor white staining, propidium iodide/Syto9 staining, scanning electron microscopy with vapour or gel fixation all gave similar frequencies of lysis. Therefore it can be concluded that imaging or staining methods did not bias the results. However, staining under conditions of osmotic stress elevated the frequency

Table 2. Comparison of Echinocandin susceptibility of *Aspergillus* spp. studied on Sabouraud agar and on PAO placed on Sabouraud plates.

Strain	Species	MIC[a] ANI		MEC[b] ANI	MIC[a] CASP		MEC[b] CASP
		E-test	PAO		E-test	PAO	
JBZ11	A. fumigatus	0.06	0.06	<0.0625	0.125	0.25	0.0625
JBZ17	A. fumigatus	0.125	0.125	0.0625	0.25	0.25	0.0625
JBZ32	A. fumigatus	0.06	0.125	<0.0625	0.125	0.12	0.125
CWZ93	A. fumigatus	<0.008	<0.0625	<0.008	8	4	0.125
CWZ1243	A. fumigatus	<0.008	0.125	<0.008	>8	>32	1
ATCC204305	A. fumigatus	<0.008	<0.0625	<0.008	0.25	0.25	0.03
CWZ59	A. terreus	<0.008	0.0625	0.0008	0.5	1	0.0625

[a]MIC values in μg/ml.
[b]MEC values in μg/ml.

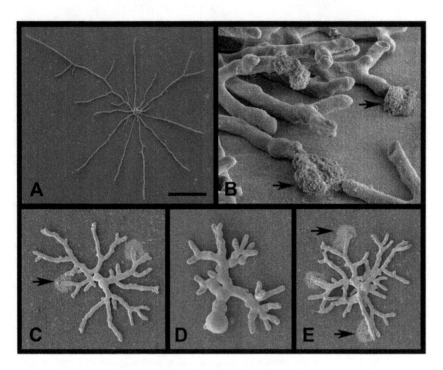

Figure 1. Imaging microcolonies of *A. fumigatus* JBZ11 grown on PAO with echinocandins by SEM. Culture was for 14 h at 37°C on Sabouraud medium with imaging from directly above unless stated otherwise. A: Microcolony without drugs. B: Growth with 0.0625 µg/ml anidulafungin. Imaging from the side at a 30 degree angle. C: Growth with 0.06 µg/ml anidulafungin. D: Growth with 2 µg/ml anidulafungin. E: Growth with 0.5 µg/ml caspofungin. Arrows indicate lysed hyphal tips. Scale bar in panel A indicates 90 µm. When applied to panel B, 8 µm; panel C, 30 µm; panel D, 15 µm; panel E, 30 µm.

of lysis to over 60% (Figure 6). This suggests that a subpopulation of cells existed that did not lyse with an echinocandin alone but which were vulnerable when the antifungal agent was combined with an osmotic shock.

Correlation between observed lysis and staining with Syto9 or propidium iodide for anidulafungin. Microcolonies of *A. fumigatus* cultured from conidia with echinocandins were analyzed using Syto9/propidium iodide staining, relating the number of lysed and intact hyphal tips per microcolony to the staining pattern. Microcolonies cultured without drugs for 14 h showed no lysis and >98% of hyphal tips (154 out of 157) stained with Syto9. The same analysis was repeated for 0.125 µg/ml anidulafungin, giving maximal lysis. Assessing 198 hyphal tips, 47% (n = 93) showed lysis and 53% (n = 105) were apparently intact. Of the 105 lysed tips imaged, >94% (n = 87) were stained with propidium iodide in preference to Syto9. In contrast, of the 93 apparently intact tips 37% were found to stain with propidium iodide (n = 34) and Syto9 (63%, n = 59). However, some intact tips that stained with propidium iodide were adjacent to lysed tips, and it is likely that many of this subpopulation were part of the same cell or hyphal compartment (Figure 4). When a higher concentration of anidulafungin was used (2 µg/ml) only low rates of lysis were observed (6% of hyphal tips, 10/163). All 10 hyphal tips preferentially stained with propidium iodide. This supports the conclusion obtained from SEM and Fun-1/calcofluor white staining methods; that whilst clearly limited in growth rate *A. fumigatus* growing on high concentrations of anidulafungin on appear less vulnerable to cell death than those growing on an intermediate concentration. In contrast, amphotericin B (at the MICs of each strain, which varied from 0.025 to 0.1 µg/ml) gave <90% propidium iodide staining of cells forthe strains of *A. fumigatus* listed in Table 1. Additionally, little

heterogeneity was observed, most microcolonies stained completely with propidium iodide.

Recovery of microcolonies of *A. fumigatus* from the effect of anidulafungin and caspofungin. PAO strips were moved from plates containing echinocandins to those lacking the drugs in order to look at the potential for recovery from echinocandins. Control experiments were first performed by moving uninoculated strips from plates containing echinocandins to drug free plates and then inoculating with *A. fumigatus* conidia. No inhibition of growth was seen, suggesting that the carry-over of drugs within the minimal volume of the PAO pores was not sufficient to influence the results. It was clear from assessment of the average microcolony diameter at different time points that after removal recovery was widespread throughout the population (Figure 7). Microcolonies cultured for 14 h on plates containing 0.125 µg/ml anidulafungin were stained with propidium iodide and Syto9 and compared with those cultured for 17 h on the same concentration or those shifted to zero anidulafungin after 14 h then grown for 3 h. In all cases (>40 observations) lysed hyphal tips stained (>90%) with propidium iodide. Furthermore, there was no evidence for extension of a lysed hyphal tip beyond the debris created by the lysis event. Similar experiments looking at recovery from 0.5 µg/ml caspofungin gave the same result. Therefore, there was no regrowth of lysed cells. Whilst microcolonies containing lysed cells clearly were recovering and growing, this was due to the growth of less damaged cells. This supports other lines of evidence (SEM, fluorescence staining) that echinocandins are fungicidal for specific cells [13]. Because of the overgrowth of lysed hyphae, it was not possible to conclude that lysed cells never recover, but it is likely that this is not a common event.

The interaction between Cyclosporin A and echinocandins. The immunosuppressive and calcineurin

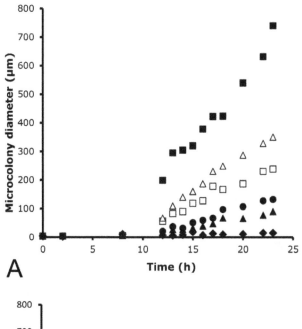

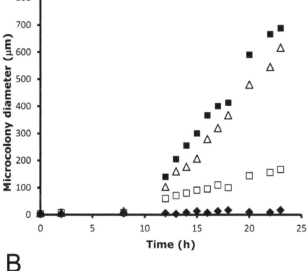

Figure 2. Growth of microcolonies of *A. fumigatus* grown on PAO with echinocandins. A: Microcolony growth from conidia of sensitive strain JBZ11 without drugs (solid squares), with 0.125 µg/ml anidulafungin (open squares) or 1 µg/ml anidulafungin (solid triangles), with 0.125 µg/ml caspofungin (open triangles) or 1 µg/ml caspofungin (solid circles) or 0.1 µg/ml amphotericin B (solid diamonds). B: As panel A, except using caspofungin resistant and anidulafungin sensitive strain CWZ1243.

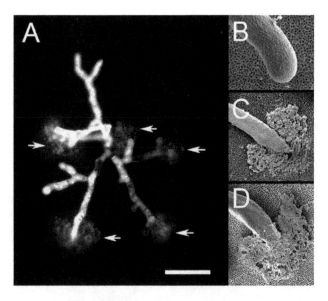

Figure 3. Lysis of microcolonies of *A. fumigatus* JBZ11 grown on PAO with echinocandins. Culture was for 14 h at 37°C. A: Microcolony cultured on 0.125 µg/ml anidulafungin. Staining was with Fun-1/calcofluor white and imaging by fluorescence microscopy. B: Imaging by SEM with vapour fixation of a hyphal tip grown without drugs. C: As panel B except growth with 0.125 µg/ml anidulafungin. D: As panel C, except growth with 0.125 µg/ml caspofungin. Arrows indicate lysed hyphal tips. Scale bar in panel A indicates 20 µm. When applied to panel B, 3 µm; panel C, 4 µm; panel D, 4 µm.

Discussion

A. fumigatus was found to germinate, grow and produce conidia on a PAO support. Conidial germination and hyphal extension and microcolony growth appeared efficient on aluminium oxide, these parameters were similar to growth on agar. The addition of echinocandins allowed the effects of these drugs to be studied at the microscopic level, in microcolonies up to several hundred microns in diameter. As with other culture methods on PAO, imaging by fluorescence microscopy after staining with fluorogenic dyes or processing for SEM was facilitated by the inert support. This has advantages over previous microscopic studies of the effect of caspofungin on cells, with the ready addition and removal of reagents without disturbing the microcolony. Additionally, effective imaging was possible on the planar surface. This permitted relatively high throughput image processing and quantification compared to previous studies, which used liquid culture to investigate caspofungin and micafungin lethality [7,8]. As far as we are aware, this is the first report of anidulafungin lethality.

Caspofungin is the only echinocandin to have been studied both in liquid culture and on PAO. Broadly, the two studies reveal similar trends, with one exception. Lysis was most apparent on PAO at intermediate concentrations of caspofungin, i.e. those close to MIC values. Higher concentrations of both echinocandins limited microcolony growth to a greater extent but resulted in reduced lysis. This may be due to more rapidly growing cells being more vulnerable to lysis. Paradoxical growth has previously been described for *A. fumigatus* treated with echinocandins, i.e. when very high doses have a lesser effect on growth compared to intermediate dosage. Explanations of paradoxical growth often invoke compensatory mechanisms for limiting cell wall damage, such as the stimulation of chitin synthesis, which reduce drug effects at high concentrations. However, both at the microcolony

pathway inhibitor cyclosporine A is known to have fungicidal activity for *A. fumigatus*. Cyclosporine has been shown to combine positively as an antifungal agent with caspofungin against some clinical isolates of *A. fumigatus* [14]. Therefore, this interaction was tested on PAO. After 14 h culture on PAO, cyclosporine A induced rounded cells staining predominantly with Syto 9 when present at concentrations >3 µg/ml. At 3 µg/ml cyclosporine had little effect on cell morphology alone but did enhance the action of echinocandins (Figure 8).

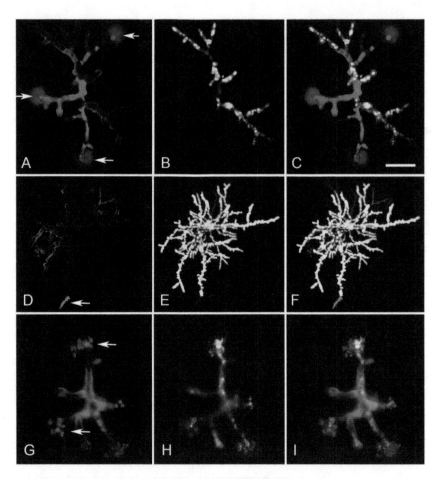

Figure 4. Examples of imaging microcolonies of *Aspergillus spp.* **grown on PAO with 0.5 μg/ml caspofungin and dual stained with propidium iodide and Syto9.** Culture in all cases was for 14 h at 37°C on Sabouraud medium. Panels A–C; CASP sensitive strain of *A. fumigatus* (JBZ11). A: Staining with PI. B: Staining with Syto9. C: Merged. Arrows in panel A indicate lysed hyphal tips. Panels D–E; as previous panels but imaging an CASP resistant strain of *A. fumigatus* (CWZ1243) showing greater growth and only limited echinocandin mediated damage (e.g. arrow). G–I; as panels A–C, but showing staining of an ANI and CASP sensitive strain of *A. terreus* (CWZ59). Scale bar in panel C indicates 20 μm for panels A–C, 45 μm for panels D–F and 15 μm for G–I.

level, and in E-tests, we did not observe classical paradoxical effects in these studies (in the sense that increasing concentrations echinocandins including those significantly above the MIC decreased the size of microcolonies). In contrast, for liquid culture increasing concentrations of caspofungin above the MIC has been reported to lead to enhanced lysis and cell death [7]. The ability to separate growth inhibition and tip lysis using microcolony imaging should allow paradoxical effects of drugs to be studied effectively in the future.

Apart from the obvious damage, two supporting lines of evidence indicated that cell death was occurring. The first is an excellent correlation between propidium iodide staining and lysis of hyphal tips. So-called "live-dead" staining is commonly used within bacteriology as an indicator of membrane permeability and therefore as an indirect measure of cell death. Whilst this staining method is not commonly used for *Aspergillus*, this dye pairing is sold for yeast viability assays (Funga Light Kit, Invitrogen). Secondly, cell debris from lysed cells adhered to PAO and this lead to a fortuitous and novel marker for the location of a lytic event. This debris was visible when stained with calcofluor white or propidium iodide and could also be imaged by SEM. The distribution of debris was a characteristic arc pattern, suggesting a single, violent lytic event at a single location. Lysed tips did not extend further,

i.e. beyond the lysis point, under conditions in which the microcolonies were growing, at least within a few hours of removal of the drug. This suggested that repair and regrowth of lysed hyphal tips did not occur within the time frame of these experiments. Previously, we have also shown that the debris resulting from *Enterobacteriaceae* lysed by trimethoprim could be imaged on PAO [11]. Taken with this work, this supports the idea that more information can be obtained from extremely damaged or highly stressed cells than by many other culture methods.

One way of thinking about the killing efficiency of the echinocandins is that these drugs are not fungicidal at the level that matters, as complete destruction of a microcolony seems very hard to achieve despite the dramatic effect on individual cells. But describing these drugs as fungistatic also seems simplistic for two reasons. The first is that growth of microcolonies never halts completely. More important is the high level of cell death, at least at specific concentrations of echinocandins (in some cases >50% cell lethality). This latter point suggests that looking for treatment regimes, or new echinocandins that are more aggressively fungicidal, may be interesting ways of improving the clinical effectiveness of this group of drugs. For example, the combination of the echinocandin FK463 with chitin synthase inhibitor nikkomycin Z has been reported to be synergistic against *A.*

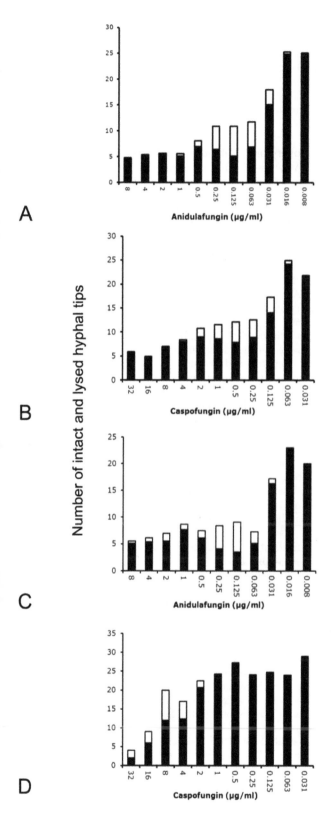

A

B

C

D

Figure 5. Quantification of the lysis of hyphal tips by echino-candins. Panels A and B: Echinocandin susceptible strain of *A. fumigatus* JBZ11. C and D: Caspofungin resistant, anidulafungin sensitive strain of *A. fumigatus* CWZ1243. Culture was for 14 h at 37°C on Sabouraud medium. A and C: Effect of anidulafungin. Black, average number of intact hyphal tips per microcolony. White, average number of lysed hyphal tips per microcolony. B and D: Effect of caspofungin. Interpretation as panel A.

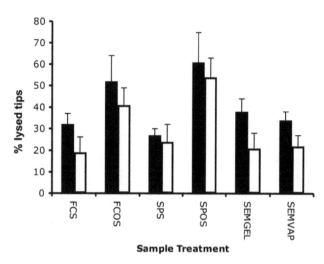

Figure 6. Quantification of the lysis of hyphal tips by echinocandins by sample preparation method. Effect of sample preparation method on percentage lysed hyphal tips. Black bars, 0.125 μg/mlanidulafungin. White bars, 0.5 μg/ml caspofungin. FCS, Fun-1 and calcofluor white staining on Sabouraud medium with imaging by fluorescence microscopy. FCOS, Fun-1 and calcofluor white staining with osmotic shock on water agar and fluorescence microscopy. SPS, Syto9 and propidium iodide staining on Sabouraud medium with assessment by fluorescence microscopy. SPOS, Syto9 and propidium iodide staining with osmotic shock on water agar and fluorescence microscopy. SEMGEL, glutaraldehyde fixation by gel method and imaging by SEM. SEMVAP vapour fixation method and SEM imaging. Error bars indicate S.D. from the mean (n = 3).

fumigatus when tested *in vitro* [13]. In our study, osmotic shock was shown to increase echinocandin-mediated lysis (Figure 6). We were also able to confirm previous work [14] indicating that the calconeurine pathway inhibitor and immunosuppressive cyclosporine A (Figure 8) enhanced the action of echinocandins. It is possible that dosing methods, complementary drugs or cofactors may be found that improve treatment with existing echinocandins. Currently, some of the clinical effectiveness of the echinocandins against *A. fumigatus in vivo* is attributed to the exposure of cell antigens to the immune system; i.e. indirect effects as well as direct inhibition of fungal growth [6]. It certainly appears possible that greater direct killing by this class of drugs is achievable.

Culture on PAO is relatively simple, with the facility to gain quantitative data on microcolony growth with changes in the environment. Here this method has been used to investigate stress and recovery of a filamentous fungus from a particular class of therapeutic agents, the echinocandins. We suggest that this approach may be more widely usable, for example to investigate fungi difficult to study in other ways because of slow growth and/ or limitations of liquid culture methods such as aggregation. It is possible to change the properties of PAO (e.g. porosity, charge, hydrophobicity, texture), which may allow a systematic investigation as to how filamentous fungi are affected by surfaces [15]. Imaging by fluorescence microscopy can detect marginal growth (Figure 1) and recovery from echinocandins (Figure 6). As gradients of drugs can be created beneath PAO supports [16], more complex drug effects can be investigated – for example issues of synergy or antagonism between multiple antibiotics – or growth under extreme stress.

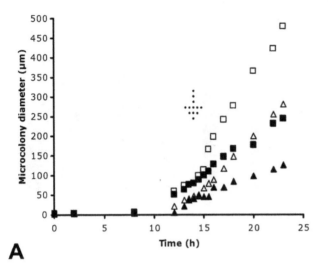

A

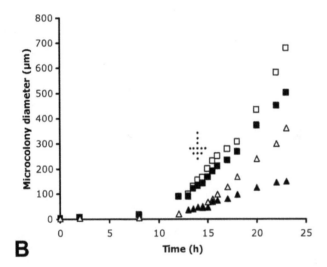

B

Figure 7. Recovery from echinocandins by sensitive strain JBZ11. A: Recovery from anidulafungin. Open squares, initial growth on 0.125 µg/ml anidulafungin then shift to Sabouraud agar with no drugs at 14 h (indicated by arrow). Open triangles, downshift from 1 µg/ml anidulafungin to none at 14 h. Solid squares, control with growth on 0.125 µg/ml throughout experiment. Solid triangles, continuous growth on 0.125 µg/ml anidulafungin. B: Recovery from caspofungin. Open squares, initial growth on 0.125 µg/ml caspofungin then shift to Sabouraud agar with no drugs at 14 h (indicated by arrow). Open triangles, downshift from 1 µg/ml caspofungin to none at 14 h. Solid squares, control with growth on 0.125 µg/ml throughout the experiment. Solid triangles, continuous growth on 0.125 µg/ml caspofungin.

Methods

Culture of *A. fumigatus* and *A. terreus* and exposure to drugs on porous aluminium oxide

Strains. All strains of *Aspergillus* species used in this study were clinical isolates or reference strains, as detailed in Tables 1 and 2.

Culture and Susceptibility Testing. PAO strips were sterilized and handled as previously described [9,10]. Dispersed preparations of *A. fumigatus* conidia [17] were inoculated onto sterile PAO (placed on the appropriate agar plate) at a density of 10–40 cfu/mm². Culture was at 37°C. PAO strips were transferred between plates using sterile parafilm. Echinocandins were delivered to fungi either at a concentration in the agar plate

or by using an E-strip (BioMerieux) aligned with the long axis of a 36 mm×8 mm PAO strip [16]. E-tests were performed directly on Sabouraud agar as recommended by the manufacturer, with 36 h incubation at 37°C. Cyclosporine A was obtained from Sigma Aldrich (NL) and was diluted from a stock solution of 25 mg/ml prepared in methanol. All experiments on growth on PAO, drug effects on PAO and conventional MIC testing were performed in triplicate.

Recovery of *A. fumigatus* from echinocandins

PAO strips were moved from plates containing echinocandins to those lacking the drugs in order to look at the potential for recovery from echinocandins. Because of the low volume of PAO strips relative to the agar beneath carry-over of drugs to plates lacking echinocandins was minimal (>1000 fold dilution).

Scanning electron microscopy (SEM)

Fixation of microcolonies for imaging by SEM was by one of two methods: 1) gel fixation was achieved by transferring the strip of PAO to an agar plate containing Sabouraud medium with 1% (v/v) glutaraldehyde for 30 min; 2) vapour fixation was performed by inverting an agar plate above the glutaraldehyde/paraformaldehyde fixative for 2 h [18]. In both cases, treatment with osmium tetroxide, ethanol dehydration, critical point drying, sputtering with tungsten and imaging by an FEI Magellan electron microscope were as previously described [11].

Staining and fluorescence microscopy

Staining of *A. fumigatus* and *A. terreus* with pairs of dyes (Fun-1/ calcofluor white or syto9/propidium iodide) was performed by transferring the PAO strip to a microscope slide covered in a thin layer of low melting point agar containing the dye. Dye concentrations and staining times were as previously described [9]. The low melting point agar was formulated with water to deliver an osmotic shock or with Sabouraud medium to stain under the same nutrient conditions as growth.

Quantification and data analysis

Image processing. TIFF format images were processing using ImageJ software (Rasband WS, 1997–2011 http://rsbweb. nih.gov/ij/). Images were inspected individually and only images with well separated microcolonies processed. Multiple images were then assembled into stacks for batch processing using an ImageJ macro that performed the operations of: [A] application of a median filter (pixel radius 3), [B] thresholding to black microcolonies on a white background and [C] using the "analyze particle" function to determine the dimensions (average diameter, area) of each microcolony within the field of view. Objects only partially within a field of view were excluded from this analysis. Datasets were exported into Microsoft Excel for further calculations.

Calculations of tip lysis and Syto9 vs propidium iodide staining. The frequency of lysed and unlysed tips was calculated from at least 50 microcolonies per condition. Hyphal tips <4 µm in length were not included in this analysis. Cells including hyphal tips were scored as Syto9 if the staining pattern was more intense than the competitor dye propidium iodide. Cells for which the converse was true were scored as propidium iodide staining.

Statistics and calculation of variance. Statistical operations (ANOVA, Student's t-test, determinations of normality) used the Vassar Statistics web server (Lowry RS, 1998–2011 http://faculty.vassar.edu/lowry/VassarStats.html). Microcolony heterogeneity was assessed using log₁₀ transformations of variance in microcolony area and diameter [19].

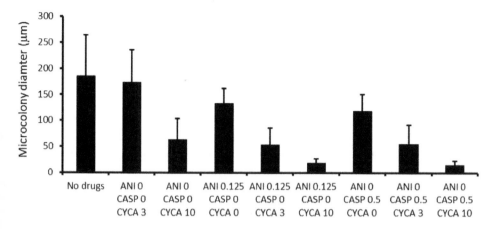

Figure 8. Effect of combining cyclosporine A with anidulofungin or caspofungin on the growth of *A. fumigatus* JBZ11. The average microcolony diameter (+/− S.D.) was determined from >100 microcolonies/data point after 16 h culture on PAO placed on Sabouraud medium containing the drugs. Drug concentrations are given in µg/ml.

Acknowledgments

Thanks to Adriaan van Aelst and Tiny Franssen-Verheijen for assistance with electron microscopy and Jacques Meis for strains. Anidulafungin was contributed by Pfizer, NL.

Author Contributions

Conceived and designed the experiments: CJI PMS. Performed the experiments: CJI. Analyzed the data: CJI PMS. Contributed reagents/materials/analysis tools: CJI. Wrote the paper: CJI PMS.

References

1. Douglas CM, D'Ippolito JA, Shei GJ, Meinz M, Onishi J, et al. (1997) Identification of the FKS1 gene of Candida albicans as the essential target of 1,3-beta-D-glucan synthase inhibitors. Antimicrob Agents Chemother 41: 2471–2479.
2. Walker LA, Gow NAR, Munro CA (2010) Fungal echinocandin resistance. Fungal Genet Biol 47: 117–126.
3. Antachopoulos C, Meletiadis J, Sein T, Roilides E, Walsh TJ (2008) Comparative *in vitro* pharmacodynamics of caspofungin, micafungin, and anidulafungin against germinated and nongerminated *Aspergillus* conidia. Antimicrob Agents Chemother 52: 321–328.
4. Hiemenz JW, Raad II, Maertens JA, Hachem RY, Saah AJ, et al. (2010) Efficiency of caspofungin as salvage therapy for invasive aspergillosis compared to standard therapy in a historical cohort. Eur J Clin Microbiol Infect Dis 29: 1387–1394.
5. Denning DW (2002) Echinocandins: a new class of antifungal. J Antimicrob Chemother 49: 889–891.
6. Perlin DS (2011) Current perspectives on echinocandin class drugs. Future Microbiol 6: 441–457.
7. Bowman JC, Hicks PS, Kutz MB, Rosen H, Schmatz DM, et al. (2002) The antifungal echinocandin caspofungin acetate kills growing cells of *Aspergillus fumigatus in vitro*. Antimicrob Agents Chemother 46: 3001–12.
8. Watabe E, Nakai T, Matsumoto S, Ikeda F, Hatano K (2003) Killing activity of micafungin against *A. fumigatus* hyphae assessed by specific fluorescent staining for cell viability. Antimicrob Agents Chemother 47: 1995–1998.
9. Ingham CJ, van den Ende M, Pijnenburg D, Wever PC, Schneeberger PM (2005) Growth and multiplexed analysis of microorganisms on a subdivided, highly porous, inorganic chip manufactured from Anopore. Appl Environ Microbiol 71: 8978–8981.
10. Ingham CJ, Sprenkels A, Bomer J, Molenaar D, van den Berg A, et al. (2007) The micro-Petri dish, a million-well growth chip for the culture and high-throughput screening of microorganisms. Proc Natl Acad Sci U S A 13: 18217–18222.

11. Ingham CJ, van den Ende M, Wever PC, Schneeberger PM (2006) Rapid antibiotic sensitivity testing and trimethoprim-mediated filamentation of clinical isolates of the *Enterobacteriaceae* assayed on a novel porous culture support. Med Microbiol 55. pp 1511–1519.
12. Hill TW, Loprete DM, Momany M, Ha Y, Harsch LM, et al. (2006) Isolation of cell wall mutants in *A. nidulans* by screening for hypersensitivity to Calcofluor White. Mycologia 98: 399–409.
13. Chiou CC, Mavrogiorgos N, Tillem E, He R, Walsh TJ (2001) Synergy, pharmacodynamics, and time-sequenced ultrastructural changes of the interaction between Nikkomycin Z and the echinocandin FK463 against *Aspergillus fumigatus*. Antimicrob Agents Chemo 12: 3310–3321.
14. Steinbach WJ, Schell WA, Blakenship JR, Onyewu C, et al. (2004) In vitro interactions between antifungals and immunosupressants against *Aspergillus fumigatus*. Antimicrob Agents Chemo 48: 1664–1669.
15. Ingham CJ, Ter Maat J, de Vos WM (2011) Where bio meets nano: The many uses of porous aluminium oxide in biotechnology. Biotechnol Advdoi:10.1016/j.biotechadv.2011.08.005.
16. Ingham CJ, Schneeberger PM (2012) The role of nanoporous materials in microbial culture. In: Hays JP, van Leeuwen W, eds. The role of new technologies in medical microbiological research and diagnosis. 1st ed. Sharjah, UAE.: Bentham Science Publishers Ltd. pp 86–98.
17. Templeton SP, Buskirk AD, Law B, Green BJ, Beezhold DH (2011) Role of germination in murine airway CD8+ T-cell responses to *Aspergillus* conidia. PLoS One 6: e18777.
18. Jones BV, Young R, Mahenthiralingam E, Stickler DJ (2004) Ultrastructure of *Proteus mirabilis* swarmer cell rafts and role of swarming in catheter-associated urinary tract infection. Infect Immun 72: 3941–3950.
19. den Besten HM, Ingham CJ, van Hylckama Vlieg JE, Beerthuyzen MM, Zwietering MH, et al. (2008) Quantitative analysis of population heterogeneity of the adaptive salt stress response and growth capacity of *Bacillus cereus* ATCC 14579. Appl Environ Microbiol 73: 4797–4804.

17

Synergistic Antifungal Effect of Glabridin and Fluconazole

Wei Liu[1], Li Ping Li[1], Jun Dong Zhang[1], Qun Li[1], Hui Shen[1], Si Min Chen[1], Li Juan He[1], Lan Yan[2], Guo Tong Xu[1], Mao Mao An[1]*, Yuan Ying Jiang[1,2]*

1 Tongji University School of Medicine, Shanghai, China, 2 New Drug Research and Development Center, School of Pharmacy, Second Military Medical University, Shanghai, China

Abstract

The incidence of invasive fungal infections is increasing in recent years. The present study mainly investigated glabridin (Gla) alone and especially in combination with fluconazole (FLC) against *Cryptococcus neoformans* and *Candida* species (*Candida albicans, Candida tropicalis, Candida krusei, Candida parapsilosis* and *Candida Glabratas*) by different methods. The minimal inhibitory concentration (MIC) and the minimal fungicidal concentration (MFC) indicated that Gla possessed a broad-spectrum antifungal activity at relatively high concentrations. After combining with FLC, Gla exerted a potent synergistic effect against drug-resistant *C. albicans* and *C. tropicalis* at lower concentrations when interpreted by fractional inhibitory concentration index (FICI). Disk diffusion test and time-killing test confirming the synergistic fungicidal effect. Cell growth tests suggested that the synergistic effect of the two drugs depended more on the concentration of Gla. The cell envelop damage including a significant decrease of cell size and membrane permeability increasing were found after Gla treatment. Together, our results suggested that Gla possessed a synergistic effect with FLC and the cell envelope damage maybe contributed to the synergistic effect, which providing new information for developing novel antifungal agents.

Editor: Joy Sturtevant, Louisiana State University, United States of America

Funding: This study was supported by National Key Basic Research Program of China (No 2013CB531602), the National Science Foundation of China (81202563, 81330083), National Science and Technology Major Project for the Creation of New Drugs (No 2012ZX09103101-003), and China posdoctoral Science foundation grant (No 2012M511141). The funders had no role in study design, data collection and analysis, decision to publish, or preparation of the manuscript.

Competing Interests: The authors have declared that no competing interests exist.

* Email: 13761575178@163.com (YYJ); anmaomao@tongji.edu.cn (MMA)

Introduction

Despite recent progress in the clinical management, invasive fungal infections are still a tricky problem and have a high mortality. *Candida* species are the fourth most important cause of hospital-acquired bloodstream infections. Besides, in developing countries systemic cryptoccocosis remains large and increasing [1]. The most common isolated *Candida* specie in clinical fungal invasive infection is *Candida albicans*, followed by *Candida tropicalis*, *Candida parapsilosis* and *Candida glabrata* [2–5]. *Cryptococcus neoformans* is the first or second most common cause of culture-proven meningitis [1]. One most common agent used in clinic is fluconazole (FLC). However, during long-time or repeated treatment, FLC resistance strains are easily developed [6]. The combination of two or more antifungal agents maybe a feasible policy to solve the problem.

Currently, researches on natural products which have potent synergisms with antifungal drugs have been raised. For example, retigeric acid B, a pentacyclic triterpenoid isolated from a lichen called *Lobaria kurokawae Yoshim*, can increase the susceptibilities of azole-resistant *C. albicans* strains in combination with azoles [7,8]. Plagiochin E, a macrocyclic bis (bibenzyl) isolated from the liverwort *Marchantia polymorpha*, has antifungal activity and resistance reversal effects for *C. albicans* [9]. Besides, berberine chloride, baicalein, allicin, pure polyphenol curcumin I, pseudo-laric acid B, eugenol and methyleugenol were also reported to have synergistic antifungal properties in combination with known antifungals [10–15].

Glabridin (Gla) [4-(8,8-dimethyl-2,3,4,8-tetrahydropyrano[2,3-f]chromen-3-yl)-benzene-1,3-diol] is a major active isoflavan isolated from *Glycyrrhiza glabra*. It has been reported that Gla had numerous beneficial properties, including antioxidant, anti-cancer, neuroprotective, anti-inflammatory activities, inhibiting fatigue or reversing learning and memory deficits in diabetic rats [16–23]. It possessed weak activity against *C. albicans*, *C. krusei*, *C. neoformans* and other filamentous fungi [24,25]. However, to our knowledge, no study to date has focused on its interaction with FLC.

In this study, synergistic antifungal effect of Gla and FLC against FLC-resistant clinical isolates of *C. albicans* and other yeast fungi (i.e. *C. neoformans*, *C. tropicalis*, *C. parapsilosis*, *C. krusei* and *C. glabratas*) and the possible mechanisms were investigated.

Materials and Methods

Strains and chemicals

25 clinical isolates of FLC-resistant *C. albicans*, and one *C. neoformans* 32609, *C. tropicalis* 2718 and *C. parapsilosis* ATCC 22019 were kindly provided by the Changhai Hospital, Shanghai, China. *C. krusei* ATCC2340 and *C. glabrata* ATCC1182 were kindly provided by professor Changzhong Wang (School of integrated traditional and western medicine, Anhui university of

traditional chinese medicine, Hefei, China). The susceptibilities of these strains to FLC were measured by broth microdilution method at advance. Frozen stocks of isolates were stored at −80°C in culture medium supplemented with 40% (vol/vol) glycerol and were subcultured twice at 35°C before each experiment. FLC (sigma Aldrich, St. Louis, MO) was obtained commercially. Gla (purity >98%) was obtained from Yuan Cheng Pharmaceutical Co. Ltd, China, and its initial stored concentration was 6.4 mg/ml in dimethyl sulfoxide (DMSO).

Antifungal susceptibility testing

The minimal inhibitory concentrations (MIC) of Gla and FLC against the yeast strains were determined by broth microdilution method as described previously [10]. The yeast at final concentration of 10^3 cells/ml in the RPMI 1640 liquid medium with serial (2×) dilutions of each drug were inoculated in 96-well flat-bottomed microtitration plates. After incubation at 35°C for 24 h or 72 h. Optical densities (OD) were measured at 630 nm with a microtitre plate reader (Thermolabsystems Multiskan MK3), and background optical densities were subtracted from that of each well. MIC_{80} was determined as the lowest concentration of the drugs that inhibited growth by 80% compared with that of drug-free wells. DMSO comprised <1% of the total test volume. The quality control strain, C. parapsilosis ATCC 22019 was included in each susceptibility test to ensure quality control. The MIC range of FLC to C. parapsilosis ATCC 22019 was from 0.5 μg/ml to 4 μg/ml, which meant this test was acceptable.

Checkerboard microdilution assay

Assays were performed on all isolates according to broth microdilution checkerboard method [10]. The initial concentration of fungal suspension in RPMI 1640 medium was 10^3 cells/ml, and the final combination concentrations ranged from 0.125 to 64 μg/ml for FLC and 1 to 16 μg/ml for Gla. The final concentration for FLC or Gla alone ranged from 0.125 to 64 μg/ml. 96-well flat-bottomed microtitration plates were incubated at 35°C for 24 h or 72 h. OD was measured at 630 nm, MIC was determined as the above.

The data obtained by the checkerboard microdilution assays were analyzed using the model-fractional inhibitory concentration index (FICI) method based on the Loewe additivity theory. FICI was calculated by the following equation: FICI = FIC A+FIC B, where FIC A is the MIC of the combination/the MIC of drug A alone, and FIC B is the MIC of the combination/the MIC of drug B alone. Among all of the FICIs calculated for each data set, the FICImin was reported as the FICI in all cases unless the FICImax was greater than four, in which case the FICImax was reported as

the FICI. Synergy was defined as an FICI value of ≤0.5, while antagonism was defined as an FICI value of >4, addition was defined as an FICI value of 0.5< FICI≤1. An FICI result between 1 and 4 (1< FICI≤4) was considered indifferent [7]. The fractional fungicidal concentration index (FFCI) was calculated the same.

Agar disk diffusion test

C. albicans 103 (one FLC-resistant isolate with a MIC of 32 μg/ml for Gla) and other yeast strains were tested by agar diffusion test [10]. 3 ml of aliquot of 10^6 cells/ml suspension was spread uniformly onto the yeast peptone dextrose (YPD) agar plate with or without 64 μg/ml FLC. Then, 6 mm paper disks impregnated with Gla alone or in combination with FLC were placed onto the agar surface. There was 5 μl DMSO in control disks. Photos were taken after incubation at 35°C for 48 h.

Time-killing test

C. albicans 103 and other yeast strains were prepared at the starting inoculum of 10^5 cells/ml. The concentrations were 4, 8, 16 μg/ml for Gla and 8 μg/ml for FLC, DMSO comprised <1% of the total test volume. At predetermined time points (0, 4, 8, 12, 16 and 24 h) after incubation with agitation at 35°C, a 100 μl aliquot was removed from every solution and serially diluted 10-fold in sterile water. A 100 μl aliquot from each dilution was spread on the sabouraud dextrose agar plate. Colony counts were determined after incubation at 35°C for 48 h. Fungicidal activity was defined as a ≥3 $\log_{10}$ reduction from the starting inoculum. Synergism and antagonism were defined as a respective decrease or increase of ≥2 $\log_{10}$ CFU/ml in antifungal activity produced by the combination compared with that by the more active agent alone [26].

Cell growth test

C. albicans 103 was prepared at the starting inoculum of 10^6 cells/ml in glass tubes. Different concentrations of Gla (2, 4, 8, 16 μg/ml) and FLC (2, 4, 8, 16, 32, 64 μg/ml) alone or the combinations of Gla (2, 4, 8, 16 μg/ml) and FLC (2, 4, 8, 16, 32, 64 μg/ml) were added into tubes. After incubation with agitation at 35°C for 24 h, pictures were taken. Aliquot was removed from each tube and serially diluted 10-fold in sterile water. A 100 μl aliquot from each dilution was spread on the sabouraud dextrose agar plate. Colony counts were determined after incubation at 35°C for 48 h.

Table 1. MICs and MFCs of Gla alone and in combination with FLC against 25 clinical FLC-resistant C. albicans.

	MIC (μg/ml)		MFC (μg/ml)	
	median	range	median	range
FLC	128	64->256	>256	>256
Gla	32	32–64	64	32–64
FLC/Gla[a]	1/4	1–1/4–4	8/16	4–16/8–16
FIC index	0.13	0.04–0.14	0.27	0.26–0.31
Interaction effect (n/%)[b]	Syn (25/100)		Syn (25/100)	

[a]MIC and MFC in combination expressed as [FLC]/[Gla].
[b]Syn, synergism. The number of strains and percentage for the interaction effect were shown.

Table 2. MICs and MFCs of Gla alone and in combination with FLC against varied yeast strains.

Yeast strains	MIC (µg/m)			FICI	MFC (µg/m)			FFCI
	FLC	Gla	FLC/Gla		FLC	Gla	FLC/Gla	
C. albicans SC5314	1	32	1/1	1.03	>64	32	16/16	0.63
C. tropicalis 2718	>64	64	4/8	0.16	>64	64	16/16	0.38
C. neoformans 32609	2	16	≤0.125/8	0.56	>64	32	8/16	0.56
C. parapsilosis ATCC22019	2	64	2/1	1.02	>64	64	32/16	0.50
C. krusei ATCC2340	>64	64	8/16	0.31	>64	64	>64/16	1.25
C. glabrata ATCC1182	>64	64	64/16	0.75	>64	64	>64/16	1.25

Cell membrane permeability

Membrane permeabilization of *C. albicans* was detected according to a previous study [27]. Briefly, *C. albicans* 103 (1×10^7 cells/ml) were incubated with 10 µM calcein acetoxymethyl ester (Fanbo biochemicals, China) for 2 h, The cells were then washed three times and *C. albicans* (1×10^7 cells/ml) was transferred to tubes. After treatment with or without FLC (64 µg/ml), MCZ (64 µg/ml) and Gla (16 µg/ml, 32 µg/ml, 64 µg/ml) for 3 h. The cells were washed three times and about 10,000 cells were acquired for flow cytometry analysis (Maflo Astrios flow cytometer). Experiments were repeated at least two times independently on separate days.

Cell wall inhibitors sensitivity test

Congored and calcofluorwhite (CFW) were incorporated into YPD agar plates at 100 µg/ml and 15 µg/ml, respectively. Yeast cells were grown in YPD medium with or without Gla (8, 16 µg/ml) for 12 h and 3 µl drops of serially diluted suspensions were inoculated into plates. After incubation for 48 h at 30°C, pictures were taken.

Results

The combination of Gla and FLC against clinical FLC-resistant *C. albicans*

The MIC values of Gla tested alone or in combination with FLC in FLC-resistant *C. albicans* were shown in Table 1. According to the interpretive breakpoints for FLC (<8 µg/ml and ≥64 µg/ml, respectively), 25 clinical FLC-resistant *C. albicans* were selected. The MICs of Gla against all tested strains ranged from 32 µg/ml to 64 µg/ml. When MIC-like assays were performed for FLC in the presence of fixed subinhibitory concentrations of Gla (4 µg/ml), the median MICs of FLC decreased from 128 µg/ml to 1 µg/ml in resistant strains (32-fold to 512-fold reductions). According to the analysis of FICI method, synergism was observed in all 25 tested strains (FICIs<0.2).

The combination of Gla and FLC against other varied FLC-susceptibility strains

We also tested antifungal effects of Gla alone or in combination with FLC in FLC-sensitive *C. albicans* and the other yeast strains (*C. neoformans, C. tropicalis, C. krusei, C. parapsilosis* and *C. glabratas*) (Table 2). In these strains, the range of MICs of Gla tested alone was from 16 µg/ml to 64 µg/ml, when in combination with FLC the MICs of Gla ranged from 1 µg/ml to16 µg/ml. In *C. tropicalis* and *C. krusei*, the MIC of FLC were reduced from >64 µg/ml to 4 µg/ml or 8 µg/ml respectively after in combination with Gla. In *C. glabratas*, no synergistic effect was observed.

In combination with FLC, Gla at lower concentrations exhibits fungicidal effect for FLC-resistant *C. albicans* by different methods

Cells from the microdilution assays after incubation with Gla, FLC or the combination of Gla and FLC at various concentrations were plated on the sabouraud dextrose agar plate to count the colony forming unit (CFU) for determination of the MFC_{100} (the minimal concentration with complete cell killing, i.e. no CFU counted). As shown in Table 1, the MFC of FLC can be much higher than the MIC and complete cell killing was not achievable. The range of the MFC of Gla was from 32 µg/ml to 64 µg/ml. When in combination with FLC (4 µg/ml or 8 µg/ml), Gla at 16 µg/ml showed fungicidal effect against all strains tested.

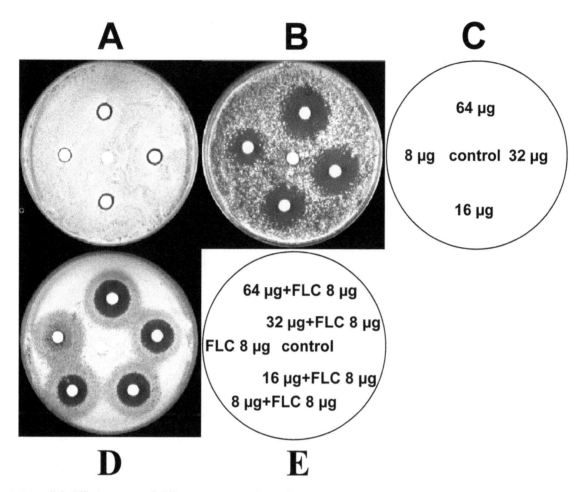

Figure 1. Agar disk diffusion assay of different concentrations of Gla combined with FLC against *C. albicans* 103. Panels A and D show agar plates, and panel B shows an agar plate containing 64 µg/ml FLC. Panel C describes the images for panels A and B, which containing 64, 32, 16, 8 µg of Gla or DMSO as control per disc. Panel E describes the image for panel D, the combination of Gla (64, 32, 16, 8 µg) with FLC (8 µg) or FLC (8 µg) alone or just DMSO as control were contained in each disc.

Further to visualize their synergistic fungicidal effect, different concentrations of Gla and the combination with FLC (8 µg/ml) were analyzed by agar disk diffusion assay. Gla alone at 64, 32, 16, 8 µg per disc had minimal fungicidal activity against the FLC-resistant *C. albicans* 103. While FLC at 8 µg per disc showed weak inhibition effect against *C. albicans*, the halo surrounding the disc was cloudy with colony (Fig. 1A). Interestingly, when FLC was combined with Gla, the halo surrounding the disc was significantly clearer. The diameters of the zones were clearer and larger than those of either drug alone on the plain agar plate, which was an indication of potent synergistic fungicidal activity (Fig. 1D). Similarly, on the agar plate (containing 64 µg/ml FLC), Gla also yielded significantly clearer and larger zones at 64, 32, 16, 8 µg per disc (Fig. 1B).

In addition, their synergistic fungicidal effect was confirmed by time-killing test (Fig. 2). Gla alone at 16 µg/ml showed fungicidal effect and led to a decrease of $3.57-\log_{10}$ CFU/ml at 24 h. No appreciable antifungal activity of FLC alone at 8 µg/ml was observed, but the combination of FLC (8 µg/ml) and Gla (4, 8 or 16 µg/ml) yielded 3.14, 3.62 or $4.10-\log_{10}$ CFU/ml reductions compared with Gla alone at 24 h (Table 3). Besides, the combination of Gla at 16 µg/ml and FLC at 8 µg/ml almost resulted in a complete cell-killing at 24 h (Fig. 2C).

In order to determine the relationship between the synergistic effect and the dosage of Gla and FLC, different concentrations of Gla (2 µg/ml–16 µg/ml) and FLC (2 µg/ml–64 µg/ml) were used in the cell growth test. Our results indicated that the synergism of the two drugs depended more on the concentration of Gla than FLC (Fig. 3). 4 µg/ml and 8 µg/ml Gla alone had no antifungal effect, while 64 µg/ml FLC had a weak antifungal activity. The antifungal effect was improved significantly after the two drugs used together at different concentrations except when FLC used at the concentration of 2 µg/ml. More specifically, 16 µg/ml Gla alone had an antifungal effect, while after combining with FLC (4 µg/ml–64 µg/ml), significantly synergistic effects were observed, and even complete cell killing activities were found when the concentration of FLC were above 16 µg/ml. Interestingly, the synergistic effects of the two drugs were unchanged when the dose of FLC declined from 64 µg/ml to 16 µg/ml, but when the combination concentration of FLC was below 16 µg/ml, the synergistic effect was lessened with the doseage of FLC decreasing.

Synergistic effect of FLC and Gla against other yeast strains

The interactions of FLC and Gla against the other yeast strains (i.e. FLC-senstive *C. albicans*, *C. tropicalis*, *C. neoformans*, *C. parapsilosis*, *C. krusei* and *C. glabrata*) were investigated by

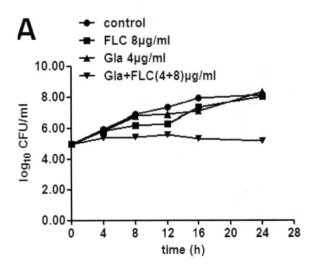

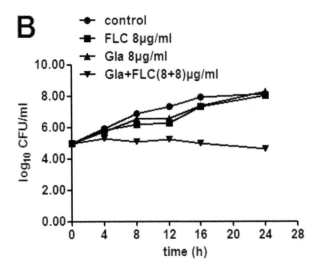

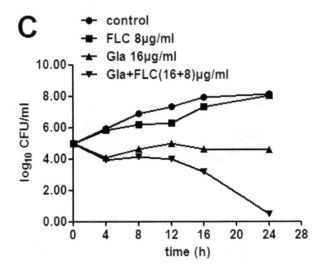

Figure 2. Time killing curves of *C. albicans* **103 treated with different concentrations of Gla and FLC.** FLC-resistant *C. albicans* 103 were treated with FLC (8 µg/ml), Gla (4 µg/ml) and FLC+Gla (4+8) µg/ml (A), Gla (8 µg/ml) and FLC+Gla (8+8) µg/ml (B) or Gla (16 µg/ml) and FLC+Gla (8+ 16) µg/ml (C) for 24 h. Aliquots were obtained at the indicated time points and serially dilutions were spreaded on agar plates. Colony counts were determined after 48 h incubation.

MFCs, agar disk diffusion assay and time-killing test. As shown in Table 2, the strains showed varied susceptibility to Gla and FLC, and there was no apparent correlation between the susceptibility towards Gla and the susceptibility towards FLC. Consistent with results from the disc diffusion assays, the *C. krusei* and *C. glabrata* were highly resistant to FLC. The range of MFC of Gla for each strain tested was from 32 µg/ml to 64 µg/ml, supporting its fungicidal property. Synergistic fungicidal interactions between

Gla and FLC were also observed in *C. tropicalis* by counting cells from the microdilution assay. The halo surrounding the discs with FLC and Gla was significantly clearer (Fig. 4) and the diameters of the zones were larger than those of either drug alone on the plain agar plate for FLC-senstive *C. albicans*, *C. tropicalis* and *C. neoformans*. Besides, the FLC+Gla combination yielded a decreased CFU compared with Gla alone in FLC-senstive *C. albicans*, *C. tropicalis* and *C. neoformans*, and even greater

Table 3. Decrease in $\log_{10}$ CFU/ml of yeast strains using different concentrations of Gla combining with FLC at 24 h.

FLC+Gla (µg/ml)	Mean (±SD) decrease in $\log_{10}$ CFU/ml compared with Gla alone			
	C. albicans **103**	*C. albicans* **SC5314**	*C. tropicalis* **2718**	*C. neoformans* **32609**
8+4	3.14 (0.08)			
8+8	3.62 (0.11)	1.51 (0.08)	1.64 (0.09)	4.42 (0.12)
8+16	4.10 (0.30)	1.82 (0.09)	3.16 (0.11)	4.50 (0.11)

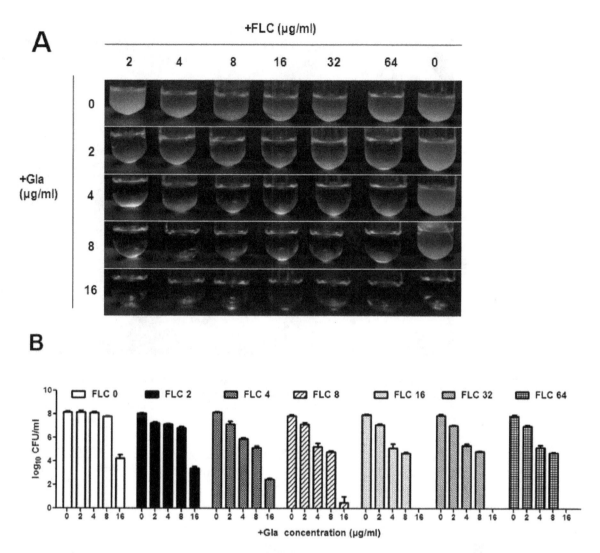

Figure 3. Growth condition of *C. albicans* 103 treated with different concentrations of Gla and FLC. Exponetially growing FLC-resistant *C. albicans* 103 were treated with or without different concentrations of Gla (2, 4, 8, 16 μg/ml) and FLC (2, 4, 8, 16, 32, 64 μg/ml) alone or the combinations of Gla and FLC in a shaking incubator. (A) Pictures of the growth condition of *C. albicans* were taken after 24 h incubation. (B) Aliquots from each tube were obtained at 24 h and serially dilutions were spread on agar plates. The number of *C. albicans* in each tube was determined by counting colonies after 48 h incubation.

reductions were observed in *C. neoformans* and *C. tropicalis* (> 3 Log$_{10}$ CFU/ml decrease, fungicidal effect can be defined) (Table 3). However, the synergistic fungicidal effect of FLC and Gla was not observed in *C. krusei* and *C. glabrata* by agar disk diffusion assay and time-killing test (results not shown), which was consistent with FFCI values of the previous tests.

Effect of Gla on the cell envelope

Flow cytometry analysis (side scatter [SSC]-forward light scatter [FSC]) showed that *C. albicans* 103 treated with FLC underwent weak cell shrinkage, while the cells treated with MCZ exhibited a significant decrease in cell size, as evidenced by the decrease in forward light scattering. Interestingly, changes in cell size were also observed after exposure to Gla, especially to Gla at the concentration of 32 μg/ml and 64 μg/ml (Fig. 5).

Calcein AM is a non-fluorescent derivative of calcein that can readily diffuse across membranes. Once within the cytoplasms of target cells, calcein AM is hydrolyzed by cytoplasmic esterases, yielding membrane-impermeable calcein which could be loaded

into intact cells. After incubation with calcein AM, the cellular fluorescence of calcein was detected and quantified by flow cytometry to evaluate the effect of Gla on the cell membrane permeabilization. Results showed that cellular calcein was markedly decreased by treatment of *Candida* cells with different concentrations of Gla (16 μg/ml, 32 μg/ml, 64 μg/ml), while by FLC a slight reduction of cellular calcein was observed (Fig. 6).

We also investigated the effect of Gla on the cell wall carbohydrates. Spot assays indicated that 16 μg/ml Gla treatment made *C. albicans* become more sensitive to both cell wall inhibitors (CFW and congored) compared with the control cells (Fig. 7). We used concanavalin A, calcofluorwhite (CFW) and specific anti-β-glucan primary antibody to stain the carbohydrates (mannan, chitin and glucan). However, fluorescence micrographs did not show obvious change in the three cell wall layers (Fig. S1).

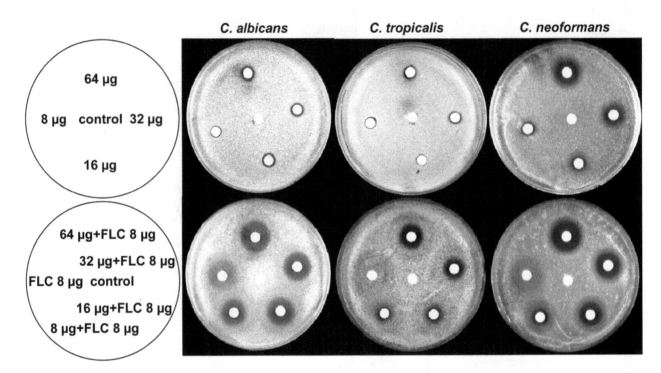

Figure 4. Agar disk diffusion assay of Gla alone or in combination with FLC against *C. albicans* **SC5314,** *C. tropicalis* **and** *C. neoformans.*
Upper agar plates of disks were impregnated with 64, 32, 16 and 8 μg of Gla or 5 μl of DMSO as control disk. In lower agar plates, disks were impregnated with FLC+Gla (64+8) μg, FLC+Gla (32+8) μg, FLC+Gla (16+8) μg, FLC+Gla (8+8) μg, FLC (8 μg) or 5 μl of DMSO as control disk. Left sketch panels describe the images for the right agar plates.

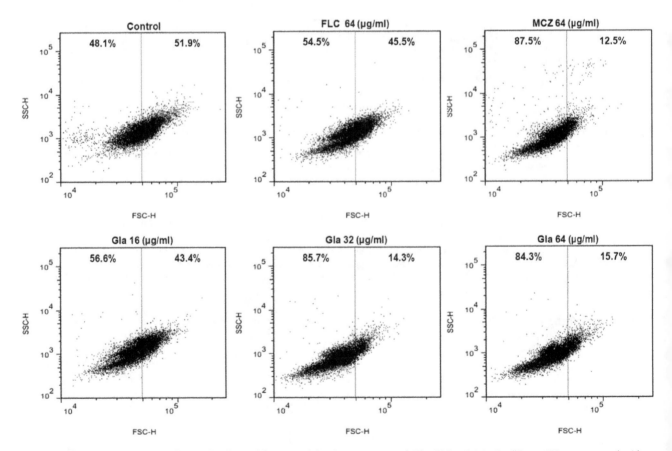

Figure 5. Changes in cell size (forward scatter-side scatter) in the presence of Gla. FLC-resistant *C. albicans* 103 were treated with or without Gla (16, 32, 64 μg/ml), FLC (64 μg/ml), MCZ (64 μg/ml) for 3 h. Then the cells were analyzed by flow cytometry.

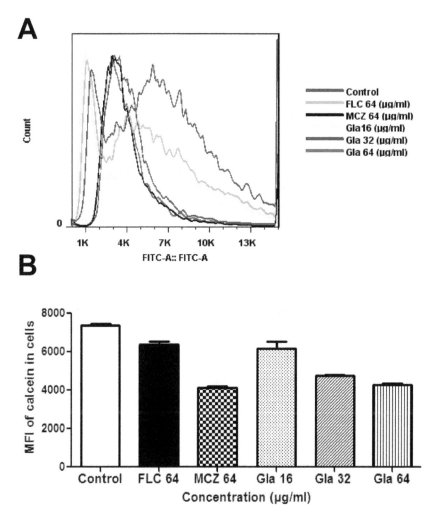

Figure 6. The influence of Gla on cell membrane permeability. *Candida* cells (1×10^7 cells/mL) were cultured in the presence of 10 μM calcein acetoxymethylester for 2 h. After washing four times, the cells were cultured in the presence or absence of Gla (16, 32, 64 μg/ml), FLC (64 μg/ml), MCZ (64 μg/ml) for 3 h. Cellular fluorescence intensities of calcein in the cells were analyzed by flow Cytometry.

Discussions

A large number of natural products from plants are reported to possess potent antifungal properties in recent years, such as terpene derivatives, flavans, nucleosides, peptides, alkaloids, saponins and sterols [28]. Gla is an isoflavan from *G. glabra* root. Previous studies have been reported that the alcohol or ethanol extract of the roots of *G. glabra* possessed weak antifungal activity against *C. albicans*, *Arthrinium sacchari*, *Chaetonmium funicola* and various filamentous fungi [24,29,30]. Synergistic activity of Gla and nystatin against oral *C. albicans* has been demonstrated previously [25]. But to our knowledge, there was no investigation on the combination of Gla and FLC against fungi.

Here, we demonstrated that Gla had weak antifungal activity against different fungi such as *Candida species and C. neoformans*, consisting with the previous reports [24,25,29,30]. When combining with FLC, Gla exerted a strong synergistic antifungal effects at lower concentrations. Different assays indicated a potent synergistic effect of Gla and FLC against FLC-resistant *C. albicans*, *C. neoformans* and *C. tropicalis*. The synergistic effects depended more on the concentration of Gla. According to Celine Messier's study on the toxic effect of Gla for oral epithelial cells (67% cell viability at 10 μg/ml), the concentration of Gla required for

reducing the MIC of FLC ranged from 1 μg/ml to 4 μg/ml, below the concentration which significantly reduced cell viability [25]. In order to determine whether this property of glabridin is specific to other isoflavans, we selected a second isoflavan equol and tested the interaction of equol and FLC against FLC-resistant *C. albicans* by checkerboard microdilution assay. The results showed a weak synergistic effect between equol and FLC. The MIC of equol alone was >1280 μg/ml, only when the concentration of equol was at 32 μg/ml a synergism was observed between equol and FLC, the concentration of FLC was reduced from >64 μg/ml to 4 μg/ml. This may suggested that there were synergistic effects between isoflavans and FLC, but the antifungal activities and synergisms of isoflavans and FLC were different for their distinct chemical structures.

The synergistic antifungal effect of Gla has not been characterized. Similar studies on the mode of synergism as follows: increasing reactive oxygen species (ROS) to accelerate apoptosis [31], inhibiting drug efflux pumps to increase intracellular drug concentration [11,32], targeting the ergosterol biosynthesis pathway to increase the fluidity for the resulted ergosterol depletion [32]. We tested the membrane sterols of *C. albicans* after treated with Gla alone or in combination with FLC by GC/MS, but none obvious sterols change was observed in Gla treatment cells (data

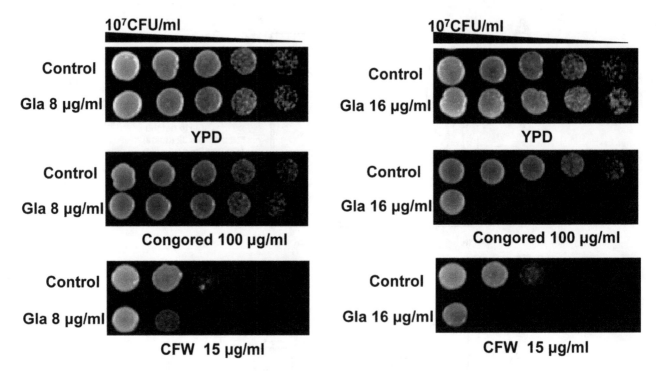

Figure 7. Susceptibilities of *C. albicans* 103 to cell wall inhibitors after treated with or without Gla. Exponetially growing FLC-resistant *C. albicans* 103 were treated with or without Gla (8, 16 µg/ml) for 12 h in YPD. Then they were diluted to 1×10^7 CFU/ml and 3 µl from 1:5 serial dilutions were spotted onto YPD agar plates containing 15 µg/ml CFW or 100 µg/ml congored.

not shown). This suggested that the synergism of Gla and FLC may not be related to the inhibition of membrane sterols synthesis. Previous study have displayed that isoflavan equol was capable of changing *Candida* cell membrane integrity by formation of membrane lesions and cell surface abnormalities against *C. albicans* [33]. Besides, flavans catechin hydrate and epigallocatechin gallate were also identified to have synergistic effect with FLC against *C. tropicalis*, cell shrinkage and plasma membrane damage were observed in the combination [34]. In our study, similar cell envelope changes were found in the *Candida* cells treated with Gla. A significant decrease in cell size and an increase of cell membrane permeability were observed in *C. albicans* after the treatment with Gla. However, *Candida* cells treated with Gla became more sensitive to cell wall inhibitors. We further stained the carbohydrate of the cell wall (mannan, chitin and glucan), while no obvious change of cell wall after Gla treatment was observed (Fig. S1).

In conclusion, the present study first demonstrated that Gla could enhance the antifungal effect of FLC, especially showed strong synergistic effect against *C. albicans*, *C. neoformans* and *C. tropicalis*. Their synergism maybe related to the effect of Gla on the cell envelope. Gla may serve as a pro-natural product for fungal infection treatment. Further studies should be carried out to identify its relationship of activity and structures.

Supporting Information

Figure S1 Fluorescence micrographs of the cell wall structures of *Candida albicans* by the treatment of Gla. Exponentially growing cells treated with or without 32 µg/ml Gla were stained by 50 µg/ml Concanavalin A alexa fluor 488 conjugate for mannan, 30 µg/ml Calcofluorwhite for chitin, or specific anti-β-glucan primary antibody and Cy3-labeled goat-anti mouse secondary antibody for glucan. Then cells were scanned under a Leica confocal laser scanning microscope and micrographs were acquired.

Acknowledgments

We thank Professor Changzhong Wang (School of Integrated Traditional and Western Medicine, Anhui University of Traditional Chinese Medicine, Hefei, China) for providing *C. krusei* ATCC2340 and *C. glabrata* ATCC1182.

Author Contributions

Conceived and designed the experiments: WL MMA YYJ. Performed the experiments: WL LPL JDZ QL HS SMC LJH LY. Analyzed the data: WL JDZ LY GTX MMA YYJ. Contributed reagents/materials/analysis tools: WL LPL JDZ QL HS SMC LJH LY. Wrote the paper: WL MMA YYJ. Final approval of manuscript: WL LPL JDZ QL HS SMC LJH LY GTX MMA YYJ.

References

1. Pukkila-Worley R, Mylonakis E (2008) Epidemiology and management of cryptococcal meningitis: developments and challenges. Expert Opin Pharmacother 9: 551–560.
2. Warnock DW (2007) Trends in the epidemiology of invasive fungal infections. Nihon Ishinkin Gakkai Zasshi 48: 1–12.
3. Morace G, Borghi E (2010) Fungal infections in ICU patients: epidemiology and the role of diagnostics. Minerva Anestesiol 76: 950–956.
4. Yapar N, Pullukcu H, Avkan-Oguz V, Sayin-Kutlu S, Ertugrul B, et al. (2011) Evaluation of species distribution and risk factors of candidemia: a multicenter case-control study. Med Mycol 49: 26–31.
5. Zirkel J, Klinker H, Kuhn A, Abele-Horn M, Tappe D, et al. (2012) Epidemiology of *Candida* blood stream infections in patients with hematological malignancies or solid tumors. Med Mycol 50: 50–55.

6. Horn DL, Neofytos D, Anaissie EJ, Fishman JA, Steinbach WJ, et al. (2009) Epidemiology and outcomes of candidemia in 2019 patients: data from the prospective antifungal therapy alliance registry. Clin Infect Dis 48: 1695–1703.

7. Sun L, Sun S, Cheng A, Wu X, Zhang Y, et al. (2009) In vitro activities of retigeric acid B alone and in combination with azole antifungal agents against *Candida albicans*. Antimicrob Agents Chemother 53: 1586–1591.

8. Chang W, Li Y, Zhang L, Cheng A, Liu Y, et al. (2012) Retigeric acid B enhances the efficacy of azoles combating the virulence and biofilm formation of *Candida albicans*. Biol Pharm Bull 35: 1794–1801.

9. Guo XL, Leng P, Yang Y, Yu LG, Lou HX (2008) Plagiochin E, a botanic-derived phenolic compound, reverses fungal resistance to fluconazole relating to the efflux pump. J Appl Microbiol 104: 831–838.

10. Quan H, Cao YY, Xu Z, Zhao JX, Gao PH, et al. (2006) Potent in vitro synergism of fluconazole and berberine chloride against clinical isolates of *Candida albicans* resistant to fluconazole. Antimicrob Agents Chemother 50: 1096–1099.

11. Huang S, Cao YY, Dai BD, Sun XR, Zhu ZY, et al. (2008) In vitro synergism of fluconazole and baicalein against clinical isolates of *Candida albicans* resistant to fluconazole. Biol Pharm Bull 31: 2234–2236.

12. An M, Shen H, Cao Y, Zhang J, Cai Y, et al. (2009) Allicin enhances the oxidative damage effect of amphotericin B against *Candida albicans*. Int J Antimicrob Agents 33: 258–263.

13. Sharma M, Manoharlal R, Negi AS, Prasad R (2010) Synergistic anticandidal activity of pure polyphenol curcumin I in combination with azoles and polyenes generates reactive oxygen species leading to apoptosis. FEMS Yeast Res 10: 570–578.

14. Guo N, Ling G, Liang X, Jin J, Fan J, et al. (2011) In vitro synergy of pseudolaric acid B and fluconazole against clinical isolates of *Candida albicans*. Mycoses 54: e400–406.

15. Ahmad A, Khan A, Khan LA, Manzoor N (2010) In vitro synergy of eugenol and methyleugenol with fluconazole against clinical *Candida* isolates. J Med Microbiol 59: 1178–1184.

16. Belinky PA, Aviram M, Fuhrman B, Rosenblat M, Vaya J (1998) The antioxidative effects of the isoflavan glabridin on endogenous constituents of LDL during its oxidation. Atherosclerosis 137: 49–61.

17. Carmeli E, Harpaz Y, Kogan NN, Fogelman Y (2008) The effect of an endogenous antioxidant glabridin on oxidized LDL. J Basic Clin Physiol Pharmacol 19: 49–63.

18. Hasanein P (2011) Glabridin as a major active isoflavan from Glycyrrhiza glabra (licorice) reverses learning and memory deficits in diabetic rats. Acta Physiol Hung 98: 221–230.

19. Shang H, Cao S, Wang J, Zheng H, Putheti R (2010) Glabridin from Chinese herb licorice inhibits fatigue in mice. Afr J Tradit Complement Altern Med 7: 17–23.

20. Tsai YM, Yang CJ, Hsu YL, Wu LY, Tsai YC, et al. (2011) Glabridin inhibits migration, invasion, and angiogenesis of human non-small cell lung cancer A549 cells by inhibiting the FAK/rho signaling pathway. Integr Cancer Ther 10: 341–349.

21. Kwon HS, Oh SM, Kim JK (2008) Glabridin, a functional compound of liquorice, attenuates colonic inflammation in mice with dextran sulphate sodium-induced colitis. Clin Exp Immunol 151: 165–173.

22. Hsu YL, Wu LY, Hou MF, Tsai EM, Lee JN, et al. (2011) Glabridin, an isoflavan from licorice root, inhibits migration, invasion and angiogenesis of MDA-MB-231 human breast adenocarcinoma cells by inhibiting focal adhesion kinase/Rho signaling pathway. Mol Nutr Food Res 55: 318–327.

23. Yu XQ, Xue CC, Zhou ZW, Li CG, Du YM, et al. (2008) In vitro and in vivo neuroprotective effect and mechanisms of glabridin, a major active isoflavan from Glycyrrhiza glabra (licorice). Life Sci 82: 68–78.

24. Fatima A, Gupta VK, Luqman S, Negi AS, Kumar JK, et al. (2009) Antifungal activity of Glycyrrhiza glabra extracts and its active constituent glabridin. Phytother Res 23: 1190–1193.

25. Messier C, Grenier D (2011) Effect of licorice compounds licochalcone A, glabridin and glycyrrhizic acid on growth and virulence properties of *Candida albicans*. Mycoses 54: e801–806.

26. Roling EE, Klepser ME, Wasson A, Lewis RE, Ernst EJ, et al. (2002) Antifungal activities of fluconazole, caspofungin (MK0991), and anidulafungin (LY 303366) alone and in combination against *Candida spp.* and *Crytococcus neoformans* via time-kill methods. Diagn Microbiol Infect Dis 43: 13–17.

27. Tanida T, Okamoto T, Ueta E, Yamamoto T, Osaki T (2006) Antimicrobial peptides enhance the candidacidal activity of antifungal drugs by promoting the efflux of ATP from Candida cells. Journal of Antimicrobial Chemotherapy 57: 94–103.

28. Di Santo R (2010) Natural products as antifungal agents against clinically relevant pathogens. Nat Prod Rep 27: 1084–1098.

29. Motsei ML, Lindsey KL, van Staden J, Jager AK (2003) Screening of traditionally used South African plants for antifungal activity against *Candida albicans*. J Ethnopharmacol 86: 235–241.

30. Hojo H, Sato J (2002) Antifungal activity of licorice (Glycyrrhiza Glabra) and potential applications to production of beverages. Foods Food Ingredients J Jpn 203: 27–33.

31. Fu Z, Lu H, Zhu Z, Yan L, Jiang Y, et al. (2011) Combination of baicalein and Amphotericin B accelerates *Candida albicans* apoptosis. Biol Pharm Bull 34: 214–218.

32. Sun LM, Cheng AX, Wu XZ, Zhang HJ, Lou HX (2010) Synergistic mechanisms of retigeric acid B and azoles against *Candida albicans*. J Appl Microbiol 108: 341–348.

33. Lee JA, Chee HY (2010) In Vitro Antifungal Activity of Equol against *Candida albicans*. Mycobiology 38: 328–330.

34. da Silva CR, de Andrade Neto JB, de Sousa Campos R, Figueiredo NS, Serpa Sampaio L, et al. (2014) Synergistic Effect of the Flavonoid Catechin, Quercetin, or Epigallocatechin Gallate with Fluconazole Induces Apoptosis in *Candida tropicalis* Resistant to Fluconazole. Antimicrobial Agents and Chemotherapy 58: 1468–1478.

18

Pesticide Exposure as a Risk Factor for Myelodysplastic Syndromes: A Meta-Analysis Based on 1,942 Cases and 5,359 Controls

Jie Jin[1,2¶], Mengxia Yu[1,2¶], Chao Hu[1,2¶], Li Ye[1,2], Lili Xie[1,2], Jin Jin[1,2], Feifei Chen[1,2], Hongyan Tong[1,2,3]*

1 Department of Hematology, the First Affiliated Hospital of Zhejiang University, Hangzhou, People's Republic of China, 2 Institute of Hematology, Zhejiang University School of Medicine, Hangzhou, People's Republic of China, 3 Myelodysplastic syndromes diagnosis and therapy center, Zhejiang University School of Medicine, Hangzhou, People's Republic of China

Abstract

Objective: Pesticide exposure has been linked to increased risk of cancer at several sites, but its association with risk of myelodysplastic syndromes (MDS) is still unclear. A meta-analysis of studies published through April, 2014 was performed to investigate the association of pesticide exposure with the risk of MDS.

Methods: Studies were identified by searching the Web of Science, Cochrane Library and PubMed databases. Summary odds ratios (ORs) with corresponding 95% confidence intervals (CIs) were calculated using random- or fixed-effect models.

Results: This meta-analysis included 11 case-control studies, all of which demonstrated a correlation between pesticide exposure and a statistically significant increased risk of MDS (OR = 1.95, 95% CI 1.23–3.09). In subgroup analyses, patients with pesticide exposure had increased risk of developing MDS if they were living in the Europe or Asia and had refractory anemia (RA) or RA with ringed sideroblasts (RARS). Moreover, in the analysis by specific pesticides, increased risk was associated with exposure to insecticides (OR = 1.71, 95% CI 1.22–2.40) but not exposure to herbicides or fungicides.

Conclusion: This meta-analysis supports the hypothesis that exposure to pesticides increases the risk of developing MDS. Further prospective cohort studies are warranted to verify the association and guide clinical practice in MDS prevention.

Editor: Shawn Hayley, Carleton University, Canada

Funding: This study was supported by grants from the Key Innovation Team Foundation of Zhejiang Province (2011R50015), major program fund of the Science Technology Department of Zhejiang Province (2013c03043-2), Zhejiang Province Fund for Distinguished Young Scholars (LR12H08001), National Public Health Grand Research Foundation (201202017) and the National Natural Science Foundation of China (No.30870914, No.81270582). The funders had no role in study design, data collection and analysis, decision to publish, or preparation of the manuscript.

Competing Interests: The authors have declared that no competing interests exist.

* Email: zjuhongyantong@163.com

¶ These authors are joint first authors on this work.

Introduction

Myelodysplastic syndromes (MDS) are a heterogeneous group of stem cell malignancies, characterized by ineffective hematopoiesis, and peripheral blood cytopenias. With disease progression, the risk of transformation into acute myeloid leukemia (AML) increased [1,2]. Despite development of new therapeutic methods in recent years, treatment of MDS is still limited and MDS remains incurable except in the case of the younger patients with good performance status, allogeneic stem cell transplantation eligibility, and adequate donor access [3]. As the whole population ages, MDS will become one of the most common myeloid malignancies. The societal impact and burden of the disease, measured in terms of the number of people affected yearly with a new diagnosis or who are living with the disease, is enormous and will continue to increase in the future. Therefore, a better comprehension of the etiology and further investigation of risk factors may significantly improve MDS prevention measures and reduce MDS incidence.

Since 1950, pesticide use has risen over 50% and pesticide toxicity has increased ten-fold [4]. Pesticide exposure is thought to increase cancer risk by promoting oxidative stress, chromosomal aberrations, cell signaling disturbances or gene mutations [5,6,7].

Over the past few decades, some epidemiological studies have analyzed the association between pesticide exposure and risk of MDS, but the findings are controversial. Five studies showed a positive association between incidence of MDS and pesticide exposure [8,9,10,11,12], and six studies illustrated no association [13,14,15,16,17,18]. Hence, the present meta-analysis was undertaken to further examine the potential involvement of pesticide exposure in MDS etiology.

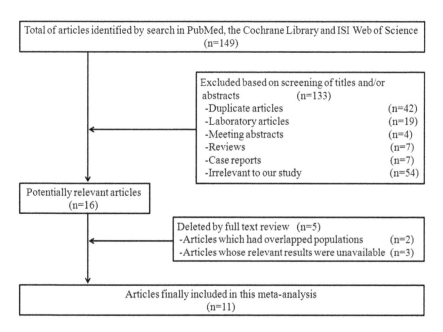

Figure 1. Process of study selection.

Materials and Methods

Literature research

A systematic literature search of the Web of Science, Cochrane Library and PubMed was executed by two independent reviewers (Chao Hu and Mengxia Yu). The following search strategy was used: (myelodysplastic syndrome OR MDS OR myelodysplastic OR myelodysplasia OR preleukemia) AND (pesticides OR herbicides OR fungicides OR insecticides). All relevant titles or abstracts were screened (see Study selection) to determine the suitability of each publication, and full-text articles were retrieved. We also checked the references from retrieved articles for additional studies not identified by database search.

Study selection

Studies included in this meta-analysis had to meet all the following criteria: (a) one of the exposures of interest was pesticide exposure; (b) one of the outcomes of interest was incidence of MDS; (c) a cohort design or case-control design; (d) providing the risk and corresponding 95% confidence intervals (CIs) or data to calculate these; (e) written in the English language. If there were multiple publications from the same study or overlapping study populations, the most recent and detailed study was eligible for inclusion in the meta-analysis.

Data extraction

Data were collected independently by two reviewers using a predefined data collection form. The following data were extracted from each study and included in the final analysis: the study name (together with the first author's name and year of publication), country of origin, gender, age, study design, source of patients, number of cases/controls, risk factor assessment, matching covariates, and adjusted covariates. We contacted the corresponding authors of the primary studies to acquire missing or insufficient data (when necessary), used group consensus and consulted a third reviewer to resolve discrepancies, and assigned scores of <7 and ≥ 7 for low- and high-quality studies, respectively, on the nine-score Newcastle-Ottawa Scale (NOS) [19,20].

Statistical analysis

To determine whether to use the fixed- or random-effects model, we measured statistical heterogeneity [21]. A fixed-effects model was used to calculate a pooled odds ratio (OR) with 95% CI when there was no heterogeneity. Otherwise, we calculated pooled ORs and confidence intervals assuming a random-effects model. The homogeneity of ORs across individual studies was quantified by the Q statistic and the I^2 score. $P>0.05$ for the Q-test was considered as a lack of heterogeneity among the studies. The I^2 values of 25%, 50%, and 75% represented mild, moderate, and severe heterogeneity, respectively [22]. Potential publication bias was assessed by using Begg's funnel plots (rank correlation method where an asymmetrical plot suggested possible publication bias) [23] and Egger's bias test (linear regression method where $P<0.05$ indicated the presence of statistically significant publication bias) [24]. Sensitivity analysis was conducted, in which the meta-analysis estimates were calculated by sequential omission of every study in turn, so as to reflect the influence of the data from individual studies on the pooled ORs and evaluate the stability of the results. Cumulative meta-analysis was also conducted by sorting the studies based on publication time. Subset analyses were performed by source of patients, disease subtype, geographic region, study quality, and type of pesticide. All of the statistical analyses were performed with STATA 11.0 (Stata Corporation, College Station, TX) using two-sided P-values, where $P<0.05$ was considered statistically significant.

Results

Literature search and study characteristics

The results of our literature search strategy and study selection process were detailed in Figure 1. We identified 11 case-control studies on the association of pesticide exposure with risk of MDS published between 1990 and 2011 [8,9,10,11,12,13,14,15,16,17,18]. A total of 1,942 MDS patients and 5,359 controls were included in the present meta-analysis. Among the 5,359 controls, 3853 persons were hospitalized patients without conditions related to hematological diseases and the remaining 1506 were recruited from healthy

74

Table 1. Main characteristics of studies evaluating the association between pesticides exposure and MDS.

Study	Country	Gender	Age	Study Design	Source of patients	Number of cases	Number of controls	Risk factor Assessment	Study Quality	Matching and Adjustments
Kokouva (2011)[13]	Greece	M/F	27–73	Case-control	Hospital-based	78	455	Questionnaire	5	Gender, age, smoking, family history
Lv (2011)[14]	China	M/F	20–88	Case-control	Hospital-based	403	806	Face-to-face Interview	6	Age, sex, anti-tb drugs, D860, traditional Chinese medicine, alcohol intake, benzene, gasoline, glues, hair dye, education, new building
Pekmezovic (2006)[8]	Serbia Montenegro	M/F	18–85	Case-control	Hospital-based	80	160	Interview	6	Age, sex
Strom (2005)[9]	United States	M/F	24–89	Case-control	Hospital-based	354	452	Mailed questionnaire	7	Age, sex, ethnicity, education, family history of hematopoietic cancer, alcohol intake, benzene, solvent, gasoline
Nisse (2001)[10]	France	M/F	NR	Case-control	Population-based	204	204	Interview	8	Agricultural workers, textile operators, health professionals, living next to an industrial plant, commercial and technical sales representatives, machine operators, oil use, smoking
Rigolin (1998)[11]	Italy	M/F	17–85	Case-control	Hospital-based	178	178	Interview and questionnaire	5	Age, sex
West (1995)[15]	UK	M/F	≥15	Case-control	Hospital-based	400	400	Interview and questionnaire	6	Age, sex, area of residence and hospital, year of diagnosis
Mele (1994)[16]	Italy	M/F	≥15	Case-control	Hospital-based	111	1161	Interview	6	Age, sex, education, and residence outside study town
Ciccone (1993)[12]	Italy	M/F	15–74	Case-control	Hospital-based and population-based	19	246	Interview	5	Sex, area of residence, age
Brown (1990)[17]	United States	M	≥30	Case-control	Population-based	63	1245	Interview	6	Vital status, age, state, tobacco use, family history of lymphopoietic cancer, high-risk occupations and high-risk exposure
Goldberg (1990)[18]	United States	NR	28–88	Case-control	Hospital-based	52	52	Interview	6	Age and sex

M: male; F: female; NR: not reported; tb: tuberculosis.

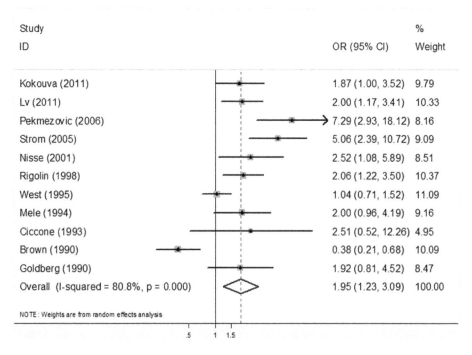

Figure 2. A forest plot illustrating risk estimates from included studies on the relationship between pesticide exposure and MDS risk.

people. A total of 1456 participants have exposed to pesticide, of whom 323 suffered from MDS. Among the chosen studies, seven were conducted in Europe, three in United States, and one in Asia. Pesticide exposure was ascertained by interview or questionnaire or both. The study quality was graded by the Newcastle-Ottawa Quality Assessment Scale, ranged from 5 to 8 (with a mean of 6). The main characteristics of the included articles were listed in Table 1.

Risk estimation

Our analysis demonstrated a significant adverse association between pesticide exposure (exposed vs. non-exposed status) and incidence of MDS (OR = 1.95, 95%CI 1.23–3.09) (Figure 2). Due to a statistically significant heterogeneity across studies ($I^2 = 80.8\%$, $P<0.001$), the summary OR were estimated using the DerSimonian and Laird random effects model [25]. A Galbraith plot identified four studies as major sources of heterogeneity (Figure 3A). After excluding these four studies [8,9,15,17], there was no study heterogeneity existed ($P = 0.999$, $I^2 = 0.0\%$) and the overall association became stronger (OR = 2.04, 95% CI 1.57–2.66).

Stratified analysis

Next, we pooled the OR estimates by patient source (population-based or hospital-based), MDS subtypes (refractory anemia (RA) and RA with ringed sideroblasts (RARS) or RA with excess blasts (RAEB) and RAEB in transformation (RAEBt)), geographic region (United States, Europe, or Asia), study quality (low or high), and type of pesticide (insecticide, herbicide or fungicide) (Table 2). When separated by patient source, the ORs (95% CI) were 2.26 (1.49–3.42) for hospital-based studies and 0.95 (0.15–6.06) for population-based studies. When stratified by MDS subtype, the associations were more positive for the RA/RARS sbutype (OR = 1.63, 95%CI 1.06–2.51) than the RAEB/RAEBt subtype (OR = 1.49, 95%CI 0.78–2.84). In the subset analyses stratified by

geographic region, a statistically significant adverse effect of pesticide exposure on MDS was observed in Europe (OR = 2.13, 95%CI 1.35–3.36) and Asia (OR = 2.00, 95%CI 1.17–3.41), but not in United States (OR = 1.52, 95%CI 0.30–7.73). Furthermore, when stratified by study quality, the relationship was more significant in high quality studies (OR = 2.19, 95%CI 1.40–3.42) than in low quality studies (OR = 1.90, 95%CI 1.09–3.33). In addition, when analyzed by type of pesticide, the ORs (95% CI) for insecticides, herbicides, and fungicides were 1.71 (1.22–2.40), 1.16 (0.55–2.43) and 0.70 (0.20–3.20), respectively.

Sensitivity analysis

We also carried out sensitivity analysis by sequentially excluding one study at a time to detect the influence of a single study on the overall estimate. The results displayed that no study disproportionately affected the summary risk estimates in this meta-analysis (Figure 3B). The eleven study-specific ORs ranged from a low of 1.72 (95%CI 1.11–2.68) to 2.27 (95%CI 1.58–3.27) via the omission of the study by Pekmezovic et al. [8] and the study by Brown et al. [17], respectively.

Cumulative meta-analysis

Cumulative meta-analysis of the relationship between pesticide exposure and risk of MDS was also implemented by sorting the studies based on publication time. Figure 4 showed the results from the cumulative meta-analysis of this connection in chronologic order. The 95% CIs became increasingly narrower with each addition of more data, suggesting the precision of each estimate was progressively increasing with the addition of more cases.

Publication bias

As reflected by the funnel plot (Begg's test, $P = 0.350$) and the Egger's test ($P = 0.113$), there was no publication bias being discovered. The data witnessed our result was statistically robust.

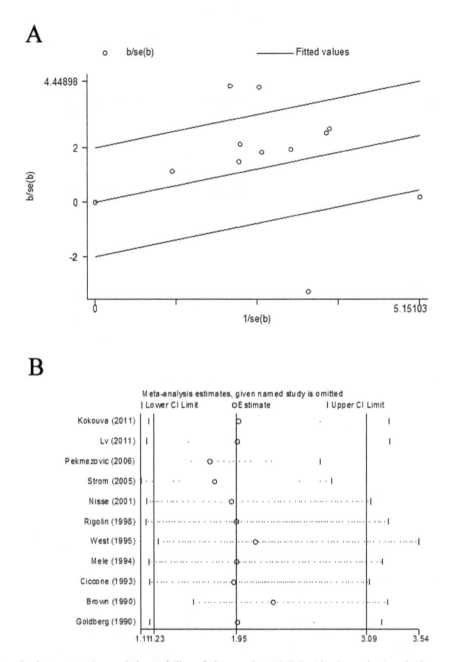

Figure 3. Evaluating the heterogeneity and the stability of the results. (A) Galbraith plot evaluating the heterogeneity; (B) Sensitivity analyses by sequential omission of individual studies in our study.

Discussion

The first myelodysplastic syndromes (MDS) case series was reported about 40 years ago [26]. Thus, the recognition of MDS is approximately 100 years behind the recognition of other hematologic malignancies. Similarly, the level of epidemiologic knowledge of MDS is far below that of other cancers. Therefore, further investigation of risk factors in MDS patients is needed to improve MDS prevention.

Pesticides are widely applied in agriculture all over the world. Three million cases of acute severe pesticide poisoning and over 200,000 deaths are reported annually [4]. Pesticides are thus considered a risk factor for some cancers. Moreover, two previous meta-analyses have been performed in hematological malignan-

cies, which indicated that pesticide exposure could increase risk of non-Hodgkin lymphoma, leukemia and multiple myeloma [27,28]. However, the result about MDS and pesticide exposure was limited. Recent epidemiological studies have examined the potential association between pesticide exposure and the risk of MDS, but none of the results has been conclusive. We attempted to clarify this possible relationship through a meta-analysis of eleven case-control studies.

To the best of our knowledge, this is the first meta-analysis assessing the relationship between pesticide exposure and MDS. Several interesting points raised by our analysis are worth discussing. Firstly, our research demonstrated a significantly positive correlation between pesticide exposure and MDS, which indicated pesticide exposure was associated with a 95% increased

Table 2. Stratified pooled odds ratios of the relationship between pesticide exposure and risk of MDS.

Variables	Number of studies	Pooled OR (95%CI)	Q-test for heterogeneity P value (I^2 score)	Egger's test P value	Begg's test P value
Total	11 (8, 9, 10, 11, 12, 13, 14, 15, 16, 17, 18)	1.95 (1.23–3.09)	<0.001 (80.8%)	0.350	0.113
Source of patients					
Population based	2 (10, 17)	0.95 (0.15–6.06)	<0.001 (92.3%)	–	1.000
Hospital based	8 (8, 9, 11, 13, 14, 15, 16, 18)	2.26 (1.49–3.42)	0.001 (71.7%)	0.098	0.266
Disease subtype					
RA/RARS	3 (9, 11, 18)	1.63 (1.06–2.51)	0.258 (25.6%)	0.413	0.734
RAEB/RAEBt	4 (9, 11, 14, 16)	1.49 (0.78–2.84)	0.005 (70.4%)	0.734	0.452
Geographic region					
Europe	7 (8, 10, 11, 12, 13, 15, 16)	2.13 (1.35–3.36)	0.006 (66.8%)	0.133	0.057
Asia	1 (14)	2.00 (1.17–3.41)	–	–	–
United States	3 (9, 17, 18)	1.52 (0.30–7.73)	<0.001 (93.4%)	0.407	1.000
Study quality					
High	2 (10, 11)	2.19 (1.40–3.42)	0.698 (0.0%)	–	1.000
Low	9 (8, 9, 12, 13, 14, 15, 16, 17, 18)	1.90 (1.09–3.33)	<0.001 (83.9%)	0.155	0.348
Type of pesticides					
Insecticides	9 (10, 11, 12, 13, 14, 15, 16, 17, 18)	1.71 (1.22–2.40)	0.009 (60.8%)	0.147	0.348
Herbicides	4 (14, 15, 16, 17)	1.16 (0.55–2.43)	0.056 (60.3%)	0.203	0.089
Fungicides	1 (17)	0.70 (0.20–3.20)	–	–	–

RA: refractory anemia; RARS: RA with ringed sideroblasts; RAEB: RA with excess blasts (RAEB); RAEBt: RAEB in transformation.

risk of MDS. Sensitivity analysis and cumulative analysis confirmed the robustness of our outcomes. In addition, subgroup analyses showed a stronger effect of pesticide exposure on RA/RARS than on RAEB/RAEBt (i.e., exposed MDS patients had 63% increased risk of RA/RARS and 49% increased risk of RAEB/RAEBt, respectively). Our study also illustrated that exposure to insecticides can the increase risk of MDS by 71%, while exposure to herbicides (OR = 1.16, 95%CI 0.55–2.43) and fungicides (OR = 0.70, 95%CI 0.20–3.20), respectively, add no risk. Our subset analysis according to geographical region noted

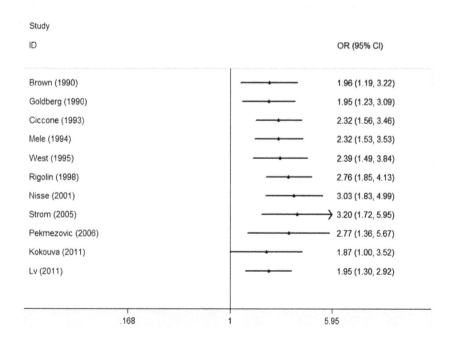

Figure 4. Forest plots showing the result of the cumulative meta-analysis.

higher risk of MDS in Europe (113%) than in Asia (100%) and the United States (52%).

The biological mechanism underlying the linkage of pesticide exposure to the pathogenesis of MDS remains largely unknown. However, several mechanisms are conceivable. Exposure to pesticides might cause overexpression of reactive oxygen species (ROS) sufficient to overwhelm antioxidant defense mechanisms and thereby lead to extensive DNA damage, protein damage, and hematopoietic irregularities [29]. On the other hand, pesticides might bind to and displace endogenous ligands of steroid nuclear receptors, including estrogen and androgen receptors, thus aberrantly activating receptor function and inducing changes in gene expression networks [30]. Recent in vitro mechanistic studies offer novel insight. For example, Boros and Williams reported that exposure of leukemic cell lines (K562) to increasing doses of an organophosphate pesticide (isofenphos) resulted in dose-dependent leukemic cell proliferation [31]. In addition, some previous studies demonstrating that pesticide exposure could induce chromosomal defects [32,33], might also suggest that pesticides could increase the risk of developing MDS. Further research is warranted to elucidate the likely biological mechanisms.

As a meta-analysis of previously published observational studies, our research has some limitations that influence interpretation of the results. First, although the present results seemed to suggest the absence of publication bias, our meta-analysis was vulnerable to publication bias, because only studies published in English were included. Limited resources prevented us from including articles published in other languages and databases. Second, no prospective studies of the association between pesticide exposure and MDS risk were available, and all included studies had a retrospective case-control design. Thus, owing to the limitations of case-control design, the possibility of undetected bias could not be excluded. Third, it is known farmers in China use large amounts of pesticides and this is an ideal population to study their effect on health. However, only one study from China was included in this meta-analysis. Fourth, too few original studies have separated biocides into insecticides, herbicides or fungicides to justify concluding the potential existence of a relationship between exposure to one or several categories of biocides and MDS. Significantly increased risks were observed, with an apparently higher increase when the exposure was to insecticides.

In summary, our findings support that pesticide exposure is associated with the increased risk of MDS, and this association varies widely across disease subtype, geographic region and specific biocide category. Larger and more rigorous analytical studies will be warranted to generate more robust conclusion to guide clinical practice in MDS prevention in the future.

Author Contributions

Conceived and designed the experiments: Jie Jin HT. Performed the experiments: CH MY. Analyzed the data: LY LX Jin Jin FC. Contributed reagents/materials/analysis tools: CH. Contributed to the writing of the manuscript: CH MY.

References

1. Kasner MT, Luger SM (2009) Update on the therapy for myelodysplastic syndrome. Am J Hematol 84: 177–186.

2. Newman K, Maness-Harris L, El-Hemaidi I, Akhtari M (2012) Revisiting use of growth factors in myelodysplastic syndromes. Asian Pac J Cancer Prev 13: 1081–1091.

3. Warlick ED, Smith BD (2007) Myelodysplastic syndromes: review of pathophysiology and current novel treatment approaches. Curr Cancer Drug Targets 7: 541–558.

4. Malek AM, Barchowsky A, Bowser R, Youk A, Talbott EO (2012) Pesticide exposure as a risk factor for amyotrophic lateral sclerosis: a meta-analysis of epidemiological studies: pesticide exposure as a risk factor for ALS. Environ Res 117: 112–119.

5. Infante-Rivard C, Weichenthal S (2007) Pesticides and childhood cancer: an update of Zahm and Ward's 1998 review. J Toxicol Environ Health B Crit Rev 10: 81–99.

6. Lafiura KM, Bielawski DM, Posecion NC, Jr., Ostrea EM, Jr., Matherly LH, et al. (2007) Association between prenatal pesticide exposures and the generation of leukemia-associated T(8;21). Pediatr Blood Cancer 49: 624–628.

7. Agopian J, Navarro JM, Gac AC, Lecluse Y, Briand M, et al. (2009) Agricultural pesticide exposure and the molecular connection to lymphomagenesis. J Exp Med 206: 1473–1483.

8. Pekmezovic T, Suvajdzic Vukovic N, Kisic D, Grgurevic A, Bogdanovic A, et al. (2006) A case-control study of myelodysplastic syndromes in Belgrade (Serbia Montenegro). Ann Hematol 85: 514–519.

9. Strom SS, Gu Y, Gruschkus SK, Pierce SA, Estey EH (2005) Risk factors of myelodysplastic syndromes: a case-control study. Leukemia 19: 1912–1918.

10. Nisse C, Haguenoer JM, Grandbastien B, Preudhomme C, Fontaine B, et al. (2001) Occupational and environmental risk factors of the myelodysplastic syndromes in the North of France. Br J Haematol 112: 927–935.

11. Rigolin GM, Cuneo A, Roberti MG, Bardi A, Bigoni R, et al. (1998) Exposure to myelotoxic agents and myelodysplasia: case-control study and correlation with clinicobiological findings. Br J Haematol 103: 189–197.

12. Ciccone G, Mirabelli D, Levis A, Gavarotti P, Rege-Cambrin G, et al. (1993) Myeloid leukemias and myelodysplastic syndromes: chemical exposure, histologic subtype and cytogenetics in a case-control study. Cancer Genet Cytogenet 68: 135–139.

13. Kokouva M, Bitsolas N, Hadjigeorgiou GM, Rachiotis G, Papadoulis N, et al. (2011) Pesticide exposure and lymphohaematopoietic cancers: a case-control study in an agricultural region (Larissa, Thessaly, Greece). BMC Public Health 11: 5.

14. Lv L, Lin G, Gao X, Wu C, Dai J, et al. (2011) Case-control study of risk factors of myelodysplastic syndromes according to World Health Organization classification in a Chinese population. Am J Hematol 86: 163–169.

15. West RR, Stafford DA, Farrow A, Jacobs A (1995) Occupational and environmental exposures and myelodysplasia: a case-control study. Leuk Res 19: 127–139.

16. Mele A, Szklo M, Visani G, Stazi MA, Castelli G, et al. (1994) Hair dye use and other risk factors for leukemia and pre-leukemia: a case-control study. Italian Leukemia Study Group. Am J Epidemiol 139: 609–619.

17. Brown LM, Blair A, Gibson R, Everett GD, Cantor KP, et al. (1990) Pesticide exposures and other agricultural risk factors for leukemia among men in Iowa and Minnesota. Cancer Res 50: 6585–6591.

18. Goldberg H, Lusk E, Moore J, Nowell PC, Besa EC (1990) Survey of exposure to genotoxic agents in primary myelodysplastic syndrome: correlation with chromosome patterns and data on patients without hematological disease. Cancer Res 50: 6876–6881.

19. Xu X, Cheng Y, Li S, Zhu Y, Zheng X, et al. (2014) Dietary carrot consumption and the risk of prostate cancer. Eur J Nutr.

20. Wells G, Shea B, O'Connell D (2009) Ottawa Hospital Research Institute: The Newcastle-Ottawa Scale (NOS) for assessing the quality of nonrandomized studies in meta-analyses. Available: http://www.ohri.ca/programs/clinical epidemiology/oxford.asp.

21. Hu ZH, Lin YW, Xu X, Chen H, Mao YQ, et al. (2013) Genetic polymorphisms of glutathione S-transferase M1 and prostate cancer risk in Asians: a meta-analysis of 18 studies. Asian Pac J Cancer Prev 14: 393–398.

22. Castillo JJ, Dalia S, Shum H (2011) Meta-analysis of the association between cigarette smoking and incidence of Hodgkin's Lymphoma. J Clin Oncol 29: 3900–3906.

23. Begg CB, Mazumdar M (1994) Operating characteristics of a rank correlation test for publication bias. Biometrics 50: 1088–1101.

24. Egger M, Davey Smith G, Schneider M, Minder C (1997) Bias in meta-analysis detected by a simple, graphical test. BMJ 315: 629–634.

25. DerSimonian R, Laird N (1986) Meta-analysis in clinical trials. Control Clin Trials 7: 177–188.

26. Saarni MI, Linman JW (1973) Preleukemia. The hematologic syndrome preceding acute leukemia. Am J Med 55: 38–48.

27. Van Maele-Fabry G, Duhayon S, Lison D (2007) A systematic review of myeloid leukemias and occupational pesticide exposure. Cancer Causes Control 18: 457–478.

28. Merhi M, Raynal H, Cahuzac E, Vinson F, Cravedi JP, et al. (2007) Occupational exposure to pesticides and risk of hematopoietic cancers: meta-analysis of case-control studies. Cancer Causes Control 18: 1209–1226.

29. Alavanja MC, Ross MK, Bonner MR (2013) Increased cancer burden among pesticide applicators and others due to pesticide exposure. CA Cancer J Clin 63: 120–142.

30. Schug TT, Janesick A, Blumberg B, Heindel JJ (2011) Endocrine disrupting chemicals and disease susceptibility. J Steroid Biochem Mol Biol 127: 204–215.

31. Boros LG, Williams RD (2001) Isofenphos induced metabolic changes in K562 myeloid blast cells. Leuk Res 25: 883–890.

32. Smith MT, McHale CM, Wiemels JL, Zhang L, Wiencke JK, et al. (2005) Molecular biomarkers for the study of childhood leukemia. Toxicol Appl Pharmacol 206: 237–245.

33. Chiu BC, Blair A (2009) Pesticides, chromosomal aberrations, and non-Hodgkin's lymphoma. J Agromedicine 14: 250–255.

19

The Impact of Selective-Logging and Forest Clearance for Oil Palm on Fungal Communities in Borneo

Dorsaf Kerfahi[1,2], **Binu M. Tripathi**[1], **Junghoon Lee**[2], **David P. Edwards**[3]*, **Jonathan M. Adams**[1]*

1 Department of Biological Sciences, Seoul National University, Seoul, Republic of Korea, **2** School of Chemical and Biological Engineering, Interdisciplinary Program of Bioengineering, Seoul National University, Seoul, Republic of Korea, **3** Department of Animal and Plant Sciences, University of Sheffield, Sheffield, United Kingdom

Abstract

Tropical forests are being rapidly altered by logging, and cleared for agriculture. Understanding the effects of these land use changes on soil fungi, which play vital roles in the soil ecosystem functioning and services, is a major conservation frontier. Using 454-pyrosequencing of the ITS1 region of extracted soil DNA, we compared communities of soil fungi between unlogged, once-logged, and twice-logged rainforest, and areas cleared for oil palm, in Sabah, Malaysia. Overall fungal community composition differed significantly between forest and oil palm plantation. The OTU richness and Chao 1 were higher in forest, compared to oil palm plantation. As a proportion of total reads, Basidiomycota were more abundant in forest soil, compared to oil palm plantation soil. The turnover of fungal OTUs across space, true β-diversity, was also higher in forest than oil palm plantation. Ectomycorrhizal (EcM) fungal abundance was significantly different between land uses, with highest relative abundance (out of total fungal reads) observed in unlogged forest soil, lower abundance in logged forest, and lowest in oil palm. In their entirety, these results indicate a pervasive effect of conversion to oil palm on fungal community structure. Such wholesale changes in fungal communities might impact the long-term sustainability of oil palm agriculture. Logging also has more subtle long term effects, on relative abundance of EcM fungi, which might affect tree recruitment and nutrient cycling. However, in general the logged forest retains most of the diversity and community composition of unlogged forest.

Editor: Andrew Hector, University of Oxford, United Kingdom

Funding: This project was funded by a National Research Foundation grant (NRF-2013-031400), Ministry of Education, Science and Technology, South Korea. The funders had no role in study design, data collection and analysis, decision to publish, or preparation of the manuscript.

* Email: david.edwards@sheffield.ac.uk (DPE); foundinkualalumpur@yahoo.com (JMA)

Introduction

Tropical forests are one of the world's most important reservoirs for biodiversity [1]. They contain an exceptional concentration of the world's species, but are being reduced in area faster than any other ecosystem [2]. Roughly half of the world's natural extent of tropical forest has been logged or converted to different land uses [3]. Some 403 million hectares of tropical rainforests have been included in timber estates and slated for selective logging [4], and between 2000 and 2010, approximately 13 million hectares of forest within the tropics were cleared for agricultural activities [5,6], including oil palm plantations [7].

Anthropogenic disturbances in tropical forests are causing a dramatic decline in global biodiversity, and in associated biological processes that maintain the productivity and sustainability of ecosystems [8]. Several studies have shown that selective logging does not drastically impact the overall species richness and diversity of tropical forest [9–12], however, it has been shown that the impact of selective logging could be anything between fairly mild and severe depending on the intensity of logging [13] with changes in the composition of species, as forest-interior specialists decline and edge-tolerant, gap specialists increase in abundance [14,15]. In contrast, the conversion of both primary and logged forest to agricultural land uses has been to shown to have a far greater negative impact on biodiversity than does logging. Conversion to agriculture results in a major reduction in biodiversity, again across a host of animal and plant taxa [16]. The conversion of primary and logged forest to agricultural plantations also results in a substantial decrease in the functional diversity of tropical ecosystems, with implications for the provision of ecosystem functions, whereas logging has lesser impacts on these metrics [17,18]. As well as affecting plants and larger animals, land use change also affects the soil biota. Land use change affects soil pH, carbon and nutrient content [19,20], causing shifts in soil microbial communities [21–23].

Fungi constitute one of the most diverse and dominant groups of organisms in soil, and they play important ecological roles in the ecosystem as decomposers, pathogens and plant mutualists [24,25]. Understanding the structure and diversity of soil fungal communities is fundamental to the understanding of their function in the ecosystem and their impact on plant communities [26]. However, while minimal work has been done in the tropics to assess the effect of land use changes on soil bacterial communities [23,27], until recently relatively little was known about the impacts of tropical land use change on soil fungal communities. Various studies have suggested that forest clearance to tree plantations or agricultural crops shifts soil fungal communities, linked to strong changes in soil properties [22,28]. However, previous studies of forest clearance to other forms of agriculture on forest fungal communities have mostly been limited to techniques that give relatively low taxonomic resolution (i.e. T-RFLP and PCR-DGGE). Nevertheless, a recent study by McGuire et al. [29], which used high throughput sequencing to analyze soil fungal communities in Southeast Asian tropical forests in west Malaysia, showed that conversion of primary forest to oil palm plantations alters fungal community composition and function, whereas primary and logged forests were more similar in composition and nutrient cycling potential. However, there is a need for further studies to understand the impacts of logging cycles on soil fungal communities and of the conversion of logged forest to agriculture, since logged forests now dominate the tropics [3] and are much more likely to be converted to agriculture than primary forests [2,6].

In this study, we also focused on the rainforests of the Sundaland region, of Southeast Asia, but some 1,800 km away in east Malaysia (Borneo). Across the Sundaland region, the primary forest has been subject to differing degrees of logging intensity. Much of the region's forest (about 50%) has never been logged, while many areas have been subject to one or two logging cycles [30]. Also, oil palm is one of the most rapidly expanding crops in this region. This provides an opportunity to study the effect of different intensities of logging on the soil fungal community and also to evaluate if conversion of forest to oil palm plantation has a stronger impact on soil fungi than logging, as is the case for numerous macroscopic taxa. Our objective here was to understand whether land use change has an impact on the structure and diversity of fungal communities in the Yayasan Sabah (YS) logging concession in Malaysian, Borneo. We compared the fungal communities in forests with different logging histories (unlogged, once-logged and twice-logged), and oil palm plantations. We examined whether there are differences in α and â-diversity, as well as community composition. These results may provide important information for soil management policies, and estimation of ecological impact of land use change in this region.

Materials and Methods

Study area

The study area is located within the Yayasan Sabah (YS) logging concession and contiguous oil palm plantation areas, in Sabah, Malaysian Borneo (4°58′ N, 117°48′ E). The forests in this area are naturally dominated by valuable timber tree species belonging to the family Dipterocarpaceae [31]. Due to logging for the wood industry and clearance for palm oil plantations, the area of forest in Borneo - as elsewhere in the tropics - has been dramatically reduced in recent decades [32].

Fieldwork was conducted in the Ulu Segama-Malua Forest Reserve (US-MFR) and adjacent oil palm estates in Sabah, Borneo. Some areas of forest were logged between 1970 and 1990, and some of these were then re-logged between 2000 and 2007. During the first logging rotation, approximately 113 m^3 per hectare (range 73 m^3 to 166 m^3) of commercially valuable trees > 0.6 m diameter were extracted. During the second logging rotation, an additional 31 m^3 per hectare (range 15 m^3 to 72 m^3) of timber were removed [10,31]. Selectively logged forest in the US-MFR is adjacent to the 45,200 ha Danum Valley Conservation Area (DVCA) and Palum Tambun Watershed Reserve, containing large areas of unlogged forest [10,16]. Oil palm plantations are situated to the north and south of the US-

MFR, with mature palms of 20 to 30 years old, planted at a density of 100 trees per hectare.

Soil sampling and DNA extraction

From September to October 2012, twenty-four transects each of 200 m in length were located across four different land uses: unlogged (primary) forests, once-logged and twice-logged forests, and oil palm plantations, with six transects per habitat. Within each habitat, distances between transects ranged from 500 m to 65 km, whereas across habitats, distances between transects ranged from 1 km to 67 km. From each transect, at 50 m intervals, approximately 50 g from the top 5 cm of soil (excluding the leaf litter layer) was taken in a sterile plastic bag using a trowel, giving five samples of soil per transect. The trowels were thoroughly cleaned with ethanol between successive transect sampling. All soil samples were then sieved (2 mm) in laboratory to homogenize the sample and stored at –20°C until DNA extraction [10]. Twenty-four soil DNA extractions (one for each transect) were performed using 0.3 g of soil, with the Power Soil DNA extraction kit (MO BIO Laboratories, Carlsbad, CA, USA) following the directions described by the manufacturer.

PCR amplification and pyrosequencing

Fungal DNA was amplified using ITS primers targeting the internal transcribed spacer (ITS) region 1 and 2. Forward primers comprised the 454 Fusion Primer A-adaptor, a specific multiplex identifier (MID) barcode, and the ITS1F primer (5′-CTTGGTCATTTAGAGGAAGTAA-3′) [33], while the reverse primer was composed of the B-adapter and ITS4 primer (5′-TCCTCCGCTTATTGATATGC-3′) [34]. Polymerase chain reactions (PCR) were performed in 50 µl reactions using the following temperature program: 95°C for 10 min s; 30 cycles of 95°C for 30 s, 55°C for 30 s, 72°C for 30 s; and 72°C for 7 min. The PCR products were purified using the QIAquick PCR purification kit (Qiagen) and quantified using PicoGreen (Invitrogen) spectrofluorometrically (TBS 380, Turner Biosystems, Inc. Sunnyvale, CA, USA). 50 ng of purified PCR product for each sample were combined in a single tube and sent to Macrogen Inc. (Seoul, Korea) for sequencing using 454/Roche GS FLX Titanium Instrument (Roche, NJ, USA).

Sequence processing

Initial quality filtering and denoising were performed following the 454 SOP in the mothur pipeline [35]. The ITS1 region was verified and extracted using the ITS1 extractor for fungal ITS sequences [36]. Putative chimeric sequences were detected and removed via the Chimera Uchime algorithm contained within mother [37]. Operational taxonomic units (OTUs) were assigned using the QIIME implementation of UCLUST [38], with a threshold of 97% pairwise identity. OTUs were classified taxonomically using the classify command in mothur at 80% Naïve Bayesian bootstrap cutoff with 1000 iterations against the UNITE database [39]. Ectomycorrhizal (EcM) fungi were determined by matching taxonomy assignments with established EcM lineages as determined by recent phylogenetic and stable isotope data [40]. The 454 sequence run has been deposited in the NCBI Sequence Read Archive under accession number SRP041467.

Statistical analysis

To correct for differences in number of reads, which can bias diversity estimates, all samples were rarified to 3,347 reads per sample. To test for effects of land use types on the OTU richness

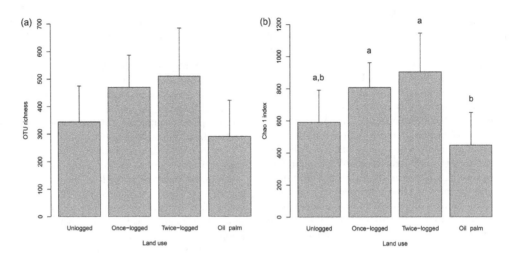

Figure 1. Diversity indices of the fungal community across different land uses in Sabah, Malaysian Borneo. (a) OTU richness and (b) Chao1 index. Pairwise comparisons are shown; different letters denote significant differences between groups at $P<0.05$.

and diversity indices, we used a linear model (LM) for normal data or generalized linear model (GLM) for non-normal data, considering land use as the major factor. We used the same procedure to test whether relative abundance of the most abundant phyla differed among different land use types. We also assessed the effect of land use on the relative abundance at the order and genus levels within those phyla that showed significant differences due to land use. Post-hoc Tukey tests were used for

pairwise comparisons. When neither a linear nor a generalized linear model fitted the data, we used a Kruskall-Wallis test to assess the effect of land use on the relative abundance of fungal taxa, with the Bonferroni correction to assess pairwise comparisons.

To test whether species composition results may have been influenced by pseudoreplication within study sites, we used a Mantel test (Mantel Nonparametric Test Calculator 2.0) [41] to compare transect matrices of fungal compositional to geographic

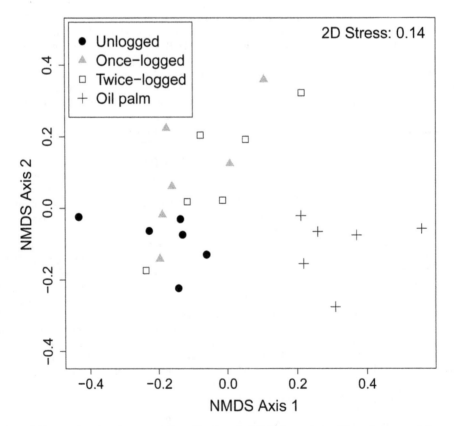

Figure 2. Non-metric multidimensional scaling (NMDS) ordination showing clustering of fungal communities among different land uses in Sabah, Malaysian Borneo.

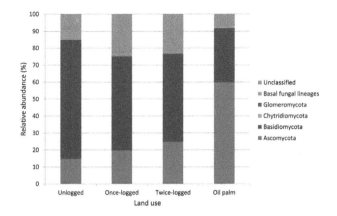

Figure 3. Relative abundance of dominant fungal phyla among different land uses in Sabah, Malaysian Borneo.

distance between pairs of transects within a site and between pairs of transects across the entire dataset [42,43].

The OTU-based community similarity was calculated by using the Bray-Curtis index [44]. Non-metric multidimensional scaling (NMDS), to visualize the change in species composition across the land use types, was conducted in Primer-E software (Version 6, Plymouth, UK). Then, we tested the difference among different land use types using an analysis of similarity (ANOSIM).

We measured β-diversity amongst land use types following Anderson et al. [45], which is defined as the variation in community structure without defining a particular gradient or direction. Therefore, we estimated true β-diversity following Whittaker [46] in [47] for every land use type. In addition, we calculated β-diversity as the average distance from each site to the group centroid [45]. The betadisper function in R was used to test if β-diversity shows any difference between land use types. True β-diversity (i.e. $S/\bar{\alpha}$) for each pair of samples within each of the four land uses was estimated by the following equation:

$$\frac{S}{\bar{\alpha}} = \frac{a+b+c}{(2a+b+c)/2}$$

Where S is the total number of OTUs in two samples, $\bar{\alpha}$ is the average number of OTUs for both samples, a is shared OTUs between both samples, b is OTUs found only in sample 1 and c are OTUs found only in sample 2. To compare true β-diversity among land uses, we used a linear model using land use as factor, and sample as random factor to control for pseudoreplication, as every sample is used in more than one comparison within each land use. Post-hoc Tukey tests were used for pairwise comparisons among different land uses.

Results

A total of 114,744 quality sequences were obtained from the 24 soil samples, with coverage ranging from 3,347–6,456 sequences. After rarifying to 3,347 reads per sample, we obtained a total of 80,328 sequences, and of these around 84% sequences were classified up to phylum level with a total of 5,327 OTUs (defined at ≥97% sequence similarity level). Fungal OUT richness (i.e. number of OTUs) was marginally significantly different across land use types ($F_{3,24}$ = 2.98, P = 0.05; Fig. 1a), with lowest levels of OTU richness observed in oil palm plantations compared to primary and logged forests. Predicted OTU richness calculated using the Chao1 estimator was significantly higher in logged forests than oil palm plantations (F3,24 = 4.74, P = 0.01; Fig. 1b), whereas Shannon index did not show any variation among land uses (F3,24 = 1.93, P = 0.15).

The NMDS plots of pairwise Bray–Curtis dissimilarities showed that fungal communities were clustered significantly across land use types (ANOSIM: R = 0.51, P<0.001; Fig. 2). Mantel tests showed no effect of distance on the composition of fungal communities across different land use types in Borneo (all P> 0.07). The majority of fungal sequences recovered in our study belonged to the Basidiomycota and Ascomycota, with relative abundances of 52% and 29%, respectively (Fig. 3). The basal

Table 1. Comparison of relative abundance of the dominant fungal orders within the phyla Ascomycota and Basidiomycota among land uses[a].

Taxa	F or χ^{2b}	df	P	Pairwise comparisons[c]
Ascomycota				
Helotiales	3.67	3, 24	0.02	Once-logged > oil palm
Hypocreales	5.54	3, 24	0.006	Once-logged/twice-logged/unlogged < oil palm
Pleosporales	6.42	3, 24	0.003	Once-logged/unlogged < oil palm
Basidiomycota				
Agaricales	10.9	3, 24	0.0001	Once-logged/twice-logged/unlogged > oil palm
Russulales	11.3	3, 24	0.0001	Unlogged > Once-logged/twice-logged/oil palm
Sebacinales	11.1*	3	0.01	Unlogged/twice-logged > oil palm
Sporidiobolales	15.1*	3	0.001	Once-logged/twice-logged/unlogged < oil palm
Thelephorales	10.0*	3	0.01	Unlogged/twice-logged > oil palm
Trichosporonales	13.3	3, 24	0.0001	Once-logged/twice-logged/unlogged > oil palm

[a]Only orders for which significant differences were found are shown.
[b]Effect of land use on relative abundance evaluated by linear or generalized linear model or by the Kruskal-Wallis test (*).
[c]Pairwise comparisons by *post hoc* Tukey test for linear/generalized linear models or P values Bonferroni-corrected for Kruskal-Wallis. Differences were considered significant at a P value of <0.05.

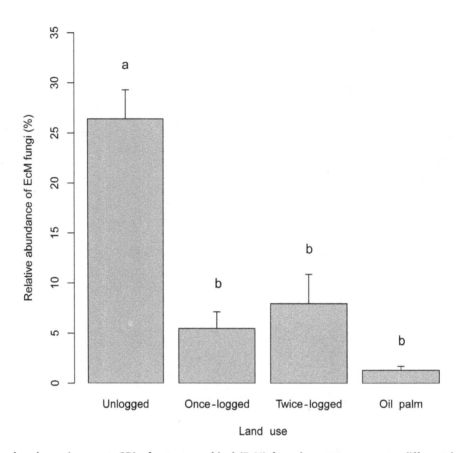

Figure 4. Relative abundance (means ± SD) of ectomycorrhizal (EcM) fungal sequences among different land uses in Sabah, Malaysian Borneo. Pairwise comparisons are shown; different letters denote significant differences between groups at P<0.05.

fungal lineages represented 2%, followed by Glomeromycota and Chytridiomycota with less than 1%, and 15% of the detected sequences were unclassified (Fig. 3). We found a significant change in the relative abundance of the two most dominant phyla: Basidiomycota ($F_{3,24}$ = 4.23, P = 0.01) and Ascomycota ($F_{3,24}$ = 4.81, P = 0.01) along different land uses. The abundance of Basidiomycota was greater in the unlogged forest, intermediate in the logged forest, and least in the oil palm. Conversely, the oil palm plantations had greater abundance of Ascomycota compared to forest soils (Fig. 3). At the order level, there were significant differences in the relative abundances of the most dominant orders (Table 1). The results revealed that the relative abundance of *Agaricales, Russulales, Thelephorales, Trichosporonales, Sebacinales* and *Helotiales* were significantly higher in the forest than oil palm plantations (P<0.05; Table 1). However, *Hypocreales, Sporidiobolales* and *Pleosporales*, were more abundant in oil palm plantations than unlogged and logged forests (P<0.05; Table 1).

A total of 11,421 sequences belonged to known groups of ectomycorrhizal (EcM) fungi, with the EcM fungi representing around 10% (11,421 sequences) of the total detected fungal sequences, with 180 OTUs. The relative abundance of EcM sequences was significantly different across land use types, with highest and lowest relative abundances observed in primary forest (mean relative abundance = 26%) and oil palm plantations (mean relative abundance = 0.5%), respectively (χ^2 = 18.04, P<0.001; Fig. 4). From our soil samples, we identified 14 genera belonging to EcM fungi with *Russula* as the most dominant genus (65% of total EcM sequences), followed by *Tomentella, Sebacina* and *Lactarius*. The relative abundance of these four dominant EcM

genera combined were significantly higher in forest soils compared to oil palm plantations (P<0.05). For *Russula* alone, we also found a difference in abundance between unlogged (greatest abundance), once-logged (less abundant), and twice-logged forest (lowest abundance of *Russula*) (P<0.05 in each case).

The β-diversity, measured as the average distance of all samples to the centroid in each land use type, did not show significant difference among land uses ($F_{3,24}$ = 2.77, P = 0.06). However, there was a significant effect of forest conversion to oil palm plantations on fungal true β-diversity (i.e. S/α; $F_{3,24}$ = 3.85, P = 0.01), with oil palm having lowest true β-diversity compared to unlogged and logged forests (Fig. 5). Logging did not produce any significant change in true β-diversity.

Discussion

Our results showed that fungal OTU richness and Chao1 index differed among land uses. Oil palm plantations had lower OTU richness, which can be attributed to the effects of anthropogenic intervention and forest conversion. This contrasts with the findings on bacterial communities of Lee-Cruz et al. [27] in the same study site, where they found that OTU richness and diversity indices did not differ among land uses, and that α-diversity was similar in forests and oil palm plantations.

We found that the structure of fungal communities differed most fundamentally between forests and oil palm plantations. This finding mirrors that of McGuire et al. [29], who found that the community of soil fungi collected across three different land uses in Malaysia differed between oil palm plantations and forests. Changes in fungal community composition in logged forest and

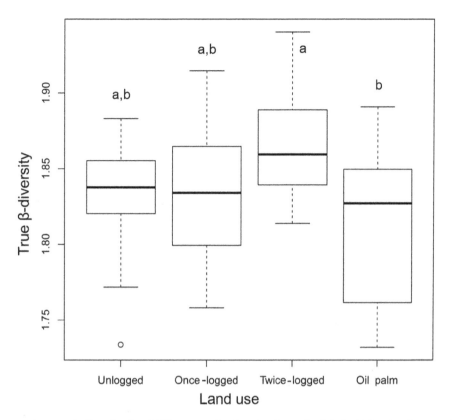

Figure 5. Fungal community true β-diversity (i.e. $S/\bar{\alpha}$) among the four land uses in Sabah, Malaysian Borneo. Boxes show the lower quartile, the median and the upper quartile. Pairwise comparisons are shown; different letters denote significant differences between groups at P< 0.05.

former forest areas could directly impact the functioning of soil communities and their ability to provide key ecosystem services, such as decomposition and nutrient recycling [22]. Given that fertilizer prices are predicted to rise dramatically in the coming decades [48], these results suggest that improvements in agricultural methods by establishing diversified farms could be necessary for sustaining vital soil biodiversity and ecosystem-service values.

The abundance pattern of fungal taxa detected here is similar to the soils of the Neotropics and elsewhere, where Basidiomycota and Ascomycota are also the most prevalent groups [49–53]. The most abundant orders of Basidomycota in the forest areas we sampled were *Russulales*, *Agaricales*, *Sporidiobolales*, *Telephorales*, *Trichosporonales* and *Sebacinales*. They all became less abundant in oil palm plantations compared to primary forest. These orders are generally reported to be the most varied and abundant groups of fungi in forests around the world [40,54,55], they have also been characterized as being lignicolous, saprobic or mycorrhizal, associated with litter decomposition, and they are known to degrade plant-derived cellulose [56]. In our forest plots, the ectomycorrhizal orders *Russulales* and *Telephorales* were the most common. This pattern has also been found in forests from the Neotropics and African tropics [57–59].

The lower abundance of Basidiomycota in oil palm soils may be due to the lack of large quantities of coarse woody debris, often derived from roots or branches of forest trees, in oil palm plantations. However, it may be worth investigating whether the change in abundance of Basidiomycota - and change in their overall community structure - with conversion of forest to oil palm may have implications for long term nutrient processing in the oil palm soils. Since Basidiomycota as a group are often able to break

down relatively recalcitrant substrates and changes in their abundance and community composition might impede nutrient recycling in oil palm soils. The converse increase in Ascomycota, may be seen in terms of the relatively lignin-poor and nutrient-rich character of most of the organic matter reaching the soil from roots and leaves of oil palm, and from the herbaceous weedy layer that grows under the palms.

We found that both a history of logging, and forest conversion to oil palm plantation, resulted in shifts in EcM fungal communities. The most drastic effects were with forest conversion to oil palm. However, logging history also had detectable effects EcM fungi relative abundance. Ectomycorrhizas are one of the most important widespread types of mycorrhiza in forests of the cool temperate and boreal latitudes [60], and they also form an important group and often the dominant ecologically and economically important minority of Dipterocarpaceae family trees in tropical Asia [61–65].

The lower abundance of EcM fungi in logged forests might be due to a thinner, more incomplete root mat following past logging disturbance. The much lower abundance of EcM fungi in oil palm plantation soils could be partly due to the much lower abundance of potential host roots (e.g. Dipterocarpaceae which are absent from the palm plantations). Such fungi have generally been found to recover slowly from disturbance even when potential host plants are present [66]. Among the detected genera in our study, we found that *Russula*, *Sebacina*, *Lactarius* and *Tomentella* showed significant impact of land use change. *Russula* was relatively the most abundant EcM genus in our samples: this genus has been found to be common on roots of dipterocarp forests, tropical and southern hemisphere angiosperm forests [62,67,68].

This study also investigated the impacts of land use change on the turnover of fungal communities across space (β-diversity), and results suggest that there is a spatial homogenization of fungal communities in oil palm agriculture compared to forest: fungal β-diversity in oil palm plantations was lower than forest. Habitat conversion to agriculture also reduces β-diversity of soil bacteria in Amazonia [69]. In contrast, across the same land use system and sites in Borneo as we studied here, there was actually an increase in β-diversity of bacteria with both logging and conversion to oil palm [27].

There are some caveats that accompany our findings. We worked in only one biogeographic region and on only one form of agriculture. It is thus important to replicate this work in other logging systems and in other key expanding crops, including soya, sugar cane, and cacao. We also did not explicitly demonstrate the impacts of changing fungal composition on soil ecosystem functions and services: this is a major knowledge gap, with critical importance to the development of sustainable logging and, in particular, agricultural systems in the tropics.

In conclusion, the conversion of both primary and logged forest to oil palm drives a change in the overall fungal community, including EcM fungi abundance, and an associated decrease in total fungal community beta-diversity. This finding invites further studies that investigate the long-term implications of such changes for agricultural sustainability. There was a more subtle long-term impact of logging on fungal communities. Most measurable features of the unlogged forest fungal community remained unchanged after logging. However, there were significant changes in EcM fungal abundance due to logging, which could have a pervasive impact since EcM fungi are thought to play a key role in tree growth and community structure. Despite this, the lack of drastic changes in the overall forest fungal community structure following logging strengthens the view that logged forest is not necessarily an irretrievably damaged and drastically altered system, and that protecting it from conversion to oil palm may still have considerable conservation benefits.

Acknowledgments

We thank Yayasan Sabah, Danum Valley Management Committee, Sabah Parks, the state Secretary, Sabah Chief Minister's Departments, and the Sabah Biodiversity Council for permission to conduct research. We also, thank the Royal Society's Southeast Asian Rainforest Research Program (SEARRP) and the Borneo Rainforest Lodge for logistical support and site access.

Author Contributions

Conceived and designed the experiments: DPE JMA. Performed the experiments: DK BMT DPE. Analyzed the data: DK BMT DPE JL. Contributed reagents/materials/analysis tools: DK BMT DPE JMA JL. Contributed to the writing of the manuscript: DK BMT DPE JMA.

References

1. Whitmore TC (1990) An introduction to tropical rain forests. Oxford: Clarendon Press. 226 pp.
2. Asner GP, Rudel TK, Aide TM, Defries R, Emerson R (2009) A contemporary assessment of change in humid tropical forests. Conserv Biol 23: 1386–1395.
3. Wright SJ (2005) Tropical forests in a changing environment. Trends Ecol Evol 20: 553–560.
4. Blaser J (2011) Status of tropical forest management 2011: International Tropical Timber Organization.
5. Food and Agriculture Organization of the United Nations (2010) Global forest resources assessment 2010 : main report. Rome: FAO. 340 p.
6. Hansen MC, Stehman SV, Potapov PV, Loveland TR, Townshend JR, et al. (2008) Humid tropical forest clearing from 2000 to 2005 quantified by using multitemporal and multiresolution remotely sensed data. Proc Natl Acad Sci U S A 105: 9439–9444.
7. Morris RJ (2010) Anthropogenic impacts on tropical forest biodiversity: a network structure and ecosystem functioning perspective. Philos Trans R Soc Lond B Biol Sci 365: 3709–3718.
8. Achard F, Eva HD, Stibig H-J, Mayaux P, Gallego J, et al. (2002) Determination of Deforestation Rates of the World's Humid Tropical Forests. Science 297: 999–1002.
9. Gibson L, Lee TM, Koh LP, Brook BW, Gardner TA, et al. (2011) Primary forests are irreplaceable for sustaining tropical biodiversity. Nature 478: 378–381.
10. Edwards DP, Larsen TH, Docherty TD, Ansell FA, Hsu WW, et al. (2011) Degraded lands worth protecting: the biological importance of Southeast Asia's repeatedly logged forests. Proc Biol Sci 278: 82–90.
11. Berry NJ, Phillips OL, Lewis SL, Hill JK, Edwards DP, et al. (2010) The high value of logged tropical forests: lessons from northern Borneo. Biodiver Conserv 19: 985–997.
12. Woodcock P, Edwards DP, Newton RJ, Vun Khen C, Bottrell SH, et al. (2013) Impacts of intensive logging on the trophic organisation of ant communities in a biodiversity hotspot. PLoS One 8: e60756.
13. Burivalova Z, Şekercioğlu ÇH, Koh LP (2014) Thresholds of Logging Intensity to Maintain Tropical Forest Biodiversity. Curr Biol 24: 1–6.
14. Hamer K, Hill J, Benedick S, Mustaffa N, Sherratt T, et al. (2003) Ecology of butterflies in natural and selectively logged forests of northern Borneo: the importance of habitat heterogeneity. J Appl Ecol 40: 150–162.
15. Cleary DF, Boyle TJ, Setyawati T, Anggraeni CD, Loon EEV, et al. (2007) Bird species and traits associated with logged and unlogged forest in Borneo. Ecol Appl 17: 1184–1197.
16. Edwards DP, Hodgson JA, Hamer KC, Mitchell SL, Ahmad AH, et al. (2010) Wildlife-friendly oil palm plantations fail to protect biodiversity effectively. Conserv Lett 3: 236–242.
17. Edwards F, Edwards D, Larsen T, Hsu W, Benedick S, et al. (2013) Does logging and forest conversion to oil palm agriculture alter functional diversity in a biodiversity hotspot? Anim Conserv 17: 163–173.
18. Baraloto C, Hérault B, Paine C, Massot H, Blanc L, et al. (2012) Contrasting taxonomic and functional responses of a tropical tree community to selective logging. J Appl Ecol 49: 861–870.
19. McGrath DA, Smith CK, Gholz HL, Oliveira FdA (2001) Effects of Land-Use Change on Soil Nutrient Dynamics in Amazônia. Ecosystems 4: 625–645.
20. Murty D, Kirschbaum MUF, McMurtrie RE, McGilvray H (2002) Does conversion of forest to agricultural land change soil carbon and nitrogen? a review of the literature. Global Change Biol 8: 105–123.
21. Cornejo FH, Varela A, Wright SJ (1994) Tropical Forest Litter Decomposition under Seasonal Drought: Nutrient Release, Fungi and Bacteria. Oikos 70: 183–190.
22. Lauber CL, Strickland MS, Bradford MA, Fierer N (2008) The influence of soil properties on the structure of bacterial and fungal communities across land-use types. Soil Biol Biochem 40: 2407–2415.
23. Tripathi B, Kim M, Singh D, Lee-Cruz L, Lai-Hoe A, et al. (2012) Tropical Soil Bacterial Communities in Malaysia: pH Dominates in the Equatorial Tropics Too. Microb Ecol 64: 474–484.
24. Orgiazzi A, Lumini E, Nilsson RH, Girlanda M, Vizzini A, et al. (2012) Unravelling soil fungal communities from different Mediterranean land-use backgrounds. PloS one 7: e34847.
25. Wu TH, Chellemi DO, Martin KJ, Graham JH, Rosskop EN (2007) Discriminating the effects of agricultural land management practices on soil fungal communities. Soil Biol Biochem 39: 1139–1155.
26. Martin F, Cullen D, Hibbett D, Pisabarro A, Spatafora JW, et al. (2011) Sequencing the fungal tree of life. New Phytol 190: 818–821.
27. Lee-Cruz L, Edwards DP, Tripathi BM, Adams JM (2013) Impact of Logging and Forest Conversion to Oil Palm Plantations on Soil Bacterial Communities in Borneo. Appl Environ Microbiol 79: 7290–7297.
28. Lupatini M, Jacques RJS, Antoniolli ZI, Suleiman AKA, Fulthorpe RR, et al. (2013) Land-use change and soil type are drivers of fungal and archaeal communities in the Pampa biome. World J Microbiol Biotechnol 29: 223–233.
29. McGuire KL, D'Angelo H, Brearley FQ, Gedallovich SM, Babar N, et al. (2014) Responses of soil fungi to logging and oil palm agriculture in Southeast Asian tropical forests. Microbial Ecol.
30. Wilcove DS, Giam X, Edwards DP, Fisher B, Koh LP (2013) Navjot's nightmare revisited: logging, agriculture, and biodiversity in Southeast Asia. Trends Ecol Evol 28: 531–540.
31. Fisher B, Edwards DP, Giam XL, Wilcove DS (2011) The high costs of conserving Southeast Asia's lowland rainforests. Front Ecol Environ 9: 329–334.
32. Gibbs HK, Ruesch AS, Achard F, Clayton MK, Holmgren P, et al. (2010) Tropical forests were the primary sources of new agricultural land in the 1980s and 1990s. Proc Natl Acad Sci U S A 107: 16732–16737.
33. Gardes M, Bruns TD (1993) ITS primers with enhanced specificity for basidiomycetes–application to the identification of mycorrhizae and rusts. Mol Ecol 2: 113–118.

34. White TJ, Bruns T, Lee S, Taylor J (1990) Amplification and direct sequencing of fungal ribosomal RNA genes for phylogenetics. PCR protocols: a guide to methods and applications 18: 315–322.

35. Schloss PD, Gevers D, Westcott SL (2011) Reducing the effects of PCR amplification and sequencing artifacts on 16S rRNA-based studies. PLoS One 6: e27310.

36. Nilsson RH, Veldre V, Hartmann M, Unterseher M, Amend A, et al. (2010) An open source software package for automated extraction of ITS1 and ITS2 from fungal ITS sequences for use in high-throughput community assays and molecular ecology. Fungal Ecol 3: 284–287.

37. Edgar RC, Haas BJ, Clemente JC, Quince C, Knight R (2011) UCHIME improves sensitivity and speed of chimera detection. Bioinformatics 27: 2194–2200.

38. Edgar RC (2010) Search and clustering orders of magnitude faster than BLAST. Bioinformatics 26: 2460–2461.

39. Abarenkov K, Nilsson RH, Larsson KH, Alexander IJ, Eberhardt U, et al. (2010) The UNITE database for molecular identification of fungi - recent updates and future perspectives. New Phytol 186: 281–285.

40. Tedersoo L, May T, Smith M (2010) Ectomycorrhizal lifestyle in fungi: global diversity, distribution, and evolution of phylogenetic lineages. Mycorrhiza 20: 217–263.

41. Liedloff A (1999) Mantel nonparametric test calculator. Version 2.0. School of Natural Resource Sciences, Queensland University of Technology, Australia.

42. Ghazoul J (2002) Impact of logging on the richness and diversity of forest butterflies in a tropical dry forest in Thailand. Biodiver Conserv 11: 521–541.

43. Ramage BS, Sheil D, Salim HM, Fletcher C, MUSTAFA NZA, et al. (2013) Pseudoreplication in tropical forests and the resulting effects on biodiversity conservation. Conserv Biol 27: 364–372.

44. Magurran AE (2004) Measuring biological diversity. Maldan, MA: Blackwell Pub. viii, 256 p.

45. Anderson MJ, Crist TO, Chase JM, Vellend M, Inouye BD, et al. (2011) Navigating the multiple meanings of beta diversity: a roadmap for the practicing ecologist. Ecol Lett 14: 19–28.

46. Whittaker RH (1960) Vegetation of the Siskiyou mountains, Oregon and California. Ecol Monogr 30: 279–338.

47. Koleff P, Gaston KJ, Lennon JJ (2003) Measuring beta diversity for presence-absence data. J Anim Ecol 72: 367–382.

48. Piesse J, Thirtle C (2009) Three bubbles and a panic: An explanatory review of recent food commodity price events. Food Policy 34: 119–129.

49. O'Brien HE, Parrent JL, Jackson JA, Moncalvo JM, Vilgalys R (2005) Fungal community analysis by large-scale sequencing of environmental samples. Appl Environ Microbiol 71: 5544–5550.

50. Wubet T, Christ S, Schöning I, Boch S, Gawlich M, et al. (2012) Differences in soil fungal communities between European beech (Fagus sylvatica L.) dominated forests are related to soil and understory vegetation. PLoS one 7: e47500.

51. Gomes NC, Fagbola O, Costa R, Rumjanek NG, Buchner A, et al. (2003) Dynamics of fungal communities in bulk and maize rhizosphere soil in the tropics. Appl Environ Microbiol 69: 3758–3766.

52. Oros-Sichler M, Gomes NC, Neuber G, Smalla K (2006) A new semi-nested PCR protocol to amplify large 18S rRNA gene fragments for PCR-DGGE analysis of soil fungal communities. J Microbiol Methods 65: 63–75.

53. Bridge PD, Newsham KK (2009) Soil fungal community composition at Mars Oasis, a southern maritime Antarctic site, assessed by PCR amplification and cloning. Fungal Ecol 2: 66–74.

54. Geml J, Laursen GA, Timling I, McFarland JM, Booth MG, et al. (2009) Molecular phylogenetic biodiversity assessment of arctic and boreal ectomycorrhizal Lactarius Pers. (Russulales; Basidiomycota) in Alaska, based on soil and sporocarp DNA. Mol Ecol 18: 2213–2227.

55. Geml J, Laursen GA, Herriott IC, McFarland JM, Booth MG, et al. (2010) Phylogenetic and ecological analyses of soil and sporocarp DNA sequences reveal high diversity and strong habitat partitioning in the boreal ectomycorrhizal genus Russula (Russulales; Basidiomycota). New Phytol 187: 494–507.

56. Kuramae EE, Hillekens RHE, de Hollander M, van der Heijden MGA, van den Berg M, et al. (2013) Structural and functional variation in soil fungal communities associated with litter bags containing maize leaf. Fems Microbiol Ecol 84: 519–531.

57. Ba AM, Duponnois R, Moyersoen B, Diedhiou AG (2012) Ectomycorrhizal symbiosis of tropical African trees. Mycorrhiza 22: 1–29.

58. Smith ME, Henkel TW, Aime MC, Fremier AK, Vilgalys R (2011) Ectomycorrhizal fungal diversity and community structure on three co-occurring leguminous canopy tree species in a Neotropical rainforest. New Phytol 192: 699–712.

59. Tedersoo L, Sadam A, Zambrano M, Valencia R, Bahram M (2010) Low diversity and high host preference of ectomycorrhizal fungi in Western Amazonia, a neotropical biodiversity hotspot. Isme Journal 4: 465–471.

60. Smith SE, Read DJ (2010) Mycorrhizal symbiosis: Academic press.

61. Brearley FQ (2012) Ectomycorrhizal Associations of the Dipterocarpaceae. Biotropica 44: 637–648.

62. Peay KG, Kennedy PG, Davies SJ, Tan S, Bruns TD (2010) Potential link between plant and fungal distributions in a dipterocarp rainforest: community and phylogenetic structure of tropical ectomycorrhizal fungi across a plant and soil ecotone. New Phytol 185: 529–542.

63. Moyersoen B, Becker P, Alexander I (2001) Are ectomycorrhizas more abundant than arbuscular mycorrhizas in tropical heath forests? New Phytol 150: 591–599.

64. Haug I, Weiß M, Homeier J, Oberwinkler F, Kottke I (2005) Russulaceae and Thelephoraceae form ectomycorrhizas with members of the Nyctaginaceae (Caryophyllales) in the tropical mountain rain forest of southern Ecuador. New Phytol 165: 923–936.

65. Natarajan K, Senthilarasu G, Kumaresan V, Riviere T (2005) Diversity in ectomycorrhizal fungi of a dipterocarp forest in Western Ghats. Curr Sci 88: 1893–1895.

66. Peay KG, Schubert MG, Nguyen NH, Bruns TD (2012) Measuring ectomycorrhizal fungal dispersal: macroecological patterns driven by microscopic propagules. Mol Ecol 21: 4122–4136.

67. Tedersoo L, Jairus T, Horton BM, Abarenkov K, Suvi T, et al. (2008) Strong host preference of ectomycorrhizal fungi in a Tasmanian wet sclerophyll forest as revealed by DNA barcoding and taxon-specific primers. New Phytol 180: 479–490.

68. Riviere T, Diedhiou AG, Diabate M, Senthilarasu G, Natarajan K, et al. (2007) Genetic diversity of ectomycorrhizal Basidiomycetes from African and Indian tropical rain forests. Mycorrhiza 17: 415–428.

69. Rodrigues JL, Pellizari VH, Mueller R, Baek K, Jesus Eda C, et al. (2013) Conversion of the Amazon rainforest to agriculture results in biotic homogenization of soil bacterial communities. Proc Natl Acad Sci U S A 110: 988–993.

20

A Brazilian Population of the Asexual Fungus-Growing Ant *Mycocepurus smithii* (Formicidae, Myrmicinae, Attini) Cultivates Fungal Symbionts with Gongylidia-Like Structures

Virginia E. Masiulionis[1]*, **Christian Rabeling**[2,3¤], **Henrik H. De Fine Licht**[4], **Ted Schultz**[3], **Maurício Bacci Jr.**[1], **Cintia M. Santos Bezerra**[1], **Fernando C. Pagnocca**[1]

1 Instituto de Biociências, São Paulo State University, Rio Claro, SP, Brazil, 2 Museum of Comparative Zoology, Harvard University, Cambridge, Massachusetts, United States of America, 3 Department of Entomology, National Museum of Natural History, Smithsonian Institution, Washington, D.C., United States of America, 4 Section for Organismal Biology, Department of Plant and Environmental Sciences, University of Copenhagen, Copenhagen, Denmark

Abstract

Attine ants cultivate fungi as their most important food source and in turn the fungus is nourished, protected against harmful microorganisms, and dispersed by the ants. This symbiosis evolved approximately 50–60 million years ago in the late Paleocene or early Eocene, and since its origin attine ants have acquired a variety of fungal mutualists in the Leucocoprineae and the distantly related Pterulaceae. The most specialized symbiotic interaction is referred to as "higher agriculture" and includes leafcutter ant agriculture in which the ants cultivate the single species *Leucoagaricus gongylophorus*. Higher agriculture fungal cultivars are characterized by specialized hyphal tip swellings, so-called gongylidia, which are considered a unique, derived morphological adaptation of higher attine fungi thought to be absent in lower attine fungi. Rare reports of gongylidia-like structures in fungus gardens of lower attines exist, but it was never tested whether these represent rare switches of lower attines to *L. gonglyphorus* cultivars or whether lower attine cultivars occasionally produce gongylidia. Here we describe the occurrence of gongylidia-like structures in fungus gardens of the asexual lower attine ant *Mycocepurus smithii*. To test whether *M. smithii* cultivates leafcutter ant fungi or whether lower attine cultivars produce gongylidia, we identified the *M. smithii* fungus utilizing molecular and morphological methods. Results shows that the gongylidia-like structures of *M. smithii* gardens are morphologically similar to gongylidia of higher attine fungus gardens and can only be distinguished by their slightly smaller size. A molecular phylogenetic analysis of the fungal ITS sequence indicates that the gongylidia-bearing *M. smithii* cultivar belongs to the so-called "Clade 1"of lower Attini cultivars. Given that *M. smithii* is capable of cultivating a morphologically and genetically diverse array of fungal symbionts, we discuss whether asexuality of the ant host maybe correlated with low partner fidelity and active symbiont choice between fungus and ant mutualists.

Editor: Nicole M. Gerardo, Emory University, United States of America

Funding: FCP and MB are grateful to CNPq and Fapesp for their financial support. VEM is a recipient of a CAPES/PEC-PG scholarship. CR was financially supported by a Junior Fellowship from the Harvard Society of Fellows and the HMS Milton Fund. TS was supported by the U.S. National Science Foundation grant DEB 0949689, the Smithsonian Institution Scholarly Studies Program, and the Smithsonian NMNH Small Grants Program. HHDFL was supported by grants from the Danish Research Council and the Carlsberg Foundation. The funders had no role in study design, data collection and analysis, decision to publish, or preparation of the manuscript.

Competing Interests: The authors have declared that no competing interests exist.

* Email: vemasiulionis@gmail.com

¤ Current address: Biology Department, University of Rochester, Rochester, New York, United States of America

Introduction

Mutualisms, symbiotic interactions between organisms in which each partner benefits, are widespread across the tree of life [1]. Many eukaryotes evolved obligate relationships with symbiotic organelles, such as mitochondria and chloroplasts, and provide stunning examples of ancient, evolutionarily stable mutualisms [2–4]. Co-evolutionary processes, reciprocal genetic changes in one species in response to changes in the partner species, shape these tight relationships, selecting for ecological specialization and resulting in co-diversification and eventually co-speciation [5–10]. Evolutionary patterns of co-speciation can be inferred secondarily from congruent phylogenetic histories (i.e., co-clado-genesis). Unfortunately, however, it is inherently difficult to study currently co-evolving organisms in order to understand the selective processes and proximate mechanisms underlying obligate interdependencies because currently observed patterns may not necessarily reflect the evolutionary interactions that shaped the symbiosis when it originated.

The complex symbiosis of fungus-growing ants with leucoco-prineaceous fungi and other associated microorganisms provides a system that is well suited for studying the evolution of mutualistic interactions and the origins of fungiculture in insects [11–17]. The fungus-gardening ants of the tribe Attini comprise a monophyletic group of more than 250 described species [15,18,19] that are distributed throughout the New World from Argentina in the south to the United States in the north [20–23]. All fungus-growing ant species rely obligately on basidiomycete fungi that they cultivate for food [21,24–26]. To enable the growth of the fungal symbionts, the ants provide nutrition to the fungus garden and prevent the growth of alien microorganisms [21,27–31].

Originating around 50–60 million years ago, the ancestral attine agricultural system, i.e., "lower agriculture," is practiced today by the majority of attine ant genera and species, which cultivate a closely related but poly- and paraphyletic group of leucocoprineaceous fungi [15,32]. Around 20–30 million years ago, a particular lower-attine ant-fungus association gave rise to "higher agriculture," which includes the well-known leafcutter ants [15]. The clade of fungi associated with higher attine ants is descended from a lower-attine fungal ancestor, and unlike the lower attine fungi, higher attine fungi are never found free-living apart from their ant hosts, suggesting strong co-evolutionary dynamics between higher attine ants and their cultivars [12,27,33–37]. The most significant morphological adaptations of higher attine fungi are the nutrient-rich hyphal tip swellings, the so-called gongylidia, which serve as the main food source for the ants and their brood [24,26,38–44].

In 1893, the pioneering mycologist Alfred Möller first described these hyphal tip swellings in the fungus gardens of Acromyrmex coronatus, which he called "Kohlrabikopf," due to their morphological similarity to cabbage turnips (Möller [24], pp. 26). Later Wheeler [25] suggested the Hellenistic version of Möller's term, "gongylidium, -a" (Greek = gongilis = turnip), for the same structure. In fungus gardens of higher attine ants, gongylidia occur in clusters, which Möller [24] termed "Kohlrabihäufchen," and Weber [38] later called "staphyla, -ae" (Greek = cluster of grapes). Chemical analyses showed that gongylidia contain glucose, mannitol, trehalose, glycan, arabinitol, and glycogen, in addition to lipids, and ergosterol [26,40,45,46], as well as free amino acids [34,45]. In contrast, the filamentous hyphae contain high protein concentrations but only low concentrations of lipids and carbohydrates [26,45]. In addition, a recent study demonstrated how the co-evolutionary adaptations of specific laccase enzymes, which are highly expressed in the gongylidia, participate in the detoxification of secondary plant compounds present in the leaf material [47].

The presence of gongylidia in the fungus gardens of higher attine ants is considered an exclusively co-evolved, mutualistic adaptation and gongylidia are not known to convey any fitness benefits to the fungus outside the ant symbiosis [13,15,21,35,37,48]. Here we report the occurrence of gongylidia in fungus gardens of the asexual lower attine ant Mycocepurus smithii from Brazil. M. smithii is distributed across tropical and subtropical habitats in Central and South America [49,50] and consists of a mosaic of sexually and asexually reproducing populations [50]. Populations from southeast Brazil were found to reproduce strictly asexually via thelytokous parthenogenesis [50,51]. Interestingly, M. smithii was previously reported to be one of the very few attine ant species to cultivate a genetically diverse array of fungi [13,32,52]. In contrast, some other lower attine species are known to be faithful to a single cultivar lineage [17,53]. In this study, we test whether M. smithii cultivates leafcutter ant cultivars or, alternatively, whether lower attine cultivars are capable of producing gongylidia. In addition, we explore the hypothesis of whether asexual reproduction in M. smithii may have enabled and/or prompted the cultivation of morphologically and genetically diverse cultivars.

Materials and Methods

Study site and field observations

During a field class taught at São Paulo State University in Rio Claro, Brazil (22.3955°S, 047.5424°W; elevation 608 m), we excavated nests of multiple fungus-growing ant species to illustrate their natural histories and the intricate symbiosis between ants, fungi, and other associated microorganisms. Nest excavations followed the methodology described earlier [51,54]. We excavated a total of three M. smithii fungus chambers, which received the following collection codes: CR110715-01, CR110715-02, and CR110718-01. Fungus-chamber CR110715-01 was found at 25 cm depth. It was 2.5 cm wide and 2 cm high and contained a pendant fungus garden hanging from the chamber ceiling (Fig. 1A), multiple workers, no brood, and no queen. The second chamber, CR110715-02, was located directly underneath the first chamber at 53 cm depth, was slightly larger with a diameter of 3 cm and contained a pendant fungus garden, a single queen, multiple workers, and no brood. We assume that these two chambers belonged to the same nest. The third M. smithii fungus chamber, CR110718-01, was excavated approximately 50 m distant from the first nest. Only a single chamber was encountered at 50 cm depth. It was 5 cm wide and 4 cm high, contained a pendant fungus garden, eleven workers, and neither a queen nor brood. The sizes of the fungus chambers and their locations in the soil are consistent with results reported from other M. smithii populations in Latin America [51,54,55]. The live ant colonies and their fungus gardens were collected with a surface-sterilized spoon and placed into a laboratory nest for further observation. The lab nest consisted of a plastic container with a plaster of Paris bottom; see reference [56] for lab nest setup. Ants were identified using a Leica MS5 stereomicroscope and voucher specimens were deposited in Maurício Bacci's Molecular Evolution Laboratory at São Paulo State University in Rio Claro. Collections were conducted under collecting permit number 14789-3 issued by the Ministério do Meio Ambiente – MMA and the "Instituto Chico Mendes de Conservação da Biodiversidade " (ICMBio).

Macro and microscopic observations

The presence of staphylae in M. smithii gardens was first noted during observations of the live colonies with a Leica MZ16 stereomicroscope and confirmed under higher magnification with a Leica DM750 bright-field microscope. The fungus garden of colony CR110715-02 was maintained in a dark room at 25°C for two days and measurements of gongylidia, which were grouped into staphylae, were taken with the Leica Application Suite V3 software package under a Leica DM750 bright-field microscope. For microscopic studies, we removed the staphylae from the fungus garden with a pair of acupuncture needles, placed them on a microscope slide, and submerged them in a drop of 15% glycerin solution. Staphylae were collected and 40 gongylidia were measured from a fungus garden of each of the following: lab colonies of Atta sexdens, A. laevigata, and Acromyrmex disciger, and a field-collected colony of Trachymyrmex fuscus. The colonies were collected on the campus of São Paulo State University in Rio Claro between October 2009 and September 2011.

The normality of the morphological measures was tested using Shapiro-Wilk's test and the homogeneity of variance was assessed with Levene's test of homoscedasticity. To test for variance of size distribution in gongylidia of different fungi, we conducted an analysis of variance (one-way ANOVA and Tukey test) with a significance level of less than 1% ($p < 0.01$) using the software package BioEstat 5.0 [57].

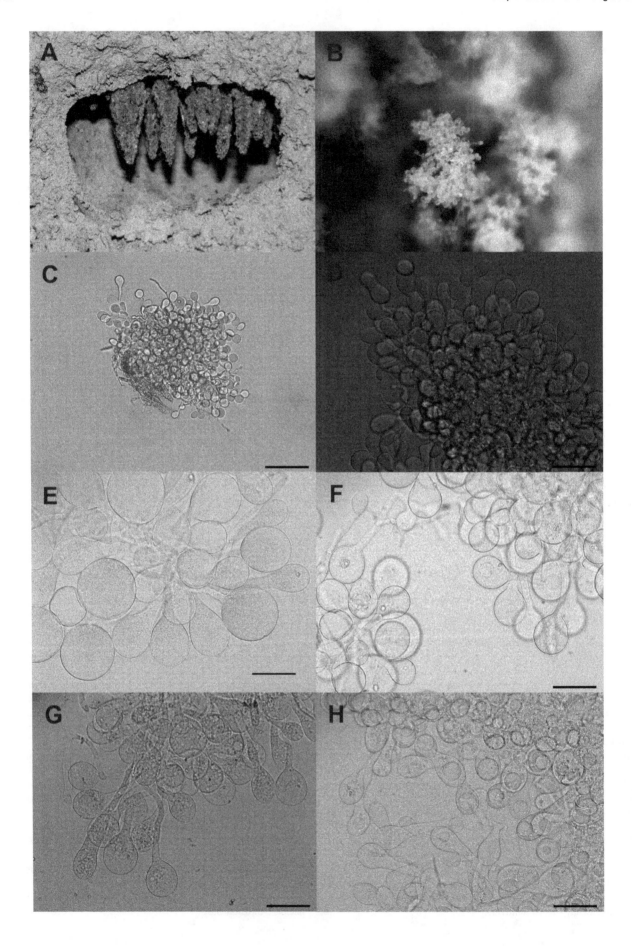

Figure 1. Cultivated fungi and gongylidia observed in gardens of *Mycocepurus smithii* **(A, B, C, D) and some species of higher Attini (E, F, G, H).** (A) Nest chamber and pendant fungus garden of *Mycocepurus smithii*; (B) gongylidia organized in staphylae in the fungus garden of *M. smithii* (8x magnification); (C) staphylae in a *M. smithii* cultivar. Gongylidia in the fungus gardens of *M. smithii* (D), *T. fuscus* (E), *Ac. disciger* (F), *A. laevigata* (G), and *A. sexdens* (H). The scale bar in figure C represents 100 µm; scale bars in figures D, E, F, G, and H represent 50 µm.

Genotyping and molecular phylogenetic analyses

To genotype the gongylidia-bearing *M. smithii* cultivar CR110715-02, we extracted genomic DNA from tissue samples of the staphylae following the methodology described in Martins Jr. et al. [58]. The ITS region was amplified according to Manter and Vivanco [59] using the forward primer ITS 5 (5'-GGAAGTAAAAGTCGTAACAAGG-3') and reverse primer ITS 4 (5'-TCCTCCGCTTATTGATATGC-3') [60]. The PCR reaction consisted of an initial 2-min incubation at 96°C, followed by 28 cycles of 46 s at 96°C, 30 s at 50°C and 4-min at 60°C. The PCR product was gel-purified using an Illustra™ GFX™ PCR DNA and Gel Band Purification Kit (GE Healthcare, UK). The same primer pair was utilized for amplification and sequencing. The sequencing reaction was prepared with 100 ng of PCR template, 6 pmols primers, 2.0 µl BigDye Terminator (Applied Biosystems), 1.0 µl buffer (200 mM Tris.HCl, 5 mM MgCl$_2$) and ddH$_2$O. Sequencing products were purified and then analyzed on an automated sequencer ABI3500 (Applied Biosystems). The consensus sequence was edited with the program BIOEDIT 7.0.5 [61] and aligned using CLUSTALW [62]. The obtained sequence was deposited at NCBI's GenBank (www.ncbi.nih.gov) as accession number JX027477 and was compared to other ITS sequences deposited in GenBank. The resulting ITS DNA sequence consisted of 606 base pairs and nucleotides at positions 76 and 514 bp were unknown. The ITS DNA sequence of *M. smithii* cultivar CR110715-02 is identical to ITS sequences of *M. smithii* cultivars from Panama, Costa Rica, and Trinidad; *M. tardus* from Panama; *Myrmicocrypta ednaella* from Panama; and a new *Mycocepurus* species from Guyana (erroneously identified as *M. cf. goeldii*), all of which were previously published in Mueller et al. [13] and Kellner et al. [63].

To determine the molecular phylogenetic placement of the gongylidia-bearing *M. smithii* fungus, we added this sample to a data matrix of attine and free-living leucocoprineaceous fungal ITS DNA sequences consisting of 305 fungal taxa and 1042 nucleotide sites (including indels), which was previously published by Mehdiabadi et al. [17]. Sequences were aligned in MAFFT v7.017 [64] using the E-INS-I algorithm, a 200PAM/k = 2 scoring matrix, a gap opening penalty of 1.53, and an offset value of 0.

Maximum likelihood analyses. Initially, a best-fit model of sequence evolution was selected for the entire unpartitioned fungal ITS alignment under the Akaike information criterion as calculated in jModel Test 2.1.1 [65]. Using the resulting model, GTR+I+G (with 6 rate categories), an initial ML "best tree" analysis was conducted in Garli 2.0.1019 [66] consisting of 100 replicates using parallel processing and default parameter values. The results of this initial analysis were used to divide the sequence data into faster- and slower-evolving character sets. This was done by evaluating character evolution on the ML tree in MacClade 4.08 [67] under the parsimony criterion and assigning characters requiring 3 or fewer steps to a "slow" partition (596 characters) and characters requiring 4 or more steps to a "fast" partition (446 characters). Each of these two partitions was fit, again using the AIC in jModel Test, to an evolutionary model. Based on the results, a second "best tree" search was conducted with two partitions, a "slow" partition under the JC model and a "fast" partition under the HKY+G model, consisting of 100 searches and deviating from the default settings as follows: topoweight = 0.01;

brlenweight = 0.002. ML bootstrap analyses in Garli, also employing the same two partitions and models, consisted of 1000 pseudoreplicates, deviating from default settings as follows: genthreshfortopoterm = 5000; scorethreshforterm = 0.10; startoptprec = 0.5; minoptprec = 0.01; numberofprecreductions = 1; treerejectionthreshold = 20.0; topoweight = 0.01; brlenweight = 0.002.

Bayesian analyses. The fungal alignment was analyzed under Bayesian criteria as implemented in MrBayes v3.2.1 [68] with the two partitions and models described above and with 10 million generations and samplefreq = 1000. All parameters were unlinked across partitions except for branch-lengths and topology. All analyses were carried out using parallel processing (one chain per CPU) with nucmodel = 4by4, nruns = 2, nchains = 8, and samplefreq = 1000. To address known problems with branch-length estimation in MrBayes, we reduced the branch-length priors using brlenspr = unconstrained:Exp(100) based on the procedure suggested in Brown et al. [69] and as applied in Ward et al. [70] and Rabeling et al. [50]. Burn-in and convergence were assessed using Tracer v1.5 [71], by examining potential scale reduction factor (PSRF) values in the MrBayes.stat output files, and by using Bayes factor comparisons of marginal likelihoods of pairs of runs in Tracer, which employs the weighted likelihood bootstrap estimator of Newton and Raftery [72] as modified by Suchard et al. [73], with standard error estimated using 1,000 bootstrap pseudoreplicates.

Results

The fungal cultivars collected from three fungus chambers belonging to at least two *M. smithii* colonies contained gongylidia-like structures that were organized in staphylae (Figs. 1B–D). Our microscopic examination at 400x magnification showed that the gongylidia of this *M. smithii* fungal cultivar were morphologically very similar to the gongylidia found in fungal cultivars of *T. fuscus*, *Ac. disciger*, *A. laevigata*, and *A. sexdens* (Fig. 1 E,F,G,H), differing only in their smaller size (Fig. 2).

The diameter of gongylidia found in the *M. smithii* garden varied between 16.3 µm and 25.41 µm, which was significantly smaller than the gongylidia diameters in the studied *Trachymyrmex, Acromyrmex,* and *Atta* species (one-way ANOVA and Tukey test, p<0.01; Fig. 2). Gongylidia diameter in the higher attine cultivars was distributed along a continuum, on which *A. sexdens* gongylidia were the smallest, ranging from 21.04 µm to 39.99 µm in diameter, and on which *T. fuscus* gongylidia were the largest, ranging between 42.01 µm and 68.26 µm in diameter. The cultivars of all five examined ant species had gongylidia diameters significantly different from one another, except for *Ac. disciger* and *A. laevigata* (Fig. 2).

Gongylidia are thought to be highly specialized morphological adaptations of higher attine cultivars. The detection of gongylidia in the fungus garden of one of the most "primitive" attine lineages therefore suggests at least three alternative explanations: (i) *M. smithii* is capable of growing leafcutter ant cultivars; (ii) gongylidia production arose independently in the cultivars of lower and higher attine ants; or (iii) gongylidia are ancestrally present in at least some lower attine fungi, including, presumably, the lineage that gave rise to higher attine fungi. To distinguish between these evolutionary scenarios, we conducted molecular phylogenetic

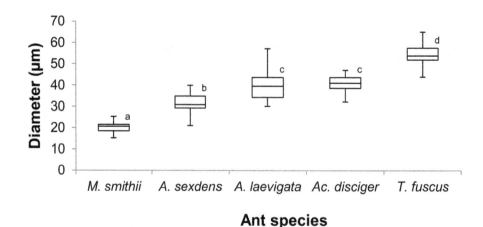

Ant species

Figure 2. Comparison of gongylidia diameter from fungi cultivated by *M. smithii* and four species of higher Attini. The diameter of each gongylidium (n = 40 gongylidia per colony per ant species) was measured at its widest point. Mean values that are annotated with different letters are significantly different from each other (one-way ANOVA and Tukey test, p<0.01).

analyses of the ITS region of attine cultivars to determine the phylogenetic position of the gongylidia-producing *M. smithii* fungus.

Both Bayesian and maximum likelihood phylogenetic analyses agree that the gongylidia-bearing *M. smithii* fungal cultivar of the Rio Claro population is embedded in the so-called "Clade 1" of the fungus tribe Leucocoprineae (Fig. 3) and is sequence identical with fungi cultivated by *M. smithii* (from Trinidad, Panama, Costa Rica), *M. tardus* (Panama), an undescribed species of *Mycocepurus* from Guyana, and *Myrmicocrypta ednaella* (Panama) (MLBP = 95, BPP = 100). Interestingly, a clade only a couple of nodes removed from the gongylidia-producing *M. smithii* cultivar contains a fungus cultivated by *T. papulatus* (Fig. 3), which is a member of the higher Attini. The highly derived higher attine fungal clade, which includes the cultivars of the leafcutter ants, is shown in our analysis to have arisen from a lower-attine fungal ancestor in Clade 1, which is consistent with earlier results [12,16].

Discussion

Although somewhat smaller in diameter, the gongylidia found in the fungus gardens of the asexual fungus-growing ant *M. smithii* occurring in Rio Claro are similar in shape to gongylidia found in the cultivars of higher attine species. This is surprising because gongylidia are thought to be a derived morphological specialization in higher attine fungi that originated as a result of the ant-fungus co-evolution that produced the higher attine ants, including the leafcutter ant genera *Atta* and *Acromyrmex* [14,21,48]. The fact that gongylidia were encountered in three fungus gardens of at least two separate nests suggests that gongylidia may occur frequently in *M. smithii* fungus gardens in this population in Rio Claro. Unfortunately, the frequency of occurrence of gongylidia in the *M. smithii* study population is currently unknown and will only be determined by a comprehensive survey across populations and seasons. It should be noted here that we have never observed gongylidia in *M. smithii* gardens previously, although we have excavated and inspected cultivars from dozens of *M. smithii* nests in Central and South America (Bacci, Rabeling, Schultz, unpublished data). One of us (TRS), however, recalls observing in 1993 what appeared to be staphylae in the garden of a lab colony of a lower attine ant, probably *Mycetosoritis hartmanni*.

In general, accounts of gongylidia-like structures in the fungus gardens of lower attines are rare in the Attini literature. Möller

[24] was the first to provide a detailed account of the fungi cultivated by leafcutter ants and by lower attine species and reported hyphal swellings from gardens of *Apterostigma wasmannii* and to a lesser extent from fungus gardens of *Cyphomyrmex strigatus*. Möller [24] clearly distinguished these hyphal swelling from the gongylidia and staphylae of the leafcutter ants, however. Paraphrasing Möller's original account ([24], pp.108), he reported that the globular and densely packed gongylidia of the leafcutter ants represented the most "perfect" food bodies, whereas the gongylidia of *A. wasmannii* were tubular in shape and less densely aggregated in the staphylae. Möller [24] further distinguished the fungal cultivars of *Acromyrmex* and *Apterostigma* experimentally, demonstrating that only leafcutter fungi continued to produce gongylidia in culture medium, whereas the *A. wasmannii* and *C. strigatus* cultivars reverted invariably to the characteristic filamentous mycelial growth. Unfortunately, we were not able to culture the *M. smithii* fungus, but its morphology was significantly different from the tubular hyphal swellings of *Apterostigma* and *Cyphomyrmex* fungi observed by Möller (compare Figs. 26 & 27 in [24]).

It remains uncertain whether the tubular swellings in *Apterostigma* and *Cyphomyrmex* gardens observed by Möller, the gongylidia of higher attine cultivars, and the gongylidia of the *M. smithii* cultivar reported here are developmentally homologous. Interestingly, the developmental origins of gongylidia remain entirely uninvestigated. One hypothesis suggests that gongylidia are modified cystidia, the hyphal swellings found in the hymenium of many free-living basidiocarps (J.A. Scott pers. obs., cited in [39]) and a hypothesis of the origin of ant fungivory suggests that "proto-gongylidia" may have served as the fungal analogue of ant-attractant elaiosomes of plant seeds, which provide a food reward in return for ant dispersal [14,24].

The mean diameter of the gongylidia in the *M. smithii* cultivar was smaller than the mean diameter of gongylidia of higher attine species collected in Rio Claro. Earlier mycological studies by Möller [24] on the fungus gardens of *Acromyrmex coronatus* and *Ac. disciger* reported relatively small gongylidia sizes (10–24 µm), which are comparable in size to those found in *M. smithii* gardens. *Ac. disciger* was also studied by us and the fungi from the Rio Claro population had significantly larger gongylidia than those from Blumenau reported by Möller [24]. Gongylidia size could differ significantly between colonies of the same ant species

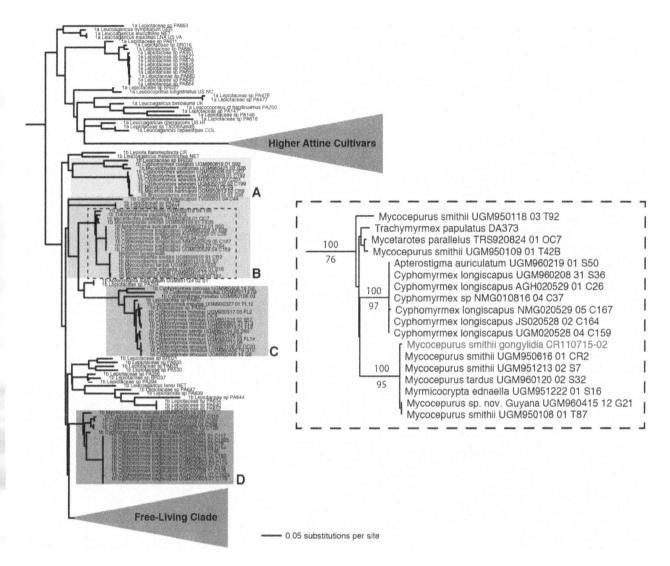

Figure 3. ITS phylogeny of Clade 1 of the fungal tribe Leucocoprineae. Clade 1 of the Leucocoprineae includes four primary clades of attine cultivars and closely related free-living fungi, here indicated as subclades A–D (sensu Mehdiabadi et al. [17]). The clade within subclade B that contains the gongylidia-bearing fungal cultivar of *M. smithii* from Rio Claro is indicated by dashed lines and is enlarged in the inset. Within the inset, the name of the gongylidia-bearing cultivar is indicated in red. The names of other cultivars of *M. smithii* in subclades A and B are indicated in red; *M. smithii* additionally cultivates fungi in Clade 2 (not shown here; see Mehdiabadi et al. [17]). The phylogram results from maximum-likelihood analyses of ITS sequence data of 305 fungal taxa. In the inset, numbers above branches are Bayesian posterior probabilities; numbers below branches are maximum-likelihood bootstrap proportions.

because the same species may utilize different cultivar species or strains in different nests and in different geographic locations. Gongylidia size could also differ in the same cultivar strain depending on the developmental status of the ant colony and the amount of nutrition provided by the ants.

The mechanisms underlying the production of gongylidia and the factors determining their size remain poorly understood. Previous studies of staphylae in higher attine nests suggested that the ants' constant pruning stimulated the formation of gongylidia [74,75]. In addition, Powell and Stradling [76] showed that the quality of the substrate, the pH, and the temperature affect the growth and size of gongylidia in cultivars of three higher attine species (*A. cephalotes*, *Ac. octospinosus*, and *T. urichi* from Trinidad) when cultivated *in-vitro* on agar plates. A recent genetic expression study added a functional dimension to our understanding of gongylidia by showing that one laccase enzyme, which is

highly expressed in the gongylidia of leafcutter ants, digests plant defensive phenolic compounds [46]. The ability to up-regulate laccase expression in the presence of fresh leaf material is thought to be a pre-adaptation of the fungus that evolved prior to the symbiosis with attine ants [46].

The phylogenetic position of the gongylidia-producing *M. smithii* cultivar in Clade 1 of the Leucocoprinae phylogeny, the uncertainty surrounding the phylogenetic position of the higher attine fungi within Clade 1, and the lack of information regarding the distribution of gongylidia in lower attine fungi leave open the question of whether gongylidia evolved repeatedly in leucocoprineaceous fungi or whether they had a single origin. The fungal phylogeny further shows that *M. smithii* cultivates a variety of genetically distant cultivars from both leucocoprineaceous Clades 1 and 2, which is consistent with earlier studies [13,17,32,52]. Fungus-growing ants and their fungal cultivars are hypothesized to have evolved largely via diffuse co-evolution, in which groups of species of ants and fungi interact [16]. However, particular ant and fungal species in the *Cyphomyrmex wheeleri* group were shown to have been exclusively associated over long periods of time, potentially for millions of years [17]. The opposite pattern is observed in *M. smithii*, which is capable of cultivating a high diversity of fungal species in both Clades 1 and 2. It has been hypothesized that this pattern of fungal association indicates that symbiont choice behavior (e.g., choosing cultivars adapted for particular microhabitats) is more important for *M. smithii* colonies than being faithful to a specific cultivar type (i.e., high partner fidelity) [63]. In the absence of a cultivar/habitat correlation, the observation that *M. smithii* cultivates an unprecedented genetic variety of cultivars is also consistent with weakened symbiont choice. *M. smithii* could only have acquired such a high diversity of cultivars by frequently domesticating free-living cultivars *de novo* and/or by acquiring novel cultivars from sympatric fungus-growing ants. A local study of Panamanian *M. smithii* colonies and their associated fungi hypothesizes that, because ants and fungi reproduce asexually and accumulate deleterious mutations (according to Muller's ratchet), the ant's frequent acquisition of novel cultivars functions as a kind of recombination, purging the ants from "deleterious fungi" and vice versa [63,77]. It remains to be demonstrated that the fungi cultivated by *M. smithii* suffer from Muller's ratchet because the large feral populations from which the cultivars are domesticated experience frequent sexual recombination [13,32].

An alternative null hypothesis is that, with regard to lower attine fungi, patterns of ant-fungus association in *M. smithii* are random. One hypothesis that would predict such randomness is that, due to the lack of genetic recombination, the ants' genomes accumulate deleterious mutations, some of which affect their olfactory receptors. Inclusive fitness theory predicts that conflict over reproduction is absent in a clonal society. Thus, nestmate recognition and mate recognition are expected to become obsolete

in an asexually reproducing species such as *M. smithii* and, if olfaction genes are not actively maintained by natural selection, these genes will accumulate non-synonymous mutations. Consequently, the ants' ability to recognize nestmates, or a specific fungal symbiont for that matter, may then deteriorate, and the ants may become unable to distinguish between cultivar types. In contrast, sexually reproducing attine species in the genus *Cyphomyrmex* have been experimentally shown to prefer the fungal cultivars they grew up with [78]. If olfactory abilities deteriorate in asexual fungus growing ants, which still needs to be demonstrated, the predicted outcome would be congruent with the pattern we observe in nature, in which an asexual fungus-growing ant cultivates a wide variety of morphologically and genetically different fungi. Future studies will first test whether the genes involved in olfaction are under relaxed natural selection in asexual populations of *M. smithii* and, second, comprehensively analyze the co-evolutionary interactions between *M. smithii* and its fungal cultivars on a global scale (Rabeling, Bacci, Schultz, unpublished data).

Acknowledgments

The authors are grateful to Dr. Silvio J. Govone for statistical advices, to both anonymous reviewers as well as the journal editor Dr. Nicole G. Gerardo for their valuable comments. A research permit to conduct fieldwork in Brazil was issued by the Ministério do MeioAmbiente and the Instituto Chico Mendes de Conservação da Biodiversidade (permit number 14789–3).

Author Contributions

Conceived and designed the experiments: VEM CR FCP. Performed the experiments: VEM CMSB. Analyzed the data: VEM CR HHDFL TS FCP MB. Contributed reagents/materials/analysis tools: FCP CR HHDFL TS MB. Wrote the paper: VEM FCP CR HHDFL TS MB.

References

1. Sachs JL, Essenberg CJ, Trucotte M (2011) New paradigms for the evolution of beneficial infections. Trends Ecol Evol 26(4): 202–209.

2. Blackstone NW (1995) A unit-of-evolution perspective on the endosymbiont theory of the origin of the mitochondrion. Evolution 49: 785–796.

3. Maynard Smith J, Szathmáry E (1998) The major transitions in evolution. Oxford: Oxford University Press. 346 p.

4. Timmis JN, Ayliffe MA, Huang CY, Martin W (2004) Endosymbiotic gene transfer: organelle genomes forge eukaryotic chromosomes. Nature Rev Genet 5: 123–135.

5. Ehrlich PR, Raven PH (1964) Butterflies and plants: a study in coevolution. Evolution 18: 586–608.

6. Benson WW, Brown Jr KS, Gilbert LE (1975) Coevolution of plants and herbivores: passion flower butterflies. Evolution 29: 659–680.

7. Janzen DH (1980) When is it coevolution? Evolution 34: 611–612.

8. Herre EA, Machado CA, Bermingham E, Nason JD, Windsor DM, et al. (1996) Molecular phylogenies of figs and their pollinator wasps. J Biogeography 23: 521–530.

9. Pellmyr O, Thompson JN, Brown JM, Harrison RG (1996) Evolution of pollination and mutualism in the yucca moth lineage. Am Nat 148: 827–847.

10. Mueller UG (2012) Symbiont recruitment versus ant-symbiont co-evolution in the attine ant-microbe symbiosis. Curr Opin Microbiol 15: 1–9.

11. Hinkle G, Wetterer JK, Schultz TR, Sogin ML (1994) Phylogeny of the Attini ants fungi based on analysis of small subunit ribosomal RNA gene sequences. Science 266: 1695–1697.

12. Chapela IH, Rehner SA, Schultz TR, Mueller UG (1994) Evolutionary history of the symbioses between fungus-growing ants and their fungi. Science 266: 1691–1694.

13. Mueller UG, Rhener SA, Schultz TR (1998) The evolution of agriculture in ants. Science 281: 2034–2038.

14. Schultz TR, Mueller UG, Currie CR, Rehner SA (2005) Reciprocal illumination: a comparison of agriculture in humans and ants. Ecological and Evolutionary Advances. In: Vega F, Blackwell M, editors. Insect-Fungal Associations. New York: Oxford University Press. 149–190.

15. Schultz TR, Brady SG (2008) Major evolutionary transitions in ant agriculture. PNAS 105: 5435–5440.

16. Mikheyev AS, Mueller UG, Abbot P (2010) Comparative dating of attine ants and lepiotaceous cultivars phylogenies reveals coevolutionary synchrony and discord. Am Nat 175: E126–E133.

17. Mehdiabadi NJ, Mueller UG, Brady SG, Himler AG, Schultz TR (2012) Symbiont fidelity and the origin of species in fungus-growing ants. Nat Commun 3: 840. doi: 10.1038/ncomms1844.

18. Schultz TR, Meier R (1995) A phylogenetic analysis of the fungus-growing ants (Hymenoptera: Formicidae: Attini) based on morphological characters of the larvae. Syst Entomol 20: 337–370.

19. Sosa-Calvo J, Schultz TR, Brandão CRF, Klingenberg C, Feitosa RM, et al. (2013) *Cyatta abscondita*: Taxonomy, evolution, and natural history of a new fungus-farming ant genus from Brazil. PLoS ONE 8(11): e80498.

20. Kempf WW (1972) Catálogo abreviado das formigas da Região Neotropical. Studia Ent 15: 1–344.

21. Weber NA (1972) Gardening Ants: The Attines. Philadelphia: American Philosophical Society. 146 p.

22. Brandão CRF (1991) Adendos ao catálogo abreviado das formigas da Região Neotropical (Hymenoptera: Formicidae). Rev Bras Entomol 35: 319–412.

23. Mayhé-Nunes AJ, Jaffé K (1998) On the biogeography of Attini (Hymenoptera: Formicidae). Ecotropicos 11(1): 45–54.

24. Möller A (1893) Die Pilzgärten einiger Südamerikanischer Ameisen. Bot Mitt Trop 6: 1–127.

25. Wheeler WM (1907) The fungus-growing ants of North America. Bull Amer Mus Nat Hist 23: 669–807.

26. Quinlan RJ, Cherrett JM (1979) The role of fungus in the diet of the leaf-cutting ants *Atta cephalotes*. Ecol Entomol 4: 151–160.

27. Weber NA (1966) The fungus-growing ants. Science 121: 587–604.

28. Quinlan RJ, Cherrett JM (1977) The role of substrate preparation in the symbiosis between the leaf-cutting ant *Acromyrmex octospinosus* (Reich) and its food fungus. Ecol Entomol 2: 161–170.

29. Currie CR, Mueller UG, Malloch D (1999) The agricultural pathology of ant fungus gardens. Proc Natl Acad Sci USA 96: 7998–8002.

30. Currie CR (2001) Prevalence and impact of a virulent parasite on a tripartite mutualism. Oecologia 128: 99–106.

31. Pagnocca FC, Masiulionis VE, Rodrigues A (2012) Specialized fungal parasites and opportunistic fungi in gardens of attine ants. Psyche 2012, Article ID 905109; doi: 10.1155/2012/905109.

32. Vo TL, Mueller UG, Mikheyev AS (2009) Free-living fungal symbionts (Lepiotaceae) of fungus-growing ants (Attini: Formicidae). Mycologia 101: 206–210.

33. Hölldobler B, Wilson EO (1990) The ants. Cambridge: Harvard University Press. 737 p.

34. Hölldobler B, Wilson EO (2010) The Leafcutter Ants: Civilization by Instinct. New York: Norton WW& Company. 160 p.

35. Mueller UG, Schultz TR, Currie CR, Adams RM, Malloch D (2001) The origin of the attine ant-fungus mutualism. Q Rev Biol 76: 169–197.

36. Mehdiabadi NJ, Schultz TR (2009) Natural history and phylogeny of the fungus-farming ants (Hymenoptera: Formicidae: Myrmicinae: Attini). Myrmecol News 13: 37–55.

37. Solomon SE, Lopes CT, Mueller UG, Rodrigues A, Sosa-Calvo J, et al. (2011) Nesting biology and fungiculture of the fungus-growing ant, *Mycetagroicus cerradensis*: new light on the origin of higher attine agriculture. J Insect Sci 11: 12.

38. Weber NA (1957) Dry season adaptations of fungus-growing ants and their fungi. Anat Rec 128: 638.

39. Littledyke M, Cherrett JM (1976) Direct ingestion of plant sap from cut leaves by the leaf-cutting ants *Atta cephalotes* (L.) and *Acromyrmex octospinosus* Reich (Formicidae: Attini). Bull Entomol Res 66: 205–217.

40. Quinlan RJ, Cherrett JM (1978) Aspects of the symbiosis of the leaf-cutting ant *Acromyrmex octospinosus* (Reich) and its food fungus. Ecol Entomol 3: 221–230.

41. Angeli-Papa J, Eymé J (1985) Les champignons cultivés par les fourmis Attinae. Ann Sci Nat Bot Biol 7: 103–129.

42. Bass M, Cherrett JM (1995) Fungal hyphae as source of nutrients for the leaf-cutting ant *Atta sexdens*. Physiol Entomol 20: 1–6.

43. Murakami T, Higashi S (1997) Social organization in two primitive attine ants, *Cyphomyrmex rimosus* and *Myrmicocrypta ednella*, with reference to their fungus substrates and food sources. J Ethol 15: 17–25.

44. Silva A, Bacci M, Bueno OC, Pagnocca FC, Hebling MJA (2003) Survival of *Atta sexdens* workers on different food sources. J Insect Physiol 49: 307–313.

45. Martin MM, Carman RM, Mac Connell JG (1969) Nutrients derived from the fungus cultured by the fungus-growing ant *Atta colombica tonsipes*. Ann Entomol Soc Am 62(6): 11–13.

46. Mônaco Furletti ME, Serzedello A (1983) Determinação de carboidratos em micélio de *Rozites gongylophora*. Rev Microbiol 14: 183–186.

47. De Fine Licht HH, Schiøtt M, Rogowska-Wrzesinska A, Nygaard S, Roepstorff P, et al. (2013) Laccase detoxification mediates the nutritional alliance between leaf-cutting ants and fungus-garden symbionts. PNAS 110: 583–587.

48. Mueller UG (2002) Ant versus fungus versus mutualism: ant-cultivar conflict and the deconstruction of the attine ant-fungus symbiosis. Am Nat 160: S67–S98.

49. Kempf WW (1963) A review of the ant genus *Mycocepurus* Forel, 1983 (Hymenoptera: Formicidae). Studia Ent 6: 417–432.

50. Rabeling C, Gonzales O, Schultz TR, Bacci Jr M, Garcia MVB, et al. (2011) Cryptic sexual populations account for genetic diversity and ecological success in a widely distributed, asexual fungus-growing ant. PNAS 108: 12366–12371.

51. Rabeling C, Lino-Neto J, Cappellari SC, Dos-Santos IA, Mueller UG, et al. (2009) Thelytokous parthenogenesis in the fungus gardening ant *Mycocepurus smithii* (Hymenoptera: Formicidae). PLoS ONE 4: e6781.

52. Rabeling C (2004) Nests, Nester, Streueintrag und symbiontische Pilze amazonischer Ameisen der Gruppe ursprünglicher Attini. Diplomarbeit, 80 pp., Universität Tübingen, Germany.

53. Schultz TR, Solomon SA, Mueller UG, Villesen P, Boomsma JJ, et al. (2002) Cryptic speciation in the fungus-growing ants *Cyphomyrmex longiscapus* Weber and *Cyphomyrmex muelleri* Schultz and Solomon, new species (Formicidae: Attini). Insectes Soc 49: 331–343.

54. Rabeling C, Verhaagh M, Engels W (2007) Comparative study of nest architecture and colony structure of the fungus-growing ants, *Mycocepurus goeldii* and *M. smithii*. J Insect Sci 7: 40.

55. Fernández-Marín H, Zimmerman JK, Wcislo WT, Rehner SA (2005) Colony foundation, nest architecture and demography of a basal fungus-growing ant, *Mycocepurus smithii* (Hymenoptera. Formicidae). J Nat Hist 39: 1735–1743.

56. Schultz TR (1993) Stalking the wild attine. Notes from Underground. Cambridge, Mass.: Museum of Comparative Zoology, Harvard University 8: 7–10.

57. Ayres M, Ayres M Jr, Ayres DL, dos Santos AS (2007) BioEstat 5.0. Imprensa Oficial do Estado do Pará. 323 p.

58. Martins Jr J, Solomon SE, Mikheyev AS, Mueller UG, Ortiz A, et al. (2007) Nuclear mitochondrial-like sequences in ants: evidence from *Atta cephalotes* (Formicidae: Attini). Insect Mol Biol 16: 777–784.

59. Manter DK, Vivanco JM (2007) Use of the ITS, primers ITS 1 F and ITS 4, to characterize fungal abundance and diversity mixed-template samples by qPCR and length heterogeneity analysis. J Microbiol Meth 71: 7–14.

60. White TJ, Bruns T, Lee S, Taylor J (1990) Amplification and direct sequencing of fungal ribosomal RNA genes for phylogenetics. In: Innis MA, Gelfand DH, Sninsky JJ, White TJ, editors. PCR Protocols: a guide to methods and applications. New York: Academic Press. 315–322.

61. Hall TA (1999) BioEdit: a user-friendly biological sequence alignment editor and analysis program for Windows 95/98/NT. Nucleic Acids Symp Series 41: 95–98.

62. Thompson JD, Higgings DG, Gibbson TJ (1994) ClustalW: improving the sensitivity of progressive multiple sequence alignment through sequence weighting, position specific gaps penalties and weight matrix choice. Nucleic Acids Res 22: 4673–4680.

63. Kellner K, Fernandez-Marin H, Ishak H, Sen R, Linksvayer TA, et al. (2013) Co-evolutionary patterns and diversification of ant–fungus associations in the asexual fungus-farming ant *Mycocepurus smithii* in Panama. J Evolution Biol 26: 1353–1362.

64. Katoh K, Misawa K,Kuma KI, Miyata T (2002) MAFFT: a novel method for rapid multiple sequence alignment based on fast Fourier transform. Nucleic Acids Research 30: 3059–3066. doi: 10.1093/nar/gkf436.

65. Posada D (2008) jModel Test: phylogenetic model averaging. Mol Biol Evol 25: 1253–1256.

66. Zwickl DJ (2006) Genetic algorithm approaches for the phylogenetic analysis of large biological sequence datasets under the maximum likelihood criterion. PhD dissertation, The University of Texas at Austin, USA.

67. Maddison DR, Madison WP (2000). MacClade v4.0. Sunderland: Sinauer Association.

68. Ronquist F, Huelsenbeck JP (2003) MrBayes 3: Bayesian phylogenetic inference under mixed models. Bioinformatics 19: 1572–1574.

69. Brown JM, Hedtke SM, Lemmon AR, Moriarty Lemmon E (2010) When trees grow too long: investigating the causes of highly inaccurate Bayesian branch-length estimates. Syst Biol 59: 145–161.

70. Ward PS, Brady SG, Fisher BL, Schultz TR (2010) Phylogeny and biogeography of dolichoderine ants: effects of data partitioning and relict taxa on historical inference. Syst Biol 59: 342–362.

71. Rambaut A, Drummond AJ (2007) Tracer v1. 5. Available: http://beast.bio.ed. ac.uk/Tracer. Accessed 2012 October 29.

72. Newton MA, Raftery AE (1994) Approximate Bayesian inference with the weighted likelihood bootstrap. J R Stat Soc, B, 56: 3–48.

73. Suchard MA, Weiss RE, Sinsheimer JS (2001) Bayesian selection of continuous-time Markov chain evolutionary models. Mol Biol Evol 18: 1001–1013.

74. Spegazzini C (1922) Descripción de hongos mirmecófilos. Ver Mus La Plata 26: 166–173.

75. Bass M, Cherrett JM (1996) Leaf-cutting ants (Formicidae, Attini) prune their fungus to increase and direct its productivity. Funct Ecol 10: 55–61.

76. Powell RJ, Stradling DJ (1986) Factors influencing the growth of *Attamyces bromatificus*, a symbiont of attine ants. Trans Br Mycol Soc 87: 205–213.

77. Himler AG, Caldera EJ, Baer BC, Fernández-Marín H, Mueller UG (2009) No sex in fungus-farming ants or their crops. Proc R Soc B doi: 10.1098/rspb.2009.0313.

78. Mueller UG, Poulin J, Adams RMM (2004) Symbiont choice in a fungus-growing ant (Attini, Formicidae). Behav Ecol 15: 357–364.

Permissions

The contributors of this book come from diverse backgrounds, making this book a truly international effort. This book will bring forth new frontiers with its revolutionizing research information and detailed analysis of the nascent developments around the world.

We would like to thank all the contributing authors for lending their expertise to make the book truly unique. They have played a crucial role in the development of this book. Without their invaluable contributions this book wouldn't have been possible. They have made vital efforts to compile up to date information on the varied aspects of this subject to make this book a valuable addition to the collection of many professionals and students.

This book was conceptualized with the vision of imparting up-to-date information and advanced data in this field. To ensure the same, a matchless editorial board was set up. Every individual on the board went through rigorous rounds of assessment to prove their worth. After which they invested a large part of their time researching and compiling the most relevant data for our readers.

The editorial board has been involved in producing this book since its inception. They have spent rigorous hours researching and exploring the diverse topics which have resulted in the successful publishing of this book. They have passed on their knowledge of decades through this book. To expedite this challenging task, the publisher supported the team at every step. A small team of assistant editors was also appointed to further simplify the editing procedure and attain best results for the readers.

Apart from the editorial board, the designing team has also invested a significant amount of their time in understanding the subject and creating the most relevant covers. They scrutinized every image to scout for the most suitable representation of the subject and create an appropriate cover for the book.

The publishing team has been an ardent support to the editorial, designing and production team. Their endless efforts to recruit the best for this project, has resulted in the accomplishment of this book. They are a veteran in the field of academics and their pool of knowledge is as vast as their experience in printing. Their expertise and guidance has proved useful at every step. Their uncompromising quality standards have made this book an exceptional effort. Their encouragement from time to time has been an inspiration for everyone.

The publisher and the editorial board hope that this book will prove to be a valuable piece of knowledge for researchers, students, practitioners and scholars across the globe.

List of Contributors

Ryan F. Seipke, Jörg Barke, Charles Brearley and Matthew I. Hutchings
School of Biological Sciences, University of East Anglia, Norwich Research Park, Norwich, United Kingdom

Lionel Hill
Metabolic Biology, John Innes Centre, Norwich Research Park, Norwich, United Kingdom

Douglas W. Yu
School of Biological Sciences, University of East Anglia, Norwich Research Park, Norwich, United Kingdom

Rebecca J. M. Goss
School of Chemistry, University of East Anglia, Norwich Research Park, Norwich, United Kingdom

Erin A. Mordecai
Department of Ecology, Evolution, and Marine Biology, University of California, Santa Barbara, Santa Barbara, California, United States of America

Shane M. Hanlon and Matthew J. Parris
Department of Biological Sciences, University of Memphis, Memphis, Tennessee, United States of America

Jacob L. Kerby
Department of Biology, University of South Dakota, Vermillion, South Dakota, United States of America

Allison McCormick, Marzena Broniszewska, Julia Beck and Frank Ebel
Max-von-Pettenkofer-Institut, Ludwig-Maximilians-University, Munich, Germany

Ilse D. Jacobsen
Department for Microbial Pathogenicity Mechanisms, Leibniz Institute for Natural Product Research and Infection Biology, Jena, Germany

Jürgen Heesemann
Max-von-Pettenkofer-Institut, Ludwig-Maximilians-University, Munich, Germany
Center of Integrated Protein Science (Munich) at the Faculty of Medicine of the Ludwig-Maximilians-University, Munich, Germany

Xinjian Zhang, Yujie Huang, Hongmei Li, Yan Ren, Jishun Li, Jianing Wang and Hetong Yang
Shandong Provincial Key Laboratory of Applied Microbiology, Biotechnology Center of Shandong Academy of Sciences, Jinan, Shandong Province, People's Republic of China

Paul R. Harvey
Shandong Provincial Key Laboratory of Applied Microbiology, Biotechnology Center of Shandong Academy of Sciences, Jinan, Shandong Province, People's Republic of China
CSIRO Sustainable Agriculture National Research Flagship and CSIRO Ecosystem Sciences, Glen Osmond, South Australia, Australia

Ting-Ting Song
Institute of Microbiology, College of Life Science, Zhejiang University, Hangzhou, Zhejiang, People's Republic of China
Horticulture Institute, Zhejiang Academy of Agricultural Sciences, Hangzhou, Zhejiang, People's Republic of China

Jing Zhao, Sheng-Hua Ying and Ming-Guang Feng
Institute of Microbiology, College of Life Science, Zhejiang University, Hangzhou, Zhejiang, People's Republic of China
Horticulture Institute, Zhejiang Academy of Agricultural Sciences, Hangzhou, Zhejiang, People's Republic of China

Anuradha Chowdhary, Shallu Kathuria, Cheshta Sharma, Gandhi Sundar and Pradeep Kumar Singh
Department of Medical Mycology, Vallabhbhai Patel Chest Institute, University of Delhi, Delhi, India,

Jianping Xu
Department of Biology, McMaster University, Hamilton, Ontario, Canada,

Shailendra N. Gaur
Department of Pulmonary Medicine, Vallabhbhai Patel Chest Institute, University of Delhi, Delhi, India

Ferry Hagen and Corné H. Klaassen
Department of Medical Microbiology and Infectious
Diseases, Canisius Wilhelmina Hospital, Nijmegen,
The Netherlands

Jacques F. Meis
Department of Medical Microbiology and Infectious
Diseases, Canisius Wilhelmina Hospital, Nijmegen,
The Netherlands
Department of Medical Microbiology, Radboud
University Nijmegen Medical Centre, Nijmegen,
The Netherlands

**Michele C. Pereira e Silva, Jan Dirk van Elsas and
Joana Falcão Salles**
Department of Microbial Ecology, Centre for Life
Sciences, University of Groningen, Groningen, The
Netherlands

Armando Cavalcante Franco Dias
Department of Soil Science, "Luiz de Queiroz"
College of Agriculture, University of São Paulo,
Piracicaba, São Paulo, Brazil

Marie-Hé lène Robin and Cé lia Cholez
Institut National de la Recherche Agronomique,
Unité Mixte de Recherche 1248 Agrosystémes et
agricultures, Gestion des ressources, Innovations
et Ruralité s, Castanet- Tolosan, France
Université de Toulouse, Institut National
Polytechnique de Toulouse, Ecole d'Ingénieurs de
Purpan, Toulouse, France

Nathalie Colbach
Institut National de la Recherche Agronomique,
Unité Mixte de Recherche 1347 Agroécologie,
Dijon, France

Philippe Lucas and Françoise Montfort
Institut National de la Recherche Agronomique,
Unité Mixte de Recherche 1099 Biologie des
Organismes et des Populations appliquöe á la
Protection des Plantes. Le Rheu, France

Philippe Debaeke and Jean-Noël Aubertot
Université Toulouse, Institut National Polytechnique
de Toulouse, Unité Mixte de Recherche 1248
Agrosystémes et agricultures, Gestion des
Ressources, Innovations et Ruralités, Castanet-
Tolosan, France

Matthew D. Castle and Christopher A. Gilligan
Department of Plant Sciences, University of
Cambridge, Cambridge, United Kingdom

Peter H. F. Hobbelen and Frank van den Bosch
Rothamsted Research, Harpenden, Hertfordshire,
United Kingdom

Neil D. Paveley
ADAS UK Ltd, High Mowthorpe, Duggleby,
Malton, North Yorkshire, United Kingdom

**Daisuke Hagiwara, Azusa Takahashi-Nakaguchi,
Takahito Toyotome, Katsuhiko Kamei, Tohru
Gonoi and Susumu Kawamoto**
Medical Mycology Research Center, Chiba
University, Chiba, Japan

Akira Yoshimi and Keietsu Abe
New Industry Creation Hatchery Center, Tohoku
University, Sendai, Japan

Ghada Abou Ammar and Reno Tryono
Institute of Agricultural and Nutritional Sciences,
Faculty of Natural Sciences III, Martin-Luther-
University Halle-Wittenberg, Halle (Saale), Germany

Katharina Dö ll and Petr Karlovsky
Molecular Phytopathology and Mycotoxin Research
Section, Georg-August-Universita¨t Go¨ ttingen,
Göttingen, Germany

Holger B. Deising and Stefan G. R. Wirsel
Institute of Agricultural and Nutritional Sciences,
Faculty of Natural Sciences III, Martin-Luther-
University Halle-Wittenberg, Halle (Saale), Germany
Interdisziplinäres Zentrum für
Nutzpflanzenforschung, Martin-Luther-Universität
Halle-Wittenberg, Halle (Saale), Germany

Yu-Mei Xiao and Yi-Hui Zhou
Department of Applied Chemistry, China
Agricultural University, Beijing, China
Laboratory of Cell Biology, Center for Cancer
Research, National Cancer Institute, NIH, Bethesda,
Maryland, United States of America

Lothar Esser, Fei Zhou and Di Xia
Laboratory of Cell Biology, Center for Cancer
Research, National Cancer Institute, NIH,
Bethesda, Maryland, United States of America

Chang Li and Zhao-Hai Qin
Department of Applied Chemistry, China
Agricultural University, Beijing, China

Chang-An Yu
Department of Biochemistry and Molecular Biology, Oklahoma State University, Stillwater, Oklahoma, United States of America

Colin J. Ingham
Department of Medical Microbiology and Infection Control, Jeroen Bosch Hospital, 's-Hertogenbosch, The Netherlands
Laboratory of Microbiology, Wageningen University, Wageningen, The Netherlands
MicroDish BV, Wageningen, The Netherlands

Peter M. Schneeberger
Department of Medical Microbiology and Infection Control, Jeroen Bosch Hospital, 's-Hertogenbosch, The Netherlands

Wei Liu, Li Ping Li, Jun Dong Zhang, Qun Li, Hui Shen, Si Min Chen, Li Juan He, Guo Tong Xu and Mao Mao An
Tongji University School of Medicine, Shanghai, China

Lan Yan
New Drug Research and Development Center, School of Pharmacy, Second Military Medical University, Shanghai, China

Yuan Ying Jiang
Tongji University School of Medicine, Shanghai, China
New Drug Research and Development Center, School of Pharmacy, Second Military Medical University, Shanghai, China

Jie Jin, Mengxia Yu, Chao Hu, Li Ye, Lili Xie, Jin Jin and Feifei Chen
Department of Hematology, the First Affiliated Hospital of Zhejiang University, Hangzhou, People's Republic of China

Institute of Hematology, Zhejiang University School of Medicine, Hangzhou, People's Republic of China

Hongyan Tong
Department of Hematology, the First Affiliated Hospital of Zhejiang University, Hangzhou, People's Republic of China
Institute of Hematology, Zhejiang University School of Medicine, Hangzhou, People's Republic of China
Myelodysplastic syndromes diagnosis and therapy center, Zhejiang University School of Medicine, Hangzhou, People's Republic of China

Virginia E. Masiulionis, Maurício Bacci Jr., Cintia M. Santos Bezerra and Fernando C. Pagnocca
Instituto de Biociê ncias, São Paulo State University, Rio Claro, SP, Brazil

Christian Rabeling
Museum of Comparative Zoology, Harvard University, Cambridge, Massachusetts, United States of America
Department of Entomology, National Museum of Natural History, Smithsonian Institution, Washington, D.C., United States of America

Henrik H. De Fine Licht
Section for Organismal Biology, Department of Plant and Environmental Sciences, University of Copenhagen, Copenhagen, Denmark

Ted Schultz
Department of Entomology, National Museum of Natural History, Smithsonian Institution, Washington, D.C., United States of America

Index